EMERGENCY RESCUE SHORING TECHNIQUES

EMERGENCY RESCUE SHORING TECHNIQUES

JOHN P. O'CONNELL

Disclaimer

The recommendations, advice, descriptions, and the methods in this book are presented solely for educational purposes. The author and publisher assume no liability whatsoever for any loss or damage that results from the use of any of the material in this book. Use of the material in this book is solely at the risk of the user.

Copyright© 2005 by
Fire Engineering Books & Videos
110 S. Hartford Ave., Suite 200
Tulsa, Oklahoma 74120 USA

800.752.9764
+1.918.831.9421
info@fireengineeringbooks.com
www.FireEngineeringBooks.com

Supervising Editor: Jared d'orr Wicklund
Production Editor: Sue Rhodes Dodd
Cover designer: Ken Wood
Book designer: Wes Rowell

Library of Congress Cataloging-in-Publication Data

O'Connell, John P., 1953-
 Emergency rescue shoring techniques / by John P. O'Connell.
     p. cm.
 Includes index
  ISBN 0-91221-259-4
  ISBN13 978-0-912212-59-3
  1. Shoring and underpinning. 2. Rescue work. I. Title.
  TH5281.O282005
  628.9'2--dc22

                                                                2004027554

Printed in the United States of America

5 6 7 8 9   27 26 25 24 23

I would like to dedicate this book to all the firefighters who have made the supreme sacrifice in the line of duty.

I dedicate this book to all the firefighters who have sacrificed their all for the people of the City of New York, before, during, and after the World Trade Center attack. I especially dedicate this book to my friends and brothers in Rescue Company #3 in the Bronx who gave their all: Al Ronaldson, Chris Blackwell, Ray Meisenheimer, Donny Regan, Tommy Foley, Gerry Shrang, Tommy Gambino, and Joe Spor. God bless you all.

I can only hope that this book may at some point prevent some of the brothers and sisters from getting injured in a rescue operation. The whole purpose of writing this text is to get the information out to all the firefighters we can and give them some more ammo in their arsenal against dangerous rescue situations.

Contents

3: Setting Up Your Shoring Operations 57

Acknowledgments

I know of no way I can possibly thank everyone who helped me on this project; so many individuals and teams contributed information that it would be impossible to document all of it. There are a few I must mention; they were the catalyst for this book and have given me unending help over the years.

To the "senior guy" of the group, a special thanks. Dave Hammond from California, your unselfish contributions to the urban search and rescue (US&R) program and this book cannot ever be repaid. I thank you for all your help and guidance through the years. I must mention three other guys from the "left" coast: Jim Hone, Mike McGroarty, and Donny Shawver. Thank you for encouraging me to keep pursuing this project and for all your expertise and help along the way.

Another group of great men I must commend are all the guys, past and present, from Rescue Company #3 "Da Bronx" who have worked with me, helped me, and corrected me when necessary in the development of this project. Thank you, brothers.

Finally, I would like to thank my beautiful children for sticking by me through thick and thin over the years: Kristen, Jennifer, Patrick, and Katelyn. I love you all and I'm very proud of you. Thanks for everything—Dad.

An Introduction to Emergency Rescue Shoring Concepts

Emergency Rescue Shoring

Emergency shoring operations for urban search and rescue incidents are defined as the temporary stabilization or resupport of any structural element that is physically damaged, missing, or structurally compromised by partial or total collapse of the structure, resulting in the danger of the structure's collapse. Shoring operations are performed in order to provide a safe and efficient atmosphere while conducting trapped victim search and rescue operations. Shoring provides a relatively safe environment of reduced risk to the victims, as well as to the trained rescue forces. Rescue shoring activities also include the stabilization of any adjacent structure or object that may be affected by the initial incident.

Rescue Shoring

Shoring for US&R is the temporary support of only that part of a damaged, collapsed, or partly collapsed structure that is required for conducting search and rescue operations at reduced risk to the victims and US&R forces.

Rescue Shoring Operations Objectives

The paramount objective of emergency shores in collapsed structures is to properly maintain the strength and integrity of any and all structurally damaged or unstable elements such as, but not limited to, beams, joists, girders, columns, arches, headers, or bearing walls.

The main objective of the rescue shoring operations is to properly and effectively receive, transmit, and/or redirect the currently unstable collapse loads. Many times, depending on the type of structure, these loads can be transferred or directed to structural elements in the remaining part of the building that are sound and capable of handling the additional collapse loads. Other times, these redirected loads cause a heavy concentrated load effect, overstressing the existing and undamaged structural elements and must be transferred ultimately to stable ground.

Basic Points

- *Shoring should be built as a system*
- *Lateral brace to prevent system from buckling*
- *Minimum level of lateral strength in vertical support should be 2%*
- *Ideally 10 %*

Concentrated versus Distributed

One of the main concepts of rescue shoring is to take the concentrated overload from debris and redirect, or redistribute, it to structural elements that can support the load. Sometimes, a collapse situation creates an overload condition on the remaining structure. This can happen when a building's contents as well as its structural elements have collapsed onto a lower floor. Depending on the type of collapse voids created, the upper floor loads are directed into a specific area, a usual occurrence in structures with a cantilever or supported lean-to, *V*-shaped or an *A*-framed collapse pattern. In these patterns, the material from the floors above is directed into specific areas. In its original state, the upper floor's structure and contents were distributed evenly throughout the space and was easily supported by the building.

However, once the material has come to rest on a lower floor in a large concentrated form, the structure's supporting elements are overloaded because the concentrated load is being supported by only a few elements, generally floor joists or a girder or both. The job of the rescue shoring officer and the structural specialist is to determine the overload and the shoring method to use for redistributing it to either the ground or to other structural elements able to support it. The following are options for redirecting overloads:

- Feed the load directly to the ground or a lower floor, normally by means of vertical shores

- Transfer the load laterally to the exterior bearing walls

There are other options available, but these two are the most common.

Basic Rescue Shoring Points

Emergency shoring: a complete system

Unlike the norm for the construction industry, shores used in buildings in an emergency after a catastrophic event has severely damaged the structure must be constructed as a complete system. When all the shores are tied together, the stability and efficiency of all the shores increases. The possibility of secondary collapse is the greatest danger at any structural collapse rescue operation. In order to minimize that risk and to maximize safety factors, all the shores must be assembled together as a complete unit.

Lateral bracing

It is very important to laterally brace all the shores in both directions. The shores must be able to withstand lateral pressures applied to the shoring system from any direction. Sudden load shifts can easily occur in unstable collapsed structures, thus applying eccentric and/or torsional loading. The minimum level of lateral strength in vertical support should be 2%; however, 10% is more desirable.

Rescue shoring is unique compared to *normal* contractor-installed shoring. Contractor shoring is generally friction-type shoring that relies on the pressure of the shore against the damaged material to keep the undamaged structure in position. A major problem with friction-type shores is that they have very little lateral stability and can be vibrated or knocked loose easily. This possibility must be avoided. The potential for secondary collapse is always present in structural collapse rescue operations. Rescuers must always prepare for it.

In rescue situations, the incident commander must have fixed shoring systems installed because of the dangerous possibility of secondary collapse. Fixed shores incorporating lateral bracing that resists forces from several directions stand up to the threat of secondary collapse much better than the friction-type shores.

Objectives

- *Maintain the integrity of all structurally unstable elements*
- *Properly transmit or redirect the collapse loads to stable ground or other suitable structural elements capable of handling the additional loads*

The Shoring System

For a shore to work properly and be considered a system, it must have four main items:

- A header, or top plate

- One or more posts or struts

- A bottom plate, or sole plate

- A lateral, or diagonal, bracing system.

Each one of these items is important for the success of the shoring system. The key to all the shores is to collect the loads from a damaged area, funnel it through the post system and redistribute the load to the ground or other suitable structural elements (see Fig. 1–1).

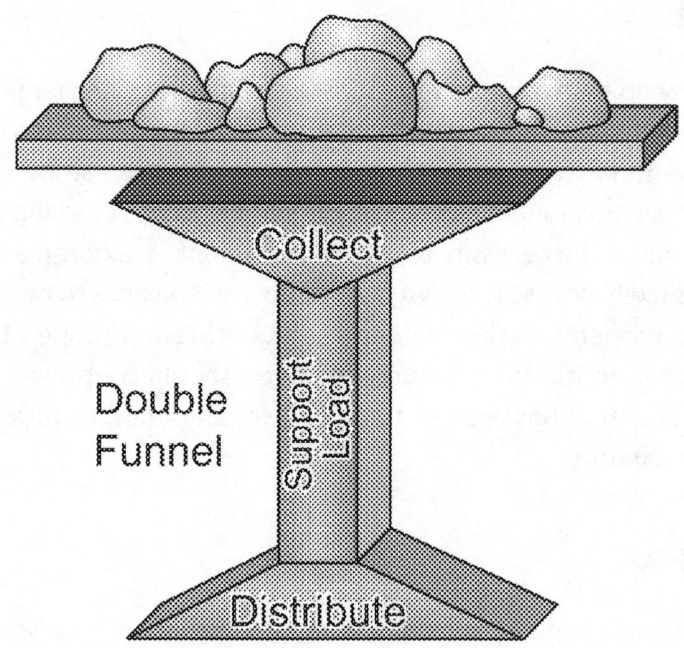

Fig. 1–1 Key shoring system principles.

Shoring Operations: Start-up Considerations

When responding to structural collapse situations, there are many things to consider: possible victims, fire problems, exposure problems, extent of the collapse, and cause of the collapse to name a few. Should rescue personnel need to enter the remains of the collapsed building, the safety of those people is paramount. Having rescuers become part of the problem instead of the solution is not acceptable. The stabilization of the structure for the protection of both victims and rescue personnel is a major concern; therefore, rescue-shoring operations are considered.

There are three main options to consider when determining whether to begin shoring the collapsed area: avoid, mitigate, or shore.

Avoid

Many times in collapse situations, especially involving larger structures, victims may be trapped in only one area; therefore, rescuers do not have to enter other unstable areas. Those areas can be blocked off, and all personnel kept out of the danger zone. Blocking off the areas negates the necessity of having to attempt an extensive shoring operation where rescue activity will not occur. The area to be avoided must be properly marked off, and safety officers stationed by it to block access to it. Also structural engineers should continually evaluate the area that has been sectioned off for any change in stability as a safety measure.

Mitigate

There will come many times when an area cannot be shored or when shoring an area will take critical time away from victim rescue operations; nevertheless, rescuers must work in or close to the area. In this case, rescuers must mitigate the situation by removing the hazard. For instance, when dealing with masonry walls or chimneys, generally it is easier and faster to remove the wall than to shore it. When the bond between mortar and bricks or blocks has separated from the joints and cracks, the entire wall section is in jeopardy of collapsing.

Unfortunately, when this situation develops, shoring an already unstable wall will take significant time and labor. Additionally, the shoring components can put pressure on the wall, creating more cracks that make the wall even more unstable. Therefore, it is much more efficient, safer, and less time-consuming to mitigate the problem by taking the wall down than to try to shore it up. However, if taking the wall down is not an option, then it must be shored up with extreme caution.

Shore

If the wall discussed in the previous section is to be shored up, there are several considerations to address. The first and foremost is the issue of the rescuers' capabilities and materials and equipment at their disposal.

- Do the current rescue personnel have the training and expertise to accomplish the mission?

- Is the necessary equipment available to do the work?

- Are the materials to accomplish the task at hand?

In order to answer the personnel questions, one must look at the type of training the rescue team has on a regular basis. How much technical training do they really do? Is it geared toward collapse, high-angle rope situations, confined space, trench rescue, or all four?

Although many teams have some basic knowledge and training in structural collapse operations, the majority of the time they don't have extensive training in collapse rescue shoring due to the relatively high cost of the training and the relatively infrequent need for its use. Many departments' check-writers don't like to spend money on specialty training like this. Unfortunately, without the proper training in this discipline, a team will have problems erecting safe and effective rescue shoring. If the team members have taken some shoring training but have not maintained the skills, they face the potential for grave danger in a difficult shoring operation.

Shoring is one discipline that requires the proper amount of training and confidence before attempting it or someone can get seriously hurt. Many basic shoring classes available today give students only some of the overall techniques. But when involved in heavily damaged structures, none of which are square, the rescuer must be able to adapt the shores to the environment. This is where experience and thorough knowledge of types of shoring and techniques are necessary. Knowing what will and what will not work for each given scenario is a must in order to ensure the safety of the operation, the rescuers, and the victims.

Shoring Size-up: Additional Information

Missing or damaged structural supports

Beams, columns, girders, and arches are all primary structural elements that support other structural elements. These items must be checked as soon as possible. The building elements they had been supporting may be under substantial additional stresses, possibly ready to fail at any time. Shoring of these items immediately resupports sections of floors that may be compromised in some fashion. It is much easier to replace or restabilize one item than to have to erect several shores. Consequently, examine these primary structural elements first before starting the shoring size-up. The shoring team must have good skills in building construction identification in order to quickly identify these particular structural supporting elements. Normally, the quickest way to safely resupport a damaged structure is to concentrate on these elements and the proper identification of what they are supporting.

Structural fire damage

Has the building sustained any previous fire damage? The structural stability of an area previously involved in a fire must be considered suspect at best. When a structure has collapsed or partially collapsed due to a fire, the safe operation in the structure depends on how badly damaged or compromised the remaining structural supports are. The shoring team must check for fire damage on the ends of the beams where the support will be and must look for any alligatoring of the lumber, which is a sure sign of loss of structural integrity of that lumber. Shoring may need to be performed throughout the damaged area if placement of equipment and personnel can be kept to a minimum in these areas.

Age of the structures

The building's age is another shoring size-up factor. The shrinkage and expansion of structural elements due to wet-dry cycles over time results in a loss of strength and the loosening of important hangers and connecting supports. If they have not been properly maintained, building elements such as wood and masonry may have dried out and become brittle and weaker with age. Older structures typically were built with lumber of larger dimensions than is in use today. This is a definite plus for shoring collapsed structures, especially those that were subject to fires. Engineering in the past was not as exacting as it is today. Larger material than necessary was used for safety reasons due to the possibility of construction errors made in the erection of and the fabrication of those structures. As a result, the older structures have what is known as *redundancy* in their construction. Basically, redundancy means that the structural elements could support much more than the design required. Therefore, when a catastrophic incident occurred, these elements tend to hold up much better than a newer, better-engineered building. So, this inadvertent situation ends up providing an advantage to the fire service.

Structure condition

The overall condition of the structure is another important part of the damage size-up. Obviously the condition of the building affects how much damage occurred and how much of the structure remains intact and or how stable the remaining section is. This is generally due to the fact that most of the joints are butted together at specific ends, leaving more surface area of those elements exposed to the weather. This usually results in these sections deteriorating first and, consequently, failing first. A structure that has received proper, continual maintenance and repair has a better chance of avoiding a collapse than a building that has fallen into a state of disrepair. The supporting elements of a well-maintained building may be used to help support and transfer the collapse load throughout the structure. However, if the building's condition is suspect, as it well may be in the case of a vacant building, the shoring team cannot assume there is sufficient structural support without first inspecting the building. The possibility of a secondary collapse in deteriorated buildings is a primary consideration.

Six-sided approach

All six sides of a structure—the top, the bottom, and all four sides—plus the area in which the shoring team is working or about to start working must be examined. To start a shoring operation, check each item or section to be shored up and determine its approximate weight. It is extremely important to survey the top of the structure to identify loose, shifting, or hanging debris or structural elements. These items may have to be moved, shored, or completely avoided, depending on the situation. The decisions made regarding these items affect the rescue operation. After the load that needs to be transferred has been calculated, rescuers must determine if the floor to receive the load will be able to hold it. If it cannot, another floor must be considered. It may take several floors to hold the weight; or in smaller structures, the load may have to be directed to the basement for initial support. Gravity is constantly working to pull the building's remains to the earth; so from a shoring standpoint, the bottom survey where the shoring will be supported generally by the ground or a substantial section of floor is extremely crucial. It should be performed simultaneously with the survey of the upper floors if possible.

Out-of-plumb walls

Immediately upon arriving at the collapse site, the shoring team should determine the building's stability by examining the walls in the collapse area to see if they are leaning. If they are partition or nonbearing walls, determine if they affect collapse operations in any way. If they are leaning or otherwise showing signs of stress, they must be shored. Walls were designed to accept the loads in an axial position for maximum stability. If the walls are out of plumb (not level in the vertical position), the chances are very good they are loaded eccentrically. This may cause a wall to be overloaded in one area or become unbalanced, both conditions potentially leading to a problem. Another issue is the fact that floor beams generally sit on top of bearing walls only a few inches. When one of these walls shifts out of plumb, only the 2–3 in. remaining on the floor joists may not be enough to keep the floor intact. If walls are observed with the naked eye to be leaning, it is a very good idea to determine how much of the upstairs floor beams remain on that wall.

Strained/stressed structural elements

Floor beams and other structural supporting elements under the main debris pile or under a victim's location are among the top priorities of sight assessment. Older, more substantial beams can withstand large amounts of stress; however, when they are stressed to their limit, very little weight is needed for them to reach their failure point. Any severely bellied, stressed, or cracked beams must be shored up before rescue personnel commit to the operation. The assessment team must examine all the structural elements affected by the collapse and look for any deflection in them. If they are strained, they must be resupported as soon as possible. When they have been strained, they lose a significant amount of their strength, a situation that must be addressed immediately. Beams that are stressed will go back into shape and still have their strength. Beams that have been strained will not revert to their normal shape after the load is removed. It is difficult for the structural assessment team to determine whether a beam will return to its original shape at the scene of the collapse.

Types of construction material

The type of materials used in the construction of a building's structural elements has a large bearing on the size and extent of the shoring material needed. If the building is made of unreinforced masonry (URM), one size may be needed; but if the building's elements are made of steel or concrete, heavier shoring material is required. Typically, the two items that determine the size and strength of rescue shoring are the weight of the building material itself and the weight of the materials plus the items inside the structure. For this reason, as soon as possible when conducting the shoring size-up, the shoring team must be able to determine the type of structure and the size and type of building materials involved. Knowing the size and weight of the structural material provides the information needed to determine the size of shoring materials needed for the operation. The bigger and heavier the structural elements are, the bigger and heavier the shoring material must be.

Another issue to consider is the amount of weight the building is supporting, complete with its contents. For this reason, it is important to identify the type of tenant(s) occupying the building. It is important to know the weights of the most common building materials and the strengths of the shoring materials you will be utilizing to stabilize that structure. To be safe, the strength of your shoring materials must surpass the weight of the materials to be stabilized. Wood and light masonry may need 4x4s; concrete and steel may need 6x6s and larger, depending on the size of the structure. It is also imperative to identify what was housed in the building because its weight must be included in the load calculation.

Types of beams

The types of beams commonly in use today are the simple, continuous, cantilever, propped, and fixed beams; each type is supported in a particular way. When the support points for these beams are compromised, the beams must be immediately evaluated and resupported. Quickly identifying a type of beam makes it easy to find the critical support points for the beams and to replace or resupport them if necessary. Thorough knowledge of building elements and their functions is imperative to determining how the beams are loaded and how the beams are supposed to be loaded. The stress, strain, compression, and tension on the beams have to be analyzed properly. The relief of the overloads and the redirection of these loads is the primary objective of shoring these beams. This action helps stabilize the remaining structure.

Types of floor construction

Again, the type and size of the material used to construct a floor determines the size and amount of shoring lumber and the type of systems needed to safely support the damage. If the floor beams are spaced 12 in., 16 in., 24 in. or more on center (from the center of one beam to the center of the other beam), the shoring material must be spaced at these same intervals. The size of the floor beams—2 in., 4 in., or larger dimension—bar joist, or trusses of concrete or steel, determines the size of shoring lumber needed for proper support. The type of flooring itself also is part of the equation. The makeup of the floor and its thickness are, for the purposes of this book, the determining factors in assessing the floor's weight. For wood flooring, the main weight factor is wood thickness. Typically, wood-flooring systems can weigh roughly 25–35 lb per sq ft. Estimating the weight of concrete and steel floors is more difficult. The weights of concrete floors vary with the thickness of the material, the type of concrete, and the amount and size of the reinforcing steel embedded in the concrete. It is helpful to consult the structural drawings of the building or to closely examine the floor sections themselves.

Proper beam connections

Another one of the more important size-up points is checking all the beam joints and connections. After the stress and strain that a structural collapse has imposed on the rest of the building, all the connections in the area affected must be examined and checked for continuity. The physical connections themselves and their supports also must be checked. Items to look for are the stability of the supports and of the anchors connecting the beams to those supports. Examining these items does not take a major effort and must be done as soon as possible. This is one of the first items that must be checked to ensure the safety of the rescue forces.

Door and window access

At the majority of structural collapses, access to the structure is hampered by debris and possibly by dangerously hanging materials. Access may be limited to windows or side and back doors whose size may limit the shoring team's access for tools and materials. Any time an existing opening is used as an access or egress way, it must be closely examined for structural defects and instability and any problems resolved before rescue teams can continually utilize the opening safely. Some mitigation of hazards may have to be done or some shoring may have to be erected. When utilizing these areas as access ways, it is necessary to constantly monitor their stability throughout the entire operation period.

Out-of-square door and window frames

An out-of-square door or window indicates to the rescue team that major structural movement has occurred. Some of the causes of this are racking of one or more walls in the structure, in a major wall, or possibly in the foundation. Whatever the reason, corrective action must be taken immediately. Generally stabilization of the walls and large sections of the affected floors helps; however, each collapse is different, and the exact cause of the problem must be identified in order to properly correct the situation. Because window and door openings are the weakest parts of a wall, structural movement occurs at or around their location first. Placing bracing and shoring in the opening restabilizes this weak area. Diagonal bracing has been used successfully on many previous occasions to help prevent a structure from racking any further. This should be one of the first options considered.

Another indicator of imminent structural collapse is a swinging door. A swinging door means that the structure has drastically shifted or settled and that some sort of structural element has failed or will fail in a very short time. The rescue teams should exit the building and reevaluate the stability of the structure before entering again to perform rescue or shoring operations.

Sagging floors and roofs

Frequently, this condition is due to overloading of the floors or roof from any number of sources. Roofs are normally overloaded by either weather conditions, such as snow or ice, or by large objects, e.g., water tanks and heating, ventilating, and air conditioning

(HVAC) equipment. As with any situation, the size-up must include examining all areas in and around the collapsed structure as well as other areas around the structure that may have contributed to the collapse. Floors sag from the weight of collapsed debris and furnishings. A sagging floor means the beams are excessively overloaded, potentially resulting in further collapse at any time. The beams must be shored, and the debris removed from the floors if feasible.

Out-of-plumb columns

Generally, columns are supporting beams, columns, or girders. If they are out of plumb, their weight-bearing capacity is diminished. The more out of plumb they are, the less effective their support strength will be. Many times the columns are under the joint of a girder or a set of beams. If this is the case, it is important that the column remains plumb because there are normally only a few inches of beam bearing on the columns. If the column gets knocked out of plumb, the bearing of one or both of the beams will be compromised, possibly causing a structural collapse.

Another important column-related issue to address is whether there is any *belly*, or *deflection* in the column. This condition also drastically diminishes the strength and stability of a column. The items the columns are still supporting must be secured even to the extent of shoring around the entire column. It is very important that these items are checked as soon as possible. The longer and thinner the column, the faster and easier it deflects, causing structural problems. Damaged columns with no deflection that are still bearing their loads must be watched closely. The damaged areas are the first place the column will fail.

Framed or unframed structure

Knowing if a building is framed or unframed helps to determine its general structure and to identify the load-bearing elements, information used to properly size-up additional collapse potential and shoring operations. In a framed structure, a skeletal-like system supports the building and the walls. Collapses are generally more localized and less extensive than those of an unframed structure. In an unframed structure, the exterior walls are the bearing walls for the structure. If a lower section of one of these walls fails, everything above it may collapse. In a collapse of an unframed structure, the damage may be more extensive than that of a framed structure.

Access to the structure

Access to the entire structure may be extremely difficult due to massive debris build-up or the danger of the remaining structure falling on rescue forces. This has to be evaluated before operations start. Bringing in tools, lumber, and equipment for the safe removal of trapped victims or for the purpose of rescue shoring may be a problem if the access space is too small or damaged. There will be many occasions in which the primary access, usually the front door, is blocked with debris or is too heavily damaged to use. If this is the case, the safest and easiest point of access should be used. Your initial point of entry is generally through the front of the structure. However, you may have to change the staging area for all your equipment so that you can get the materials you need closer to the point of use.

Bulging walls

Identifying any bulged or heavily damaged walls is very important. If the bulge is in a bearing wall, the wall may be compromised and can fail at any moment. It is necessary to determine the total extent of the damage and the amount that the wall is bulged. As with any structural component, if a wall is not loaded through its axis, it can become unstable. Masonry walls are especially susceptible to instability due to the nature of the material itself. The main shoring operation that can take place in this instance is the shoring, or stabilization, of the floors that these walls support. Normally, that entails the erection of vertical shores under the floor beams, effectively replacing the damaged wall. With exterior walls, raker shoring may have to be erected in order to stabilize the wall section itself. This keeps the bulged wall from falling and causing a secondary collapse situation. The definite possibility of having to shore and stabilize these compromised walls must be considered.

Cracked walls

As in the case of bulging walls, walls must be examined for cracks that can indicate foundation failure and wall compromise. However, even if there are cracks in the wall masonry, the wall may not be structurally compromised to the point of failure. If, for instance, a wall that is 10 ft high and 30 ft long has a hairline crack that is 3 ft long, it does not necessarily have a structural integrity problem. Small cracks like this are not uncommon in masonry construction. However, a much larger or longer crack that has opened a space in the masonry indicates a potential problem. Another indicator of a structural problem is a traveling crack and an *X*-type crack in a wall. The *X* crack suggests that there has been movement in two separate directions—a definite problem. The source of the damage must be identified and resolved.

Separating walls

A separating wall occurs when a building starts to twist, and the walls spread apart. A check of the joints at the corners can quickly show if major movement has occurred. As the structure starts to rack, the interior walls begin to pull apart. In most cases, this situation is easy to spot because the tops of the corners of the walls are peeling apart. One approach to determining if a wall separation exists is to look at the corners from the doorway as you enter a room.

Vibration potential

Another concern to address when conducting a shoring size-up is vibration potential. By eliminating these sources of vibration and by checking all joints, connections, and precariously hanging structural members, the shoring team can make the collapse area safer. It is important to remember that everything rescue workers do has the potential to create some sort of vibration. Every tool in the collapse rescue arsenal causes vibrations when in use. Therefore, it is crucial to be aware of any adverse reactions the tools may cause.

Trusses

The problem for the fire service that has developed over the use of these highly engineered items is well known. If any part of the truss fails, the entire truss fails. The collapse usually occurs very suddenly. When and if you have to shore a truss, you must always shore the top chord. If the bottom chord is shored and one of the shores elements fails, there still may be failure of the remaining part of the truss. When the top chord is shored, the weight above is supported and held in position, safely stabilizing it.

Types of void access

If there are numerous voids with victims trapped in them, some type of shoring lumber is needed. Generally, cribbing size lumber roughly 24 in. long works well. In some collapses, much larger voids may exist. In these situations, longer lumber sizes are needed; therefore, you must provide an access large enough to accommodate the larger materials.

Bearing wall stability

The most important structural elements in any unframed building are its bearing walls. They support the majority of the structure's weight and any loads in it. In a collapse situation, failure of any part of any of these walls can cause extensive damage and further collapse. The walls must be checked for the presence of the anomalies previously described in this chapter: bulges, bellies, cracks, leaning, or any type of possible deflection or abnormal deformity. Equally important is determining if there are any sections of the wall damaged or missing. If a wall is damaged, it may no longer have its full load-bearing capacity, making that area a weak point in the structure. If a section of wall is missing, additional stress is being applied to the floor beams above and to the adjoining, remaining sections of wall. This is a very dangerous situation. If there is any real or suspected structural instability, the shoring officer must decide where and how much to shore. The foundation should also be checked if you do find any of the previously described problems.

Rules of thumb

These are a few general rules of thumb that can easily and quickly be applied on the rescue site. Even though they hold true on most occasions, the rescuers must bear in mind that each structural collapse situation is unique.

There are four basic rules of thumb to keep in mind when using existing floors in the damaged structure for support of unstable walls, debris, or other floors.

- It takes one undamaged, wood-framed floor to support one damaged, wood-framed floor.

- It takes one undamaged, steel-framed floor to support one damaged, steel-framed floor.

- It takes two undamaged, reinforced concrete floors to support one damaged, concrete floor.

- The thickness of any debris on the damaged floors must also be taken into account when calculating the amount and type of shoring needed.

The length-to-diameter ratio for all shoring material should be no more than 50 times the diameter and ideally should be in the range of 25–35 times if at all possible.

Lumber

It is critical to the collapse rescue response to have quick access to lumber for shoring or cribbing operations. One way of accomplishing this is by contacting a local lumberyard and prearranging a quick delivery of specified lengths, sizes, and types of lumber. Another option is to contact the local department of public works or a large local contractor who has shoring materials on hand and arrange for the needed materials. By preplanning for lumber needs, the shoring team will have prompt access to the type of lumber needed for shoring operations. However, the easiest and only way a team can be positive that the needed materials will be delivered to the sight is by bringing them there. There are many variables involved when depending on others. For instance, what if the collapse operation occurs after working hours? How will the lumberyard supply the need? Or what if a catastrophe occurs on a lumber company's busy day, and the company has no trucks available to deliver the materials? These problems could cause unnecessary delays that can result in tragedy.

Types of lumber

There are several types of lumber available to a rescue team that the majority of lumberyards stock. The following is a list of some of the more common types utilized in the construction industry and should be readily available:

Hardwoods and softwoods. There are two major categories of wood: hardwoods and softwoods. These names really tell more about the type of tree the lumber comes from than the wood itself. For example, balsa wood, the easily cut and lightweight wood many children play with as airplane models, comes from a hardwood tree. These names do not necessarily mean that hardwoods are hard or that softwoods are soft.

Hardwoods come from broad leaf trees that lose their leaves during the winter months. The wood is generally heavy, close-grained, generally expensive, and not well suited to shoring projects. Oak and maples are two examples of the common types available.

Oak. A tough, hard, coarse-textured, high-density wood that is native to temperate climates, oak is used for both structural and decorative applications, framing timbers, flooring, molding, and plywood. The two most common varieties are the white and red oak.

Softwoods come from trees with needle-like or scale-like leaves that stay on the tree all year, for example, Christmas trees. The most popular species of softwoods are Douglas fir, western hemlock, white fir, and spruce. Pound for pound, Douglas fir is one of the strongest woods available. It resists warping, cupping, and twisting and is normally available at lumberyards throughout the country. For these reasons, Douglas fir is the lumber chosen for use by collapse rescue teams. It is strong, readily available, not expensive, and stores well—ideal for shoring use.

Yellow pine. Yellow fir is a strong, medium-density, medium-to-course-textured softwood. It is widely used for plywood and dimensional lumber and timber in a variety of building construction situations. It is well suited to rescue operations; however, in real situations, rescue teams use whatever they can find in the shortest amount of time. However, they must bear in mind that different types of lumber may have lower supporting strengths.

Each piece of lumber delivered by a reputable mill should have a grade stamp. This stamp is to certify that the piece of material meets quality control standards set by the lumber grading associations. The grades to look for are No. 1, No. 2, stud grade, and construction grade. Utility grade should not be used; it may not be strong enough in some situations.

Lumber storage

One of the biggest problems shoring rescue teams encounter is where to store lumber supplies since they are not used everyday. The biggest enemy to storage is moisture. Lumber materials must be stored in a dry, well-ventilated area if they are to last. Moisture can be taken out of lumber by two accepted methods, kiln drying and air drying. Both methods produce quality seasoned lumber. Seasoned lumber means that the moisture content of the lumber is normally 19% or less.

Lumber having higher moisture content is called green lumber. The average shrinkage of a Douglas fir structural member from green to kiln dried is approximately 7.6% in width and 4.1% in thickness. This adds up to more than a ½ in.-reduction in width for a 2x12 member. You should be aware of the changes that occur when the moisture content in the lumber changes. If green lumber that dries too quickly is used for shoring, checks, cracks, and splits develop. Green, unseasoned lumber, especially when improperly stored, can also warp, twist, and shrink. For this reason, shoring rescue teams should avoid using it. Checks are the separations in wood that normally occur across or through the rings of a tree's annual growth. They are usually the result of seasoning and generally occur at the ends of the lumber. Splits, separations of the wood, occur when the wood cells rip or tear apart.

Use of existing lumber

In the majority of collapse situations, the tendency of many teams is to grab any available material and go to work. If a team does not carry its own lumber, the firefighter uses whatever is at hand. This leaves only the material that was involved in the construction of the collapsed structure. Evaluation of any material to be used must be made before team personnel commit to utilizing it for rescue operations.

There are several areas that should be looked into before using existing lumber for any rescue shoring. Some of these areas are age, type, condition, and the amount of stress to which the material has been subjected. The age of the building is a good indication of the condition of the lumber. In newer buildings, the lumber should be in good shape; much older structures may have weathered material that can be fatigued and unsuitable for reuse.

The type and size of the lumber also are issues. Smaller lumber such as 2x4s cannot be used as main bearing members without nailing them together. Cedar and redwood decks cannot be torn up and used for structural bearing members because the wood is too soft. The single most important determining factor in the decision whether to use lumber from the collapsed building is its condition. The wood must be thoroughly examined. If the lumber is too dry or brittle, it will split and crack easily and will not stand up under any type of stress. If it shows signs of rot or is extremely wet, its strength will also be suspect and, therefore, the wood cannot be used.

Finally, examine the lumber for any twists, bows, cracks, or splitting. If any of these conditions are present, the lumber should not be utilized for rescue operations.

Common lumber sizes used

The following is a list of lumber that, if at all possible, should be carried on the collapse apparatus. Some of the more common uses for each are also listed.

2x4–This size lumber can be used in box cribbing, as diagonal bracing for interior rake shores, cross-bracing for laced posts, interior and exterior raker shores, various size cleats, filler blocks and diagonal bracing for the vertical shore.

2x6–This size can be used for diagonal wall braces, diagonal braces for the vertical shore, interior and exterior raker shores, box cribbing, cleats for raker shores, cross-bracing and horizontal bracing for raker shoring, horizontal struts for the split-sole raker and for the flying raker shore.

2x8–Although not commonly used, 2x8 lumber is excellent for sleepers or mudsills when shoring is being erected on soft ground. It can also be used for diagonal wall braces.

4x4–The most common size of shoring lumber is the 4x4. It is used for box cribbing, T-shore, window shore, door shore, laced posts, vertical shore, horizontal shore, and interior and exterior raker shores.

4x6–Generally used in larger buildings or in buildings needing substantial holding power, the 4x6 can be used as door and window shores if heavy loads are anticipated or as interior or exterior raker shores.

6x6–This size is normally used in heavy constructed buildings where the loads are great such as an all-concrete or concrete and steel structure. It is good for use as box cribbing, raker shores, vertical shores, laced post shores, and as return blocking for a series of raker shores.

3/4-in. plywood–This plywood can be used for numerous items—gusset plates, cribbing spacers, wall plating for raker shores, in trench rescue, as work platforms and saw horses.

Length-to-diameter Ratio

One of the most critical areas that must be addressed when doing a shoring size-up is the amount of weight to be supported. The main supporting elements in most shoring are the posts or struts. These may need additional support or tensioning accomplished by the use of lateral bracing. Calculating the length-to-diameter (LD) ratio of the posts is a way of determining this.

LD Ratio

The length-to-diameter ratio of all shoring material is critical—the strength of shores depends on keeping it within accepted limits

Basically, Euler's law of columns comes into play with all shoring systems. Leonhard Euler, a Swiss mathematician who lived in the 1700s, proved that a thin strut or column submitted to an axial compressive load does not remain straight. It bends out suddenly or buckles at a specific value of the compressive load called its *critical value*. Since any element that comes under compression acts as a column (whether vertical, diagonal, or horizontal), all shoring systems come under this condition. Limiting the length of posts or struts keeps the strength at the most efficient levels. That is one of the reasons to brace shoring systems. The longer and thinner the element is, the less weight it can support. By center bracing or lateral bracing these elements, shoring teams get the best use of the strength of the lumber. In most cases, the ratio of the posts should be kept at a maximum of 50-to-1.

For example, a 4x4 actually measures 3.5 in. If you multiply 3.5 in. by a factor of 50, the answer is 175 in., which equates to roughly 14 ft. This would be the maximum length to use to shore up something lightweight. In today's typical buildings in this country, shore rescuers want to keep the ratio closer to 25 in order to keep the lumber's shoring capabilities near their maximum strength. If you multiply the same 4x4 by a factor of 25, we use 8 ft or 96 in. as a benchmark, which is the preferred length to use as a guideline for lateral bracing of shoring systems. Of course, since each collapse situation is unique, the main consideration in determining the length-to-diameter ratio of shores is the amount of weight the shore needs to support. As a rule of thumb, no more than 50-to-1 and, ideally, roughly 25-to-1 is the most efficient use of the shoring material strength.

LD Ratio

Maximum— *50 times*

Ideal— *25 times*

Nails

Nails are classified according to their use and form. They are designated by the term *penny*, which for the purposes of this book, is abbreviated by the letter *d*. The term penny came from the market places of the 1400s; a penny was the price of 100 of a particular size nail. Nowadays, the term refers to the length of the nail regardless of the wire gauge. Nails come in various sizes from 2d up to 60d, or from 1 to 6 in. in length. The most common type of nail shoring teams encounter is fabricated from steel wire.

A nail should be at least three times as long as the thickness of the lumber it is to hold. Two-thirds of the length of the nail should pass into the second piece of lumber. The nails can be driven at a slight angle toward each other to maximize hold and to keep them from pulling apart. There are several types of nails in use today—finish, box, common, galvanized, threaded, ring shanked, duplex, and resin coated, to name a few. Rescue teams most frequently use the common nail for rescue situations.

In training scenarios, it is a good idea to use the duplex head nails. The best technique is to drive the nail to the first head. For ease of removal, the second head stands up out of the wood and can easily be pulled using a crowbar, nail puller, or steel-handle hammer. When you use these nails, the lumber is not as severely damaged when it is taken apart and can be reused several times, making your training economically feasible. The nail sizes normally used, unless an engineer specifies another type, are 16d for nailing all dimensional lumber together and 8d nails for nailing plywood gusset plates or toe nailing wedges.

Pneumatic nails for use in pneumatic powered nailers come either in strips or coils, depending on the type of nailer used. Most of these nails come with a resin coating for better holding power; however, one item that a rescue team should insist upon is the use of full head nails. Some of the nailers use nails with one-half or three-fourths of the nail head. In rescue situations, the nails should have their full heads for better holding power. The rescue team can decide which type of nail it wishes to use.

Nail patterns

In order to get the proper holding power with the use of these common nails, certain nailing patterns should be followed. The most important application requiring the nailing pattern is the fastening of plywood gusset plates and 2x4 or 2x6 cleats. In the construction of raker shores, the nail patterns on these cleats and gussets are very important, as it is the number of nails that provides the proper holding power. In the use of plywood gusset plates that are normally 3/4 in. thick, 8d nails are generally sufficient. In the 2x4 or 2x6 cleats used on the raker shores, 16d nails should be used. Duplex, or *double-header nails* as they are sometimes called, can be used for training purposes. These nails are easily extracted from the material with little damage, enabling the rescue team to utilize the same lumber over again, thereby making hands-on training more economically feasible.

When utilizing a 2-ft cleat made of 2x4s, the shoring team needs (17) 16d nails. They can be staggered or nailed in a 5-nail pattern that delivers 17 nails. When using a 2x4 cleat that is 3 ft long for raker shore angles more than 45°, the team needs (26) 16d nails to have the proper holding power required to hold the rake from sliding up the wall plate. When using a 2 ft-long 2x6 for a cleat on a 4x6 or 6x6 raker shore, the shore rescue team needs (26) 16d nails. The boards can be nailed in a staggered pattern of three rows. Using the same nailing patter, (38) 16d nails are needed to nail the same width cleat that is 3 ft long.

Wedges

The proper use of wedges is one of the more important factors to consider for a successful collapse rescue operation. Wedges are normally used in pairs; and when properly joined or married together, they are excellent tools for filling gaps and transferring collapse loads. They are easily adjustable and can be tightened just enough to transfer loads without lifting them. Moving unstable loads can have serious consequences in a collapse operation. It is imperative that all personnel are properly trained in the correct usage of wedges. This may sound trivial, but it is extremely important.

The use of wedges is extensive in all types of shoring operations—interior, exterior, as well any type of void shoring and stabilization.

A good wedge that fits properly and marries together snugly is one in which the length of the wedge is only five or six times as long as its thickness. Wedges constructed with too sharp an angle do not hold properly and can easily slip out. The width of the wedges should be the same thickness as that of the materials being supported by those wedges. This makes for a much smoother operation.

Wedges that are too large hinder the installation of any bracing, and too small wedges may make the shore slightly unstable. The following is a list of some of the more popular size wedges to construct and use:

2 in. high x 3½ in. wide and 12 in. long

1½ in. high x 3½ in. wide and 9 in. long or 12 in. long

1½ in. high x 3½ in. wide and 12 in. long

3½ in. high x 3½ in. wide and 18 in. long

3½ in. high x 3½ in. wide and 24 in. long

3½ in. high x 5½ in. wide and 18 in. long

The wedges can be premade and carried on the rescue apparatus, or they can be cut in the field. Even though it takes some time to cut the wedges, it is advisable to have a preset cache of wedges on hand. The lumber can be cut with a small chain saw or with a 10¼-in. circular saw. This latter size saw is required to cut a 2x4 or 4x4 in one pass. Using a smaller diameter circular saw requires two passes. Frequently, the cuts do not line up, rendering the wedges almost useless, as they do not fit together properly, nor do they tighten up sufficiently to be effective.

The Use and Training of Rescue Personnel in Shoring Operations

The Shoring Team

The installation of rescue shoring should always operate according to a team concept. This concept works very well in the stressful and confusing situation that results from a serious structural collapse. To keep the scope of supervision to a reasonable size, the team consists of six firefighters supervised by an officer. The team's primary function is to erect the specific shores designated by the shoring officer in conjunction with the structural engineer. As with any rescue team, its scope of function may suddenly change if an unforeseen situation develops.

Each firefighter has a specific assignment and function to perform called a role. Once each team member receives an assignment, he or she knows the related tool assignment and work responsibilities. The faster the members can ready themselves for their functions, the quicker the shores can be erected. Although we try not to rush things, speed is essential for the swift completion of the rescue shores needed to stabilize the remains of a building. A well-trained team can erect shores in a relatively quick fashion; the smoother the operation goes, the faster the shores go up.

The standard team consists of the following positions:

- Shoring team officer

- Measuring firefighter

- Shoring firefighter

- Layout firefighter

- Cutting firefighter

- Tool and equipment firefighter

In most cases when using the team concept, you can break the team up into two squads: the shore assembly squad and the cutting squad in order to concentrate the team's efforts. The assembly squad erects the shores, and the cutting squad gathers and cuts the materials then supplies them to the assembly squad (also called the shoring squad). These are, of course, individual guidelines.

It must be pointed out that team assignments must remain flexible. As the rescue progresses, many things can change, sometimes at a moment's notice. All rescue personnel must recognize that their job assignments could change at anytime throughout the operation.

The Shoring Team

Shoring Squad	**Cutting Squad**
• *Shoring officer*	• *Layout FF*
• *Measuring FF*	• *Cutting FF*
• *Shoring FF*	• *Tool and equip FF*

Shoring (assembly) squad

Ideally, the shore assembly squad consists of the following fire-fighters:

- One officer

- One measuring firefighter

- One shoring firefighter

In most instances, these three firefighters will be able to effectively erect the required number of shores.

Initially one shoring squad should start working in a good (or safe) area then progress into the bad (or damaged) area. The officer directs the two men and makes sure they receive the supplies they need quickly. The squad erects one shore at a time, always working from a safe area. The men are replaced as deemed necessary by the shoring officer.

The structural engineer and the team's shoring officer should establish a plan for erecting the shores in a specific order. Generally, the most critical areas are stabilized first. The squad estimates its tool and lumber needs before beginning work so that the other personnel have time to gather the necessary materials and have them deployed for use.

Shoring officer. In charge of both the cutting team and the shore assembly team, the shoring officer has full responsibility for the shoring operation. The shoring officer performs a constant size-up and makes decisions based on experience and the fire department's shoring guidelines. The shoring size-up does not end until all shoring is installed and secured. To be in direct charge of the crew at all times, the officer's role must be flexible because he may have to operate in different locations at different times to coordinate the operation properly. The shoring officer must stay one step ahead of the team. When a series of shores is being erected, the shoring officer must ensure that the area is clear for each consecutive shore and that the lumber necessary to erect the required number of shores is available.

The shoring officer must also consult with the structural specialist on the scene. The two positions must be in full agreement as to the size, design, and placement of any shoring systems to be erected. The officer is also responsible for ensuring that all shoring conforms to accepted practices and that they are properly secured together and anchored to the structure. Consulting the structural specialist is helpful to carrying out that responsibility.

Another one of the responsibilities of the shoring officer is to select personnel for each team role. At the response to the incident—preferably on the way to the collapse—the officer selects team positions. In order to do that properly, the officer must have thorough knowledge of the abilities of all the personnel under his or her command and know the frequency and extent of their training. He must also be cognizant of the technical background of team members, such as skills acquired outside the fire service. These skills might be experience obtained as construction workers, carpenters, mechanics, engineers, or any related field that would give the firefighters a decided advantage when needed to construct the rescue shoring. Personnel with the best carpentry skills should be utilized for the measuring and erecting of the shores. The firefighters laying out and cutting the shoring materials must also have some background in the handling of tools and equipment used to do the work.

Based on this information, the officer decides which firefighter to assign to what role. Some team members may feel more comfortable in some positions than in others. For example, someone may not be proficient in the use of various types of cutting saws, such as the chain saw or the 10¼-in. circular saw. If not, the shoring officer should not force those firefighters to use them. The shoring officer should discuss project assignments with team members and make sure they are willing to accept their assignments and can do the work properly and safely.

In conjunction with the team, the shoring officer decides where the tool and cutting station should be located. A few of the issues they must consider are the safety of the set-up area, its size, and access to the tools and materials. One rule that must be enforced absolutely is that the cutting station must be set up outside the secondary collapse zone.

The officer also supervises the step-by-step building of each shore, whether a single shore or numerous shores are being erected. Additionally, the officer must always be accessible in order to answer questions from any of the rescue personnel. If, and only if, the structural specialist has been trained in rescue shoring techniques, the shoring officer may enlist the specialist's help as necessary. Remember, the officer must stay several steps ahead of the shore assembly squad, or the operation will slow down, causing problems and delays.

If possible before shoring installation begins, the officer should ensure that the area where all the shores are to be erected is cleared and prepared. Firefighters other than those on the erecting squad can assemble tools and materials. In this way, your trained and experienced personnel can be properly utilized for the technical tasks necessary for the safe completion of the shoring operation. The shoring officer must communicate to the incident commander the need for additional personnel if they are needed.

It is the shoring officer's duty to provide the proper relief to personnel on a regular basis because a tired firefighter will get hurt. At the first signs of fatigue, he must replace rescue personnel. The officer must also be able to determine if the use of more manpower is needed or justified. If more than one shoring project needs to be conducted simultaneously and the multiple teams can work safely without impacting other operations, the shoring officer can make a request to the incident commander for another shoring squad to be placed in service.

At all times, the officer must make sure that a safe means of access and egress is available to the crew in case of any unforeseen problems. An access/egress size must be at least 4–6 ft wide and free of all obstructions in case team members must exit rapidly from the area. Providing and maintaining this opening is one of the officer's primary concerns, for this clear passageway can rapidly become cluttered with tools, materials, and workers.

Measuring firefighter. The measuring firefighter leads the two-member shore assembly team and is in direct contact with the layout firefighter on a secondary radio channel (see the layout firefighter section later in this chapter for a description of that role). After confirming with the officer the exact location and type of shoring to be constructed, the measuring firefighter takes all the measurements needed for the shoring. Using a portable radio, the measuring firefighter relays all measurements to the layout firefighter who measures the material at the cutting site.

In performing the measuring function, the measuring firefighter must take into account the structure that needs to be supported, the space available in which to do the work, and whether wedges are to be used. Any space that must be deducted from posts or struts for wedges should be deducted before the lumber sizes are relayed to the layout firefighter. The number that the measuring firefighter calls to the cutting crew is the size of material that the shoring team will receive.

The cutting team should do no deductions or subtractions of material sizes. Sticking to this rule avoids confusion between the measuring and cutting teams. For this reason, the measuring firefighter must plan carefully and double-check measurements before calling out the numbers to the layout firefighter. As a rule of thumb, the measuring firefighter deducts the thickness of one wedge from the posts or struts to be used. Doing so provides room to fit the set of wedges properly during the final shore adjustments.

It is important for the measuring firefighter to write down all the measurements he takes before sending them to the cutting squad. Doing so helps keep confusion to a minimum—by the way, confusion always occurs. Another very important procedure is to call out the measurement that you want then repeat the information. Also, to make sure there is no confusion, ask the layout firefighter to repeat the information.

To make things easy, many times it's a good idea to draw a simple picture of the shore to be erected, mark the measurements on the picture; one of the firefighters, usually the runner, can take it to the cutting squad. This can then be utilized as a reference point if any confusion arises and is especially helpful when numerous shores are being erected. To keep things less confusing, the measuring firefighter should call out all measurements in inches—experience is that this leads to fewer problems.

When the measuring is complete and the shoring materials have been deployed, the measurer's job is to assist the shoring firefighter with assembling the shores.

Shoring firefighter. The third member of the shoring team is the shoring firefighter. One of that role's main functions is to prepare the area to be shored by clearing away debris and other obstructions and leveling the area. The area should be at least 3 ft wide and 3 ft longer than the shore itself in order to have enough space to install the shore and to adjust it if necessary. Any debris should be cleared down to either floor level or, working in the basement, to ground level. It is always a good idea to bring a shovel into the work area, preferably a square-faced, small D-handled type because it is easy to manipulate it in tight areas.

Frequently, the shoring firefighter also helps the measuring firefighter take the proper measurements—usually limited to holding the dummy end of the tape measure. When the shoring material is delivered to a work area, the shoring firefighter is the one who nails the material. Therefore, before materials are brought in, it is important for the shoring firefighter to ensure that there is a sufficient quantity of nails to do the job at hand. Making sure that the proper number of wedges, gusset plates, and hand tools needed to assemble the shores are available is also the responsibility of the shoring firefighter. This verification must be taken care of while the measuring firefighter is relaying his information to the cutting station.

The cutting squad

The cutting squad consists of three firefighters: the layout firefighter, the cutting firefighter, and the tool and equipment firefighter. The first responsibility of these firefighters is to secure an area as close as possible to the collapse operation (but outside the collapse danger zone) so as to minimize the number of personnel needed to relay the materials to the shoring team. This area should be determined in conjunction with the officer in command and the shoring officer.

It is not unusual for several companies to be employed moving lumber and tools to the collapse area. As a matter of course, plan on this being the case. The squad must clear an area of debris large enough to accommodate the tools, equipment, and lumber needed for the particular operation. Although each operation is unique, any shoring operation needs a specific number of tools and equipment whether it constructs one shore or ten shores. An area 12 ft wide by 24 ft long would not be considered large. In fact, for safety reasons, this size is considered a minimum size. An area that size allows enough space to place the necessary layout and cutting tools, to set up a cutting station, and to place some lumber in position to be cut and marked. For safety reasons, the cutting station should be plainly marked, and a minimum number of personnel allowed to operate in the area.

These are some of the basic tools that are necessary to get the cutting station started and in full operation:

- 16–25-ft tape measure with a 1 in.-thick blade

- Carpenters pencils, markers, lumber crayon

- Speed square and framing square

- Utility knives

- Gas chain saw, electric chain saw, 10¼-in. circular saw

- Chalk line and straight edge tool

- Saw horses

- Premade angle templates

- Power supply for electric tools

- Lights if necessary

Layout firefighter. The layout firefighter is in charge of setting up the cutting station and preparing the materials to be cut. He is the lead firefighter of the cutting squad and stays in contact with the shoring squad at all times. The layout firefighter notifies the equipment firefighter of the sizes, lengths, and amount of lumber needed, based on the information he receives from the measuring firefighter (member of the shore assembly squad). Being in direct contact with the measuring firefighter helps reduce the possibility of miscommunication. Generally the layout firefighter is on the same radio channel as the measuring firefighter. If he has to, the layout firefighter can contact the shoring officer by switching to the primary radio channel. The layout firefighter measures the lumber, marks it for cutting, and lays it so the cutting firefighter can cut it to the correct lengths and at the correct angles.

When laying out all the measurements received from the measuring firefighter, the layout firefighter must be very certain to understand the numbers and sizes called by the measuring firefighter.

Double checking and repeating the information given helps eliminate any confusion or misunderstandings that can cause errors. Cutting the wrong size lumber drastically reduces the efficiency of the operation, costing precious time delays in the erection of the shoring and possibly in the rescuing of victims.

Next, the material is cut to the proper size. It is easier and far more efficient not to rush but to take the time to make the proper sized item on the first effort. An old tried and true axiom used in the construction trades for decades—which, by the way, works perfectly in rescue shoring operations—is to measure twice, cut once. If you keep this in mind when working the cutting station, you will run a smooth and accurate operation.

Communication is the key to safe and successful operations. The layout firefighter should always tell the cutting firefighter the sizes he needs and explain what and why he is laying out a particular piece of lumber. It is also important for the layout man to tell the cutter on which side of a mark on the wood he wants the cutter to saw. One way to clearly indicate the correct side is for the layout firefighter to place a *V* mark at right angles to the cutting. The side of the line the *V* is on is the piece to be used.

Shore Assembly Team	Cutting Team
• *Shoring OIC*	• *Cutting OIC*
– *Measuring FF*	– *Layout FF*
– *Shoring FF*	– *Feeder*
– *Shoring FF*	– *Cutting FF*
– *Safety FF*	– *Tool and equip FF*
– *Runner*	– *Runner*

Cutting firefighter. The cutting firefighter's responsibilities include setting up the cutting station, cutting the shoring material, and safely operating, maintaining, and handling the cutting tools. The cutting firefighter also must ensure that all blades are sharp and all equipment is in proper working order. He works directly with the layout firefighter both in setting up the cutting area and in cutting the shoring material. Small gas- or electric-powered chainsaws can be used; however, electric saws are preferred since they operate more quietly and may be slightly easier to handle. Electric circular saws can also be used, but to cut 4x4 lumber in one pass, a 10¼-in. blade is essential.

The person that is picked to be the cutting firefighter should have thorough knowledge of handling saws and tools and must have experience in cutting lumber. A structure collapse site is not the place to start to learn how to properly cut building materials. Without the proper precautions, the cutting function can be a very dangerous operation. The cutting firefighter should use the type of saw he is most accustomed to using and with which he is most comfortable. Using these tools will make the job much easier for him and the cuts should be more accurate.

It is also important that the cutter and the layout firefighter are in full agreement on how the lumber is to be laid out and cut. For each cut, the cutting firefighter must know on which side of the line on the material indicating the place to cut the cut is to be made. Although we are not making pianos, the more accurate the cuts, the better the rescue shoring fit; and a good fit is our primary concern. It is important that the cutter makes sure all the cuts are square and neat. This is imperative. Sloppy and out-of-square cuts are not acceptable and will make the shores unstable, ineffective and dangerous. Take your time making the cuts, and make sure you are comfortable. Also review the space in which you will be working to ensure that there is enough room to work safely and that there is enough lighting. If you are utilizing an electric circular saw, a guide for the saw can be used to make the cut more accurate. A 14- or 15-in. power miter saw will also make the cuts very accurately, but keep in mind that it takes quite a bit of power to supply these tools. The cutting firefighter is a very important role; therefore, it is a good idea to assign one of your better-qualified personnel to the position.

Tool and equipment firefighter. The tool and equipment firefighter supervises the removal of tools and equipment from the apparatus to the cutting station or the shore assembly squad. This job generally goes to the apparatus driver/operator. Because so many tools are involved, the help of one or two companies goes a long way toward getting equipment off the rig and to the right location quickly. Remember, the apparatus may be some distance from the collapse area. The equipment firefighter directs his assistants as to the tools and equipment needed and where they are to be taken. He also keeps an inventory checklist or log sheet to be referenced at the conclusion of the operation when the equipment is retrieved from the site.

Another major responsibility of the tool and equipment firefighter is to make sure lumber gets to the cutting station in a timely manner. The lumber is sometimes not the easiest thing to procure and get to the site. Once the material is on site, the equipment firefighter has to make sure it gets to the cutting station as soon as possible. Other firefighters on the scene can be enlisted to get the material to the cutting station as well as to the shoring squad. This can be a labor-intensive situation, and numerous manpower units will be necessary to accomplish this task.

It is also imperative that the tool and equipment firefighter track the location and use of all tools and equipment. Since we don't have an unlimited supply of all of the tools, this is a very important activity. In an emergency situation, a specific tool may be required. It must be accessed immediately when the rescue team requests it. The tool and equipment firefighter must know the area, location, and personnel using the tool so it can be quickly transferred to another place if necessary.

Tool & Equipment Checklist

* *Firefighter's Name* ___________
* *Squad or Team* ___________
* *Squad Leader* ___________
* *Building Location* ___________
* *Time Out* ___________
* *Time Returned* ___________
* *Tool* ___________
* *Tool* ___________
* *Tool* ___________
* *Tool* ___________

Large Operations

In some large collapse operations, a single shoring team may not be able to operate effectively on its own. If this is the situation, then several shoring teams can be deployed. The shore assembly team is composed of six firefighters, and the larger cutting team has an additional six firefighters. Committing additional personnel helps the operation proceed more effectively.

The multiple teams may work in the same general area or be located in areas remote from each other such as on separate floors. When you are going to utilize more than one shoring team, a separate cutting team should be established to make the operation continue smoothly. A general rule of thumb is to establish one cutting team for every three shoring teams. Since each rescue shoring operation is different, adjustments may have to be made. Keep a close watch on the progression of the shoring operation and make location adjustments as warranted. One item that may need to be changed is the proximity of the cutting station to the shoring operations. If the two functions are located very close to each other, the cutting team may borrow some of the shoring personnel to assist with the cutting and delivery of the cut materials to the shoring team. If the shoring teams are remote from the cutting station, the cutting personnel may have to not only cut the materials but also deliver them to the shoring team. Doing double duty like this will limit the number of personnel doing the actual cutting and laying out.

When the operation starts, keep an eye on the availability of the materials to the shoring teams and check with each team to see if the materials are arriving when they need them. If they are not, adjust the operation to maintain material supplies and to speed up delivery times. Remember, team positions are flexible; they can be adjusted at any time to maximize efficiency. In order for the shoring operation to succeed safely and properly, it must progress smoothly and quickly. Any slow down in the process must be addressed and resolved immediately.

The six-member shoring team

This six-member team is called the shore assembly team. This team consists of the following positions:

- Shoring officer

- Measuring firefighter

- Two shoring firefighters

- One safety/assembler

- One runner

All these positions are very flexible, so each team member can move into any one of the positions as needed. As the situation changes at an operation, the positions can be adjusted accordingly.

The shoring officer. The shoring officer's responsibilities and role remain basically the same as those of a shoring officer in the three-member shoring squad. However, rather than being in charge of both the shore assembly team and the cutting squad, the shoring officer of a six-member team supervises the six members of the shoring team only. He may very well be supervising the erection of several shores at once, as well as working with the measuring firefighter to determine the proper positioning of new shores. It is very important for a shoring officer of a six-member team to constantly remain one step ahead of the shoring teams. As the firefighters are erecting shores, they must have the next shoring system identified and planned out. They must always anticipate the next moves the team will be making throughout the completion of the shoring operation.

The measuring firefighter. The job of the measuring firefighter is generally the same as that of the shore assembly squad. This role works with the shoring officer in determining the position of the shores for layout measurements. The officer and the measuring firefighter, after conferring with the structural specialist on the scene, determine the type and size of the shores necessary to stabilize the remains of the structure. The measurer gets the necessary numbers together for the shores' material lengths and relays them as soon as possible to the layout firefighter of the cutting team. It is advisable to have the layout man repeat the numbers to minimize mistakes.

Next, the measurer works in conjunction with the shoring firefighters. He tells them the location in which to place each measured item and whether it's a header, soleplate, or post. Communication is critical as with all rescue operations, and the shoring firefighters must know how the measurer has laid out the shore. Then the measurer moves on to the next shore and repeats the process. The operation cannot be slowed down at any point; therefore, keeping the measurements flowing to the layout crew will keep the operation on track.

The shoring firefighters. These two firefighters do the brunt of the shore assembly. One shoring firefighter places the shoring together so that the assembler firefighter can anchor it together. However, like all shoring crew positions, shoring firefighters are flexible and must assist other members of the team as necessary. For example, one firefighter may be needed to assist the measuring firefighter with clearing the area for a shore and getting the proper measurements. Or, the runner may need help delivering equipment and lumber for the operation, especially if the cutting station is a significant distance from the shoring site. The two shoring firefighters are responsible for doing whatever is necessary to speed the shoring operation, while remaining under direct supervision of the shoring officer. They must anticipate their next moves and be able to work around any obstacles that they may encounter along the way.

The safety/assembly firefighter. The safety/assembly firefighter nails and anchors the shoring systems together. If using a pneumatic or gas-operated nailer, only one firefighter is necessary to accomplish the job. This is the preferred method of anchoring any shore. However, if the nailing has to be done by hand, at least one of the shoring firefighters will have to help the assembly firefighter nail the shore together.

If your crews are not utilizing a power nailer and are anchoring the shoring material by hand, they need to carefully consider the possibility of movement or vibration occurring when they drive the nails into the lumber. The best way to reduce vibration when anchoring two pieces of lumber together is for one person to hold the joint together while the other firefighter nails it in place. Holding the lumber tightly together minimizes movement between the materials.

Also responsible for the safety of the shoring crew, this firefighter monitors the safety conditions at and around the shoring site while waiting for the tools and materials to arrive at the shoring area. The safety/assembly firefighter is primarily looking at the instability of the areas in which the team is operating. This person can look at the structural elements of the areas as well as at connection points for structural compromise and should also watch out for any other activities taking place in the area that may impact the group's working situation. Safety is a major concern and should be everyone's priority at any operation. All personnel should be constantly looking for any signs of changing conditions that could affect the stability of the structure. If the firefighter notices a potential danger, he should immediately contact his officer. While the rest of the team is doing prep work, the safety/assembly firefighter should constantly be observing the surroundings and listening to any radio transmission that could affect the building's condition.

The runner firefighter. The main responsibility of the runner firefighter is to get the tools, supplies, and lumber needed to assemble the shores. This person determines the location of the tool staging area and is responsible for transferring tools and materials to the shoring area. He also works with the cutting team's runner. Depending on the size of the shoring operation, they may need additional help. The amount of help required depends on three primary factors:

- The amount of shoring needed

- The volume of lumber required

- The distance between the cutting station and the shoring operation

When multiple teams are operating, it is a good idea for the runners to submit in writing to the cutting station the size of the lumber pieces they need. This action limits errors, something a rescue effort can't afford.

The six-member cutting team

In a larger operation, additional personnel may be needed to construct and erect the shoring quickly and effectively. A six-member cutting team consists of the following personnel:

- Cutting team officer

- Layout firefighter

- Feeder

- Cutting firefighter

- Tool and equipment firefighter

- Runner

At major operations where numerous shores have to be assembled and erected simultaneously, there is a need for one or more shoring teams. When this is the case, a six-member cutting team can be instituted. Criteria for determining the need for a specific cutting team include the following:

- Location of the cutting station

- Number of shoring teams

- Location of the shores in the structure

- Location of the storage of the shoring materials

If the cutting station is located outside the collapse zone, which in most cases is the safest place for its location, the distance between it and the shoring site will be a distinct disadvantage to speedy assembly. Obviously, the farther away the cutting station is, the more time and manpower it takes to get the tools and materials into the structure. It may be necessary to assign several additional personnel to the task of transferring shoring lumber and tools to the shoring teams.

When you have more than two shoring teams in operation, it is a safe practice to implement at least one additional cutting team. Usually, one cutting team can accommodate the needs of up to two or three shoring teams, depending on the types of shoring being implemented and the proficiency of the assembly crew. When the shoring is located throughout a structure on several floors, additional manpower is needed to deliver the material to the various areas. Additional personnel must be assigned as runners as soon as possible to deliver the needed materials to the cutting and shoring stations.

The six-member cutting team's positions remain relatively the same as those of a three-member team with the following exceptions.

Cutting team officer. The cutting team officer is in charge of all personnel on the team. He picks the team positions for personnel and always takes everyone's specific expertise into account when deciding on assignments. The officer supervises the team whose members may be spread out over a sizable area. Coordinating them as conditions constantly change—as they normally do in a collapse situation—is a complicated task. Keeping the cutting operation safe and the work continually flowing is his primary concern.

Along with the layout firefighter and the cutting firefighter, the cutting officer coordinates the cutting area set up. This officer must make certain there is sufficient space for the safe operation of tool and equipment movement, as well as for specific cutting operations. When cutting tools are being used, safety is the greatest concern, and having more than adequate room to work safely is as important to the safety of personnel as it is to completing the shoring operation as quickly as possible.

One other function of the cutting team officer in a six-member cutting team is to maintain a sufficient supply of tools, equipment, and materials. This officer must be able to quickly anticipate the need for specific tools and lumber sizes before the inventory runs out. The cutting team officer sets in motion a procurement procedure for meeting those needs with little or no delay. It is important to avoid work stoppage caused by negligent monitoring of supply levels. Therefore, the cutting officer must stay in constant contact with all shoring operations and their officers in order to know of any changes or updates in the shoring team's location, types of shores being erected, lumber sizes, and tools requirements.

Layout firefighter. The layout firefighter's responsibilities include staying in direct radio contact with the measuring firefighter of each shoring team. These two firefighters are responsible for the properly sized cut material being sent to the collapse site. When two or more shoring operations are in progress, the layout firefighter may have difficulty staying in contact with the measuring firefighters of both operations. To prevent confusion so that the proper materials get to the correct shoring area, the layout firefighter must ensure that every radio transmission is distinct and that the caller identifies for which shoring site he is speaking. To do this with little confusion, the layout firefighter marks each piece with the size and the shore assembly team's physical designation. When transferring the cut material from the cutting area to the shoring site, the layout firefighter directs the runners, telling them the exact piece of lumber to deliver, and the exact team to whom it should be delivered.

Feeder. The firefighter that is assigned the feeder role is responsible for the stacking and moving of lumber at the cutting station. This firefighter interacts with all other firefighters in the cutting team. This role's main function is to place the lumber on the cutting table ready for the layout firefighter to measure. The feeder, with help if necessary, lays the lumber in separate stacks by size, e.g., 4x4s in one stack and 2x4s in another stack, with the longer lumber on the bottom to ensure proper balance of the load. Immediately after the lumber has been cut to specified sizes, the feeder firefighter is responsible for clearing the scrap lumber from the cutting area so that it does not create a trip hazard. It is advisable to have a large container in the cutting area to hold the scrap. The feeder firefighter enlists the help of the runners to help clear the cutting and shoring areas of scrap materials.

Cutting firefighter. The responsibilities of the cutting firefighter on a six-man cutting team are essentially the same as those of the standard cutter role. The cutting firefighter works closely with the layout firefighter. Using a six-man team means that the size of the collapsed structure is large; therefore, there are multiple shoring teams working simultaneously. As the number of individuals in the cutting area increases, so should the size of the cutting area. The volume of cutting will also increase; therefore, the volume of waste material grows, making clean up a larger task.

More people working at a faster pace, creating more useable and waste lumber, calls for a greater focus on safety. For the cutting firefighter, safety is a priority. He must always be aware of his position in relation to others on his team, especially when the saw is operating. Losing sight of where another person is working while bringing lumber and materials in and out of the cutting station area can easily result in injury.

Tool and equipment firefighter. The tool and equipment firefighter on a six-man team has more responsibility than his counterpart on a standard team. Because the shoring operation on a large collapse site employs numerous cutting and shoring stations, the demands on this role are very high. The tool and equipment firefighter must anticipate the need for a great many tools and materials. In a large shoring operation, saws are working constantly and must be serviced frequently. It is the tool and equipment firefighter's responsibility to make certain saws keep working to supply the lumber required to build the shores. In a large operation, the useful life of the department's tools and the equipments' capacity may be overwhelmed rather quickly. This firefighter arranges for increased inventories of tools, equipment, and parts to be maintained onsite near the cutting stations so that out-of-service equipment does not cause operating delays.

Runner. Taking direction from the layout firefighter, the runner delivers lumber from the cutting team to the shoring teams. Accuracy and speed are this person's main objectives. He takes the correct pieces of lumber to the correct shoring site as quickly as possible. Depending on the location in the structure where the shore assembly teams are in relation to the cutting area, this firefighter may have to be relieved on a regular basis.

Training for the Rescue Shoring Firefighter

As with any facet of technical rescue, training is an absolute must. When the need arises for the use of a rescue shoring team, the team must jump into action immediately. As a result, the team must have all its ducks in a row, so to speak. Team personnel must be proficient in the proper size-up techniques, as well as in proper shore construction. All possible variables that may occur should be considered, and mitigation methods set in motion before another crisis occurs. Your team's training must be comprehensive. It has to encompass every possible facet of rescue shoring, including all phases of building construction, lumber qualities, carpentry skills, load transfer, and engineering practices. There is a list at the end of this chapter of suggested training that all rescue firefighters engaged in shoring operations should complete.

When the need for shoring stabilization comes, the shoring rescue team has to enter the unstable building and restabilize it. Therefore, your shoring team must be able to properly and safely handle itself in any type of situation that may develop while erecting the shores. Team members must be trained in all types of technical rescue they may need to use to handle any kind of mishap

There are several acceptable training guidelines that have been established for operating in collapse incidents. Shoring teams should be thoroughly familiar with the curriculum of the Federal Emergency Management Agency (FEMA) rescue-specialist training. It typically consists of 80 hours of training in three main disciplines: shoring, breaching and breaking of concrete, and lifting and moving of concrete and other debris. The National Fire Protection Association (NFPA) has guidelines, *NFPA 1670* and *1006*, for team operations in structural collapse situations. Other technical rescue disciplines with which shoring teams must be familiar are trench rescue, confined space rescue, and rope rescue techniques. Basic knowledge in water rescue and vehicle rescue are also important.

The following paragraphs present recommended training classes, with content description and duration for each, for personnel engaged in structural collapse and rescue shoring operations.

Basic structural collapse operations—8 hours

This course provides specialized training in the realm of building collapse rescue including the following:

- Warning signs
- Collapse causes
- Void identification
- Safety precautions
- Search techniques
- Team operations

- Building construction awareness

- Initial fire department operations

- Review and analysis of several case studies

Basic building construction for the fire service— 8 hours

There are many types of structures in this country, and every firefighter regardless of his job should be able to recognize them and be thoroughly familiar with all facets of the construction techniques used for each. The transferring of loads in the structure is one of the main concerns of any collapse rescue operation. Team members must be able to identify the supporting elements in any building. This is especially true in a collapse scenario because the early identification of the building elements is very important to the safe completion of any collapse rescue operation.

Building construction related to building failure— 8 hours

This class covers the many facets of structural erection of buildings, including the examination of the following:

- Identity of the main structural elements and the way they are erected

- Techniques for transferring loads

- Definitions

- Stress management of the buildings

- Construction techniques

- Construction terminology

- Common ways buildings collapse

- Tabletop exercise to reinforce the concepts discussed

Void search and rescue concepts—8 hours

This class presents a comprehensive discussion of the many facets of a safe and successful void rescue operation. Void rescue is one of the most dangerous operations that any fire department emergency response team undertakes. Class content includes the following:

- Team concepts

- Safety precautions

- Void identification

- Hazard abatement

- Shoring techniques unique to void searching

- Victim packaging

- Review and analysis of several case studies

- Tabletop exercise to enhance understanding of the dangers involved in void rescues

Hands-on void search and rescue—16 hours

This is a two-day class with intensive, hands-on training in actual void search conditions. The students are subjected to operations in several void simulators. Team concepts, proper shoring techniques, safety operations, and size-up are examined. Several actual rescue simulations are conducted. The students actually extricate victims throughout the two-day course.

Lifting and moving of objects in US&R operations—16 hours

This class is designed to inform and demonstrate to the rescue responder the basic techniques in the lifting and moving of heavy objects, especially by the simple methods utilizing leverage. The different classes of levers are discussed thoroughly as well as the basics of mechanical advantage, fulcrums, and pulley systems. The proper handling of slings and rigging equipment is also part of the course as well as working with crane operations.

Emergency rescue shoring concepts—8 hours

This class covers the many concepts and principles of the proper erection of emergency building shores. The proper size-up, placement, and types of shores are described in detail. The proper engineering concepts are explained, and the fabrication procedures for each shore are examined thoroughly. Several case studies are reviewed, and a tabletop scenario of proper placement, size-up, and mitigation of a specific a shoring operation are conducted. Additionally, the theories of emergency building shoring are covered.

Engineering concepts for rescue shoring—4 hours

This class is designed to instruct the students in the basic and advanced engineering concepts that are an extremely important part of the design of the shoring systems. The students must understand the concepts behind the different angles, the specific nail patterns that are necessary, and the proper loading of the shores in order to work effectively. When you know exactly why and how the shoring systems are constructed properly, then you will be able to deviate slightly from the usual fabrication procedures. Knowing the proper concepts and procedures allows your team to adjust the shoring systems to any possible situation with the necessary positive results. This four-hour engineering concept course is a *must* for operations- and technician-level responders.

Hands-on interior shoring—16 hours

This is a two-day class designed to present extensive hands-on, step-by-step erection of the proper exterior shoring techniques needed to safely construct all the types of exterior shores. This course is a *must* to conduct safe rescue operations and stabilize the exterior of the structure. A lecture section concentrates on the proper size-up and the correct step-by-step erection of the various shores designed for installation on the interior of collapsed structures. This is a basic class showing the most common types of shores a rescue team normally erects in a major collapse scenario. This is an intensive, hands-on course; and the students are tested on the proper erection of the various shores.

Hands-on exterior shoring—16 hours

This is a two-day class designed to present an exhaustive discussion of the proper exterior shoring techniques needed to safely construct all the types of exterior shores generally needed to conduct safe rescue operations and stabilize the exterior of the structure. A lecture section concentrates on proper size-up and the correct step-by-step erection of the various shores designed for installation on the exterior of collapsed structures. The proper stabilization techniques for larger areas of the building and the support of exterior bearing elements are examined. This is an intensive, hands-on course; and the students are tested on the proper erection of the various shores covered.

Advanced emergency shoring operations—32 hours

This course is the culmination of all the previous training, plus the presentation of several shores previously not shown. It also includes an additional lecture class supporting the necessary concepts on those shores. This class has extensive hands-on erection of the new shores as well as several examples on the variations of shoring systems that could occur. The shoring of raked openings and sloped floors is covered in this class.

Setting Up Your Shoring Operations

Long before your department gets involved in emergency rescue shoring operations, you must have a complete working knowledge of the shoring world. This not only includes knowing building construction, tool operations, and carpentry skills, but it also includes knowing the principles of rescue shoring. In addition to knowing where and when to install rescue shoring and the capacities of each shore, these principles also include the dos and don'ts of properly cutting and installing the shoring material. Because this information is very important to the success of your operation, this chapter addresses this aspect before going into the step-by-step procedures of erecting specific shores. This chapter covers the basics for the following:

- Nailing patterns

- Gusset plates

- Cleats

- Cutting tables

- Lumber cutting

- Wedge placement

- Header and post options

- Criteria for determining angles and the procedure for cutting them

Your team must know these items in order to succeed in a rescue shoring operation.

Classifying Your Shores

Shoring systems can be broken down into three classes: one, two, and three. The class indicates the stability of the shore or shore systems, with class one being the least stable and class three the most.

Class one

Class-one shores are one-dimensional shores, basically the least stable shores and include a single-post T-shore and the flying-raker shore. These are all quick-to-erect, temporary shores designed to give your team some degree of safety while installing and erecting the main shoring systems. These are only temporary shores—never rely on them alone for a rescue attempt.

Class two

Class two shores are two-dimensional shores, much more stable than the one–dimensional, class one shores described previously. For the most part, these are shores with at least two posts, a soleplate, and header. This would include the double T-shore with post spacing of at least 18 in. The two-post, vertical shore, window and door shore, horizontal shore, and the flying shore are all examples of class-two shores.

Class three

This is the class in which all the shores are three-dimensional. It is the strongest and most stable shore system. The laced post is a perfect example of this type. Also, any two shores tied together as a system with cross bracing is considered a class-three shore. For example, a set of raker shores or sloped-floor shores tied together is considered three-dimensional.

Nails

For the sake of simplicity, the shoring discussed in this chapter is designed to be fastened together with the use of two sizes of nails: the 8d nail and the 16d nail. There are numerous sizes and styles of nails at rescue shoring operations, but the information on nails in this section is limited to these two sizes. The 8d nail is 2½ in. long, and a 16d nail is 3½ in. long. The single-headed, or common, nail is the one utilized on most occasions. You can use duplex nails, also known as double-headed nails, or scaffold nails. When in training, you will use double-headed nails. They have two heads and are designed to be easily pulled out. Another type of nail is a green sinker. It is a common nail that has a rosin coating so it holds very well. The 8d nails are always utilized when nailing plywood or nailing plywood to dimensional lumber (1½ in. or thicker lumber). The 16d nails are used only when nailing dimensional lumber, whether face nailing or toe-nailing the lumber together.

Nails

- **8d** – All plywood, including gusset plates
- **16d** – All dimensional lumber, 2x4, 4x4, etc.

Gusset plate nailing

In many shores constructed, the use of gusset plates is required. There are two main reasons for installing gusset plates on your shoring systems. The most important reason is to lock a connection point—the point at which two shore elements are nailed together. This can include, but is not limited to, posts to headers or sole plates, rakers to wall and sole plates, and wall plates to sole plates. In certain cases, the gusset plates must be installed on both sides of the connection points. This action locks the joint tightly, stopping it from separating or rotating and causing problems. The other reason to utilize the gusset plate is to help the shore stay together in an earthquake situation. The aftershocks can cause your shores to shake and twist. By gusseting one side of the nailed posts, you can help prevent the nailed connection from coming apart. The following figures show several variations of the gusset plates that we can use in a number of different situations.

The nailing pattern on most gussets is an 8d, 5-nail pattern. The 5-nail pattern uses an approximate 6-in. spread. This will be the proper nail sequence for a 12x12-in. gusset plate. Note the 5-nail pattern and the nail spacing. This nail pattern can change, depending on the use of the gusset plate. The rule of thumb is to place the 8d nail spread on the longer lumber joint, generally a header or wall plate.

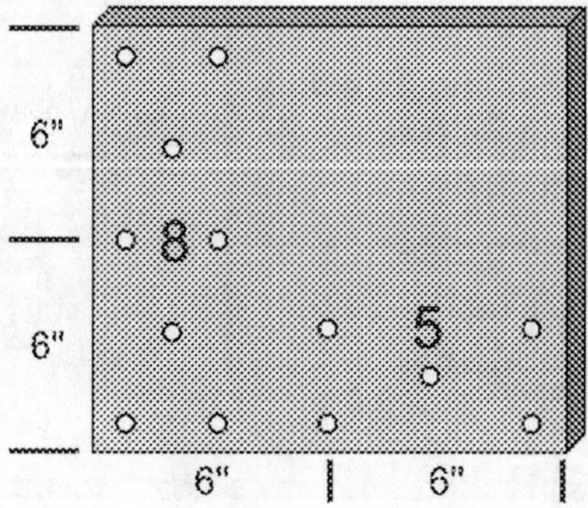

Fig. 3–1 12x12-in. gusset plate nail pattern for header or wall plate.

In some cases, especially in the interior of a building, your team may opt for a triangular gusset plate. In the following graphic, the triangle gusset is in place on one side only. Its only use is to keep the toenailed joint from being dislodged. The use of the triangle gusset is warranted in an area where there will be personnel passing through and possibly contacting the shore. The triangle has much less profile jutting past the shore's supporting elements.

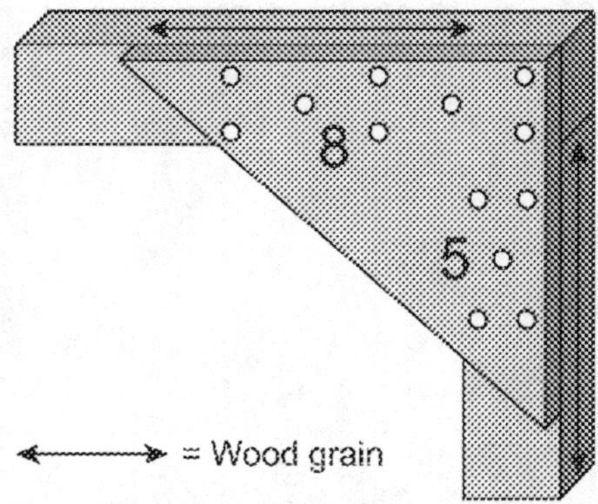

Fig. 3–2 Triangle gusset plate nail pattern.

Figure 3–3 depicts the 12x12¾-in. plywood gusset used for a top gusset for a one post T-shore. The 8-nail pattern is along the header, and the 5-nail pattern is anchored into the post. This is the most common size gusset plate used.

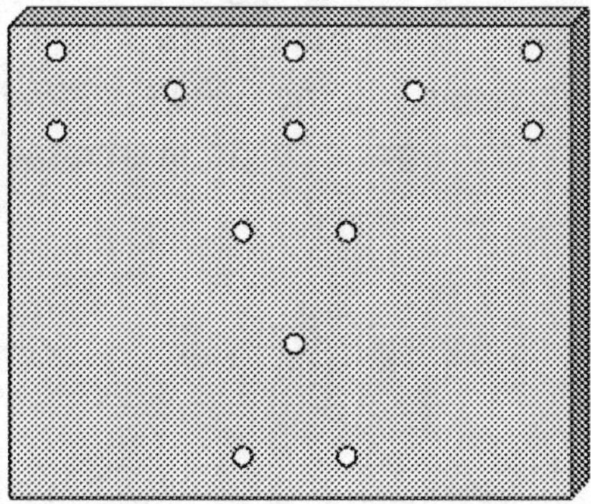

Fig. 3–3 12x12-in. gusset plate nail pattern for T-shore top plate.

If you decide to utilize a gusset plate that is 18x18 in. square, the nail pattern uses (11) 8d nails along the header and (9) 8d nails along the post, all in a 5-nail pattern. Use the same nail pattern if you use bigger plates. Your gusset plate should, for the most part, be constructed of ¾-in. thick plywood. Figure 3–4 shows a top gusset plate for a one-post T-shore.

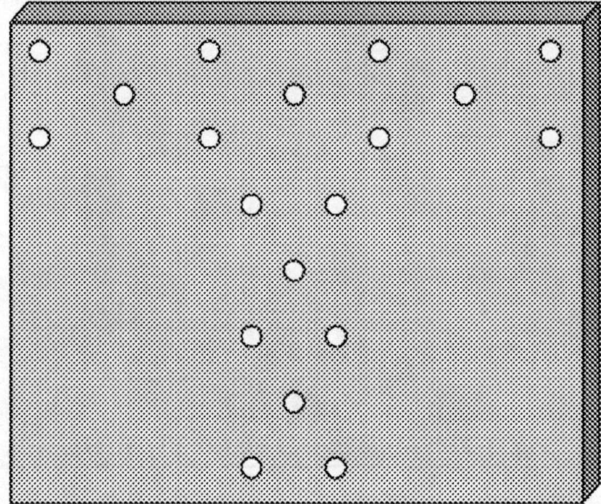

Fig. 3–4 18x18-in. gusset plate nail pattern for T-shore top plate.

Figure 3–5 shows a 12x24-in. gusset plate laid out for a two-post T-shore with the two posts 18 in. apart, outside to outside. Again, the longer nail pattern is placed along the header. There are (17) 8d nails anchored into the header. Along both posts, keeping the nailing pattern and spacing consistent, there are (5) 8d nails anchored into each post. Notice that the gusset plate overlaps 3 in. on each post, which is fine.

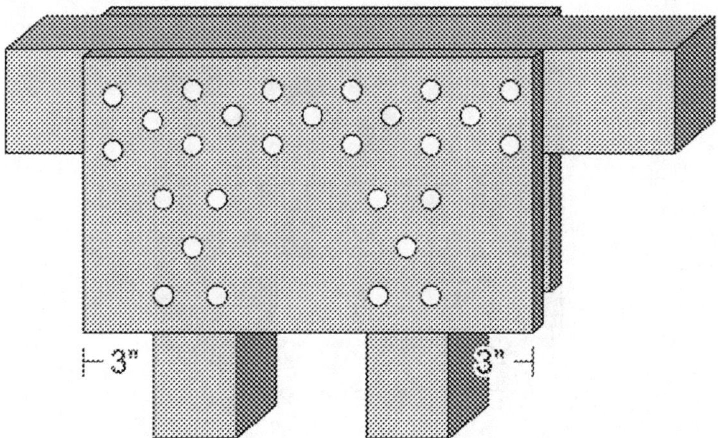

Fig. 3–5 12x24-in. gusset plate nail pattern for 18-in. two-post T-shore top plate.

The 12x24-in. gusset plate would be the largest size used. Again, it is utilized as a top gusset plate for a two-post T-shore. Figure 3–6 shows the posts spaced 24 in. apart. This pattern results in (17) 8d nails anchored into the header and 8 nails into each post, with no gusset plate overhang.

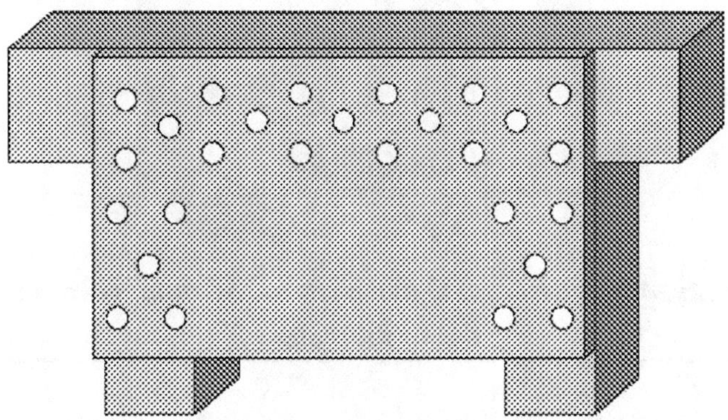

Fig. 3–6 12x24-in. gusset plate nail pattern for 24-in. two-post T-shore top plate.

Figure 3–7 depicts an option that your team has for locking the two posts of the double T-shore together. You can use this 12x24¾-in. plywood gusset plate on one side and anchor it to the posts with (8) 8d nails in the 5-nail pattern. Nail the gusset in the center of the posts.

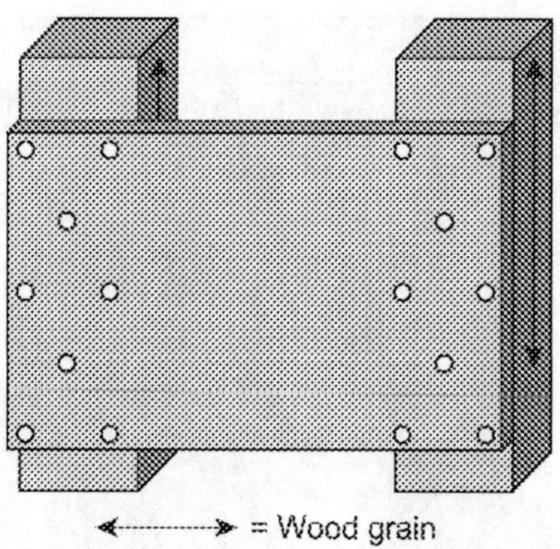

Fig. 3–7 12x24-in. gusset plate nail pattern for two-post T-shore center plate.

Figure 3-8 shows the option for locking the center of the two posts of the 18-in. spaced double T-shore. The 12x24-in. gusset plate overhangs each post 3 in. Use the 5-nail pattern, and (8) 8d to nail each post.

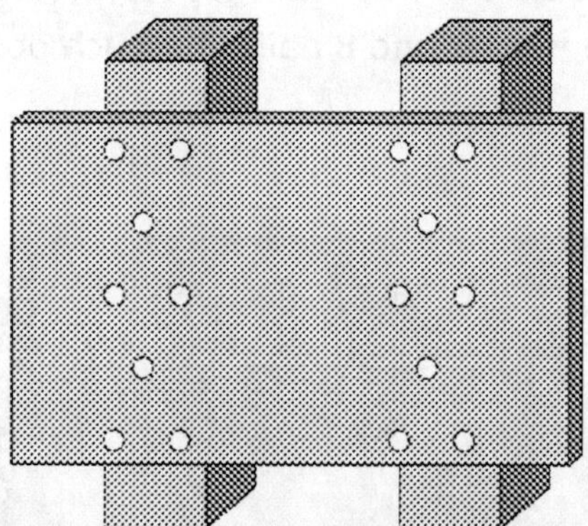

Fig. 3–8 12x24-in. gusset plate nail pattern for 18-in. spaced two-post T-shore top plate.

Nail patterns for 2x4 and 2x6 cleats

In some instances, you will use 2x4 or 2x6 cleats in your shoring situations. They are generally used as thrust blocks or cleats, designed to hold another piece of dimensional lumber from moving. In these cleats, assemblers utilize the 5-nail pattern; however, in most cases, the nail pattern spread is roughly 5 in. apart. As with any dimensional lumber, the nails to use on these shores are 16d.

2x4 nail patterns. The most common cleat layout used is a 24 in.-long 2x4. As Figure 3–9 illustrates, the 2-ft section of lumber has five 5-nail patterns, roughly 5 in. apart. This will give you a cleat with 17 nails in it.

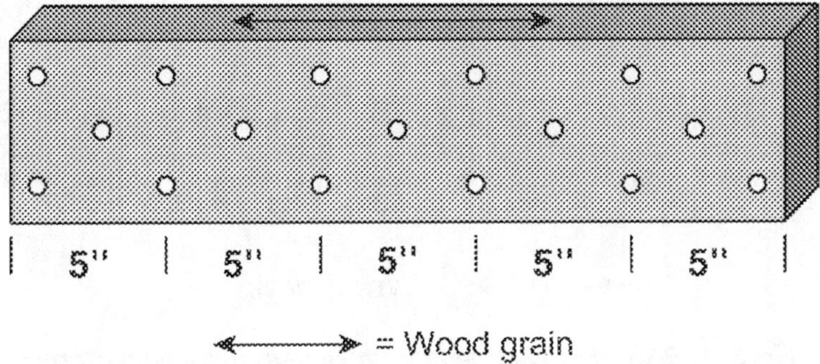

Fig. 3–9 24 in.-long 2x4 cleat with 17 nails in 5-nail pattern.

Figure 3–10 shows a cleat that is 3 ft-long 2x4. This is used mostly on raker shores that are at a greater angle than 45°. Keeping the same 5-nail pattern and using a 4½-in. nail spread instead of a 5-in. spread, you wind up with (26) 16d nails.

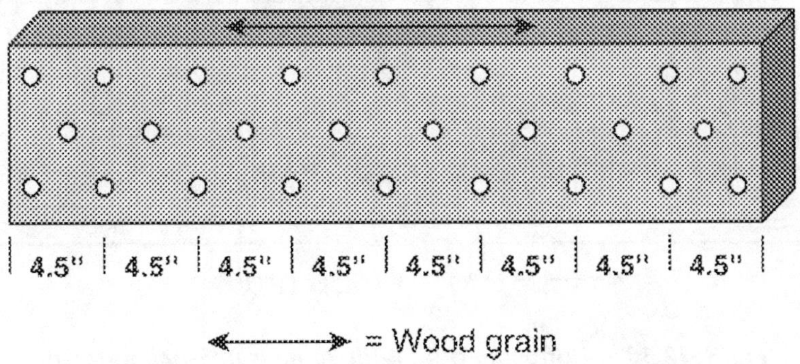

Fig. 3–10 36 in.-long 2x4 cleat with 26 nails in 5-nail pattern.

2x6 nail patterns. When using larger dimensional lumber for shores, generally 4x6 and or 6x6, your shore assemblers should utilize 2x6s for cleats. Figure 3–11 shows a typical layout for the cleat, the 5-nail pattern with 26 nails.

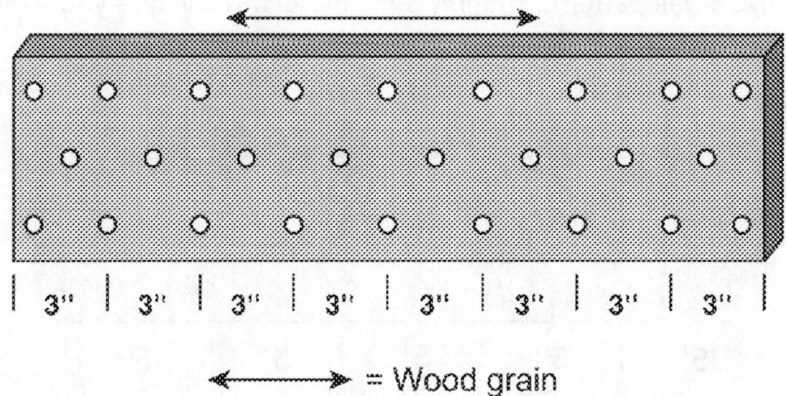

Fig. 3–11 24 in.-long 2x6 cleat with 26 nails in 5-nail pattern.

At 3 ft long, this 2x6 now has 38 nails placed in it (see Fig. 3–12). This gives the cleat enough shear resistance to hold the larger lumber you are using for your shores.

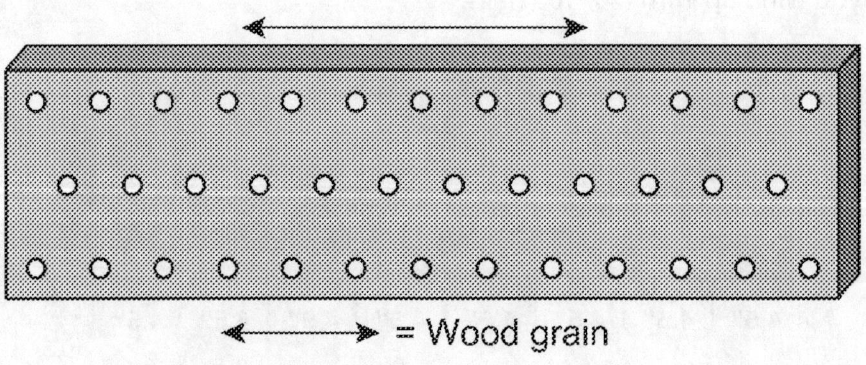

Fig. 3–12 36 in.-long 2x6 cleat with 38 nails in 5-nail pattern.

Whenever you nail a 2x4 or 2x6 bracing, follow this nail pattern system. On the 2x6 there will be five nails and on the 2x4 there will be three nails. Again, use the 5-nail pattern, and these will be 16d nails at all times.

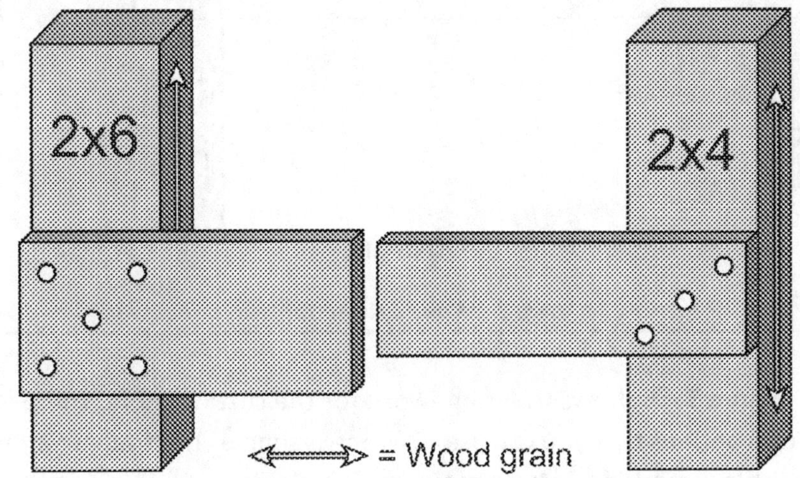

Fig. 3–13 2x4 and 2x6 nail patterns for braces.

Cutting Table

The cutting table shown in Figure 3–14 is 4 ft by 8 ft and is designed for easier cutting of the lumber needed in shoring operations. It is designed with safety in mind. If you place the material in the spaces provided on the table, it does not have to be held nor will it kick back during cutting, possibly causing injury. Basically, there are four 2x4s anchored onto a sheet of ¾-in. plywood. Hint: leaving a space of 3½ in. at the edge gives you a spot to put your tape, knife, pencil, or any other items without interfering with items on the rest of the table. The top of your table should be no more than 36 in. off the ground. Some firefighters prefer a 32 in. or 34 in. height.

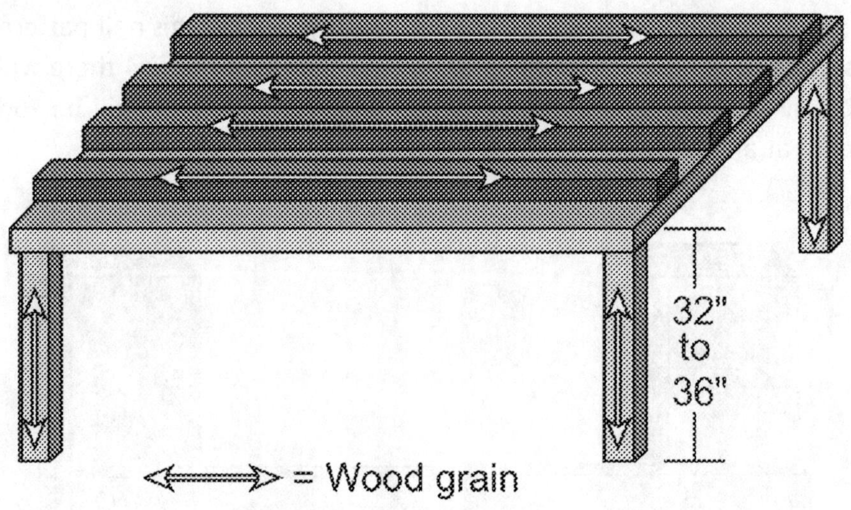

Fig. 3–14 4x8 cutting table.

Figure 3–15 shows the framework for the cutting table. There are several options that you can use. This one is simple. The framework for the table is made of 2x4s, two pieces 93 in. long and two pieces 48 in. long. The 48-in. pieces go on the outside edge of both 93-in. sections and are nailed flush. This gives you a framework that is 96 in. by 48 in. outside to outside. Place a 4 ft x 8 ft sheet of plywood on top and nail it down with 8d nails. Do this after the legs have been secured.

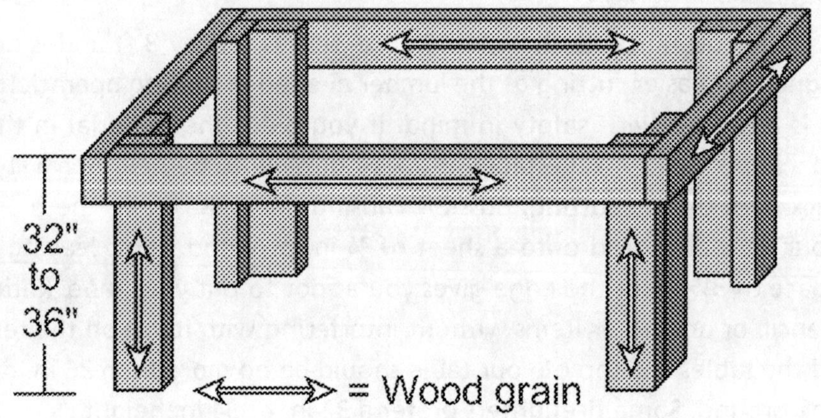

Fig. 3–15 Cutting table framework.

There are several ways to construct the legs. The following is a simple method. Utilizing 2x6s, at your preferred height (say 34 in.), nail two together. Make them at right angles to each other and nail with 16d nails every 6 in. along the face. Place them flush with the top of the 2x4 frame (as shown in Fig. 3–16) and secure them to the frame, nailing (5) 16d nails into each 2x6.

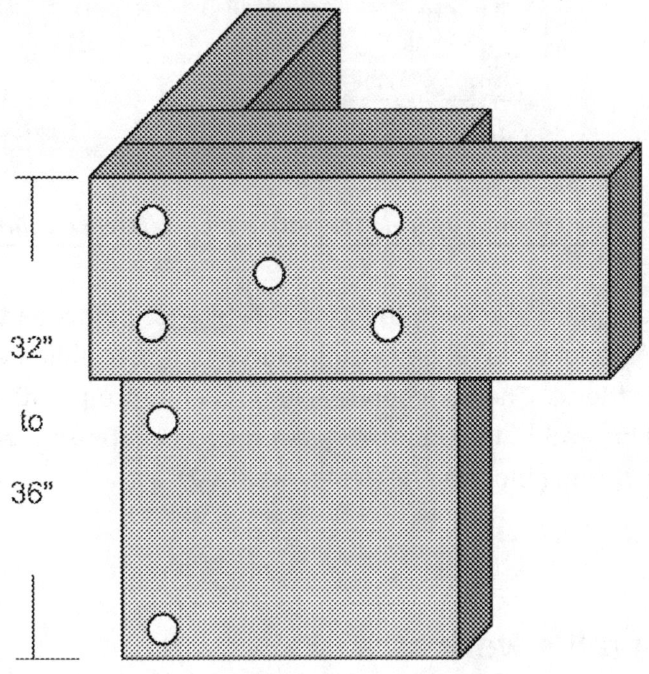

Fig. 3–16 Cutting table leg construction.

After you have nailed down the 4x8 plywood to the frame, you are ready to place the 2x4 guides on top. You then nail down one 2x4, leaving a space of 3⅝ in. before placing down the next 2x4. Using this gap, you can cut 4x4s and 2x4s on the flat. The way to make the gap's spacing is to place a 2x4 between the first and second 2x4s. Then, place a nail in the gap. This gives you just enough extra opening so the lumber won't stick between the guide pieces. The third section of 2x4 is anchored down at about 1⅝ in. distance. This gap makes cutting 2x4 wedges easy. To cut the wedge, place the 2x4 on edge (1½ side). The last space to create on the table is for cutting 2x6s, 4x6s, or 6x6s, leaving a 5⅝ in. gap.

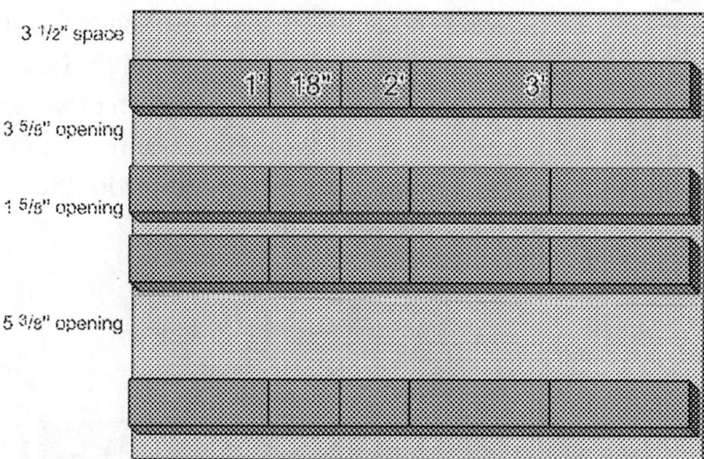

Fig. 3–17 Common spacing for cutting table reference points.

You can place marks on the 2x4s as reference points on the table for the material you will be cutting. Run a dark pencil line across all the guide blocks. The common spacing of reference points is 12 in., 18 in., 24 in., and 36 in. These are great for quick reference when you need to cut some blocking or cleats (Fig. 3–17).

Cutting table work area

Using the cutting table is a very efficient way of cutting the proper sized pieces of lumber; however, this operation must be done safely. With the operation of several cutting tools and saws and with people working beside those tools in a small area, safety becomes a major concern. By keeping a radius of 8 ft clear of anything—lumber, tools, and people—you can maintain some degree of safety in the cutting station area. It is also important to constantly clean up the scrap material. Do not let the shorts accumulate around the table. They will become a trip hazard. Remove them out of the area frequently so that the cutter can't trip on these pieces while working with tools.

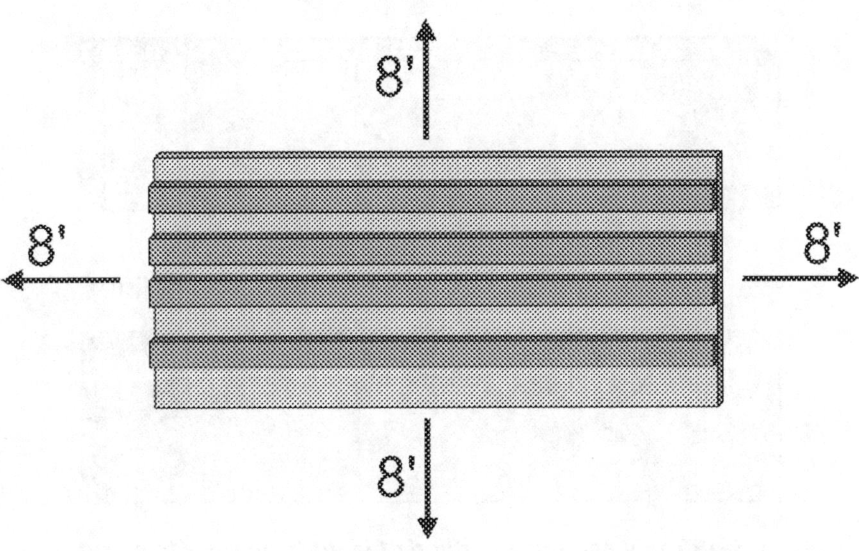

Fig. 3–18 Keep an 8-ft radius clear of debris and personnel for a safe work area.

Marking Lumber for Cuts

When things get hectic and numerous pieces of lumber are being cut and marked by different individuals, there must be some consistency and order around the cutting station. If there is not, there is a good chance that lumber will be cut to the wrong size, or worse, that someone may get injured.

There are two ways to identify the side to be cut: the *V* and the *X* method. In the *V* method, the piece of lumber has a *V* pointing to the side to be cut. Saw to the right of the line to cut the right length of material. In the *X* method, an *X* is on the wood piece to cut. Following the same procedure as used in the *V* method, cut on the right side of the line to produce the proper sized piece.

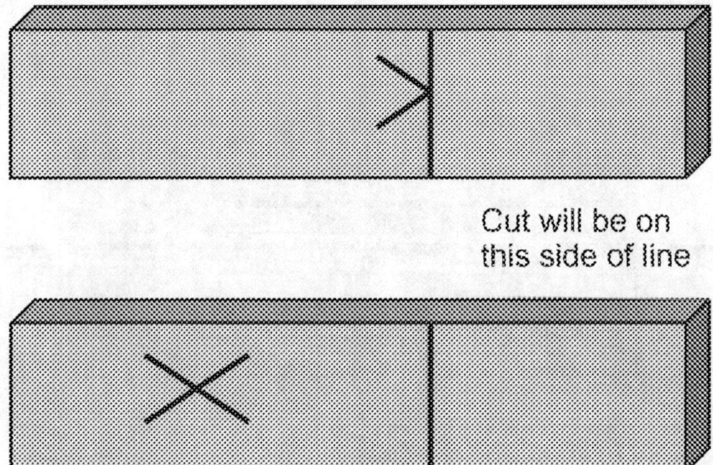

Fig. 3–19 The V and the X method of identifying sides to be cut.

Multiple Lumber Cuts in One Pass

Often when you have to cut multiple pieces of lumber the same size, you can cut all pieces in one motion. To make this action easier, align several pieces of lumber together so that you have to measure only once and make only one long cut.

Here are pieces of lumber laid out so they can be cut in one pass. By doing things this way, you can save some precious time. This method means that you need to measure and cut only once. First, set up the number of boards you want to be cut in a straight line, make sure the boards touch each other and are pushed together snuggly. Line up the ends by eye, keeping them as even as possible.

The next step is to square up all the pieces of lumber together. If you don't do this, the cuts may be off. Use the framing square to give yourself an accurate mark. You now can get your proper length and mark all the boards in one pass. While firmly holding the boards down and stopping them from shifting, you can cut all the pieces in one pass with the circular saw.

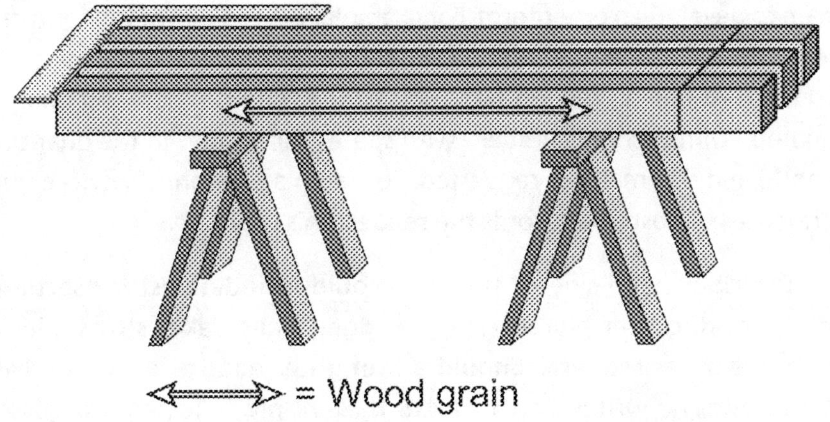

Fig. 3–20 Cutting multiple pieces of lumber the same size.

You can also use this multiple-piece cutting method to saw stacks of lumber the same size. Line up several pieces of lumber in layers. Flush up the ends. Use a framing square to square up the stacks to each other. Mark a cutting line along the top and the side of the stacks to make for a more accurate cut. Cut along the lines. When you use a chain saw to cut the wood, this method saves multiple moves and measuring.

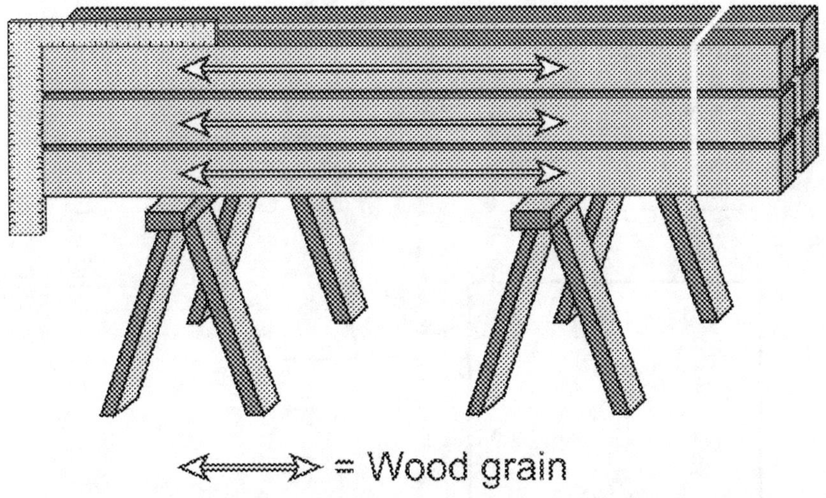

Fig. 3–21 Cutting stacked lumber of the same size.

In order to expedite communications, it is a good idea for the cutting team to develop an order form. For example, the top line of the form can be for the identity of the location and the squad requesting the material. On the next line on the left-hand side of the form can be a section designated "Dimensional Lumber" with spaces for writing in the quantity, length, and size material requested. So, when a firefighter wants eight, 8 ft-long 4x4 posts, he records the request this way: "8 8-ft 4x4s."

The right-hand side of the form could include a list for sections of plywood, gusset plates, cleats, wedges, nails, raker struts, plus a space for other requests. Should a firefighter need a section of 4x8-ft plywood, he writes down "1 4x8 ft ¾-in. ply." To request gusset plates, write "4 12 in. x 12 in. ¼-ply." Another example: to request four 24 in.-long 2x4 cleats, write down "4 24 in. 2x4s." For wedges, write "12 12-in. 2x4s"; for nails, write "20 lb 16d common nail"; for a raker strut, write "1 45° 170 in. 4x4."

The form can also have a section in which to sketch the shore to be constructed. Using a form like this makes it easy for the cutter to determine the exact amount of material needed.

Shoring Material Order Form

Rescue Squad: _______ Location: _______ Time: _______ Contact: _______

Dimensional Lumber				Plywood	Quantity	Size	Dimension
Quantity	Length	Size					

Gusset Plates

Cleats

Wedges

Shoring Sketch

Nails

Raker Strut

Other

Fig. 3–22 The cutting team order form.

Wedges

When assembly firefighters run out of the store-bought variety of wedges, the cutting team must make them on the spot. Generally, two different widths of wedges are made: the 2x4 and the 4x4. For most interior shoring, it is best to use the thinner 2x4 wedges. Those wedges make for a much neater and tighter shore. For exterior use with big lumber, use the larger 4x4 wedges.

When you have to lay out wedges in the field, you can do the layout as illustrated in Figure 3–23. Each diagonal cut produces a pair of wedges; some teams try to use matched pairs on a shore. The most popular sizes for wedges for shoring are 12 in.-long 2x4s and 18 in.-long 4x4s. The 24 in.-long 4x4s generally are used for cribbing.

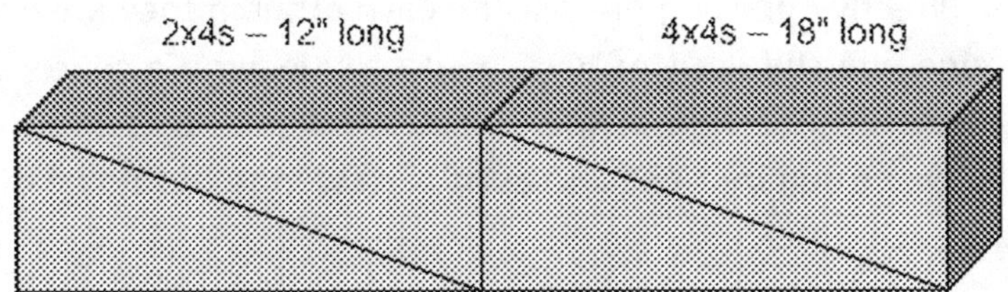

Fig. 3–23 Laying out wedges in the field.

To get the ideal spacing for the wedges, begin with the height of the post then subtract the thickness of just one of the wedges. If using 2x4s for wedges, deduct 1½ in. from the post height to have enough space for the pair of wedges to fit under the post properly without any spaces. If your posts are too short, as the one post shows in Figure 3–24, the wedges won't be as effective or stable. The wedges must have full and flush contact with the posts. You cannot over pressurize the wedges or the shoring will be ineffective. Using 4x4s for wedges provides an effective spread of up to 6 in. When using the 2x4s for wedges, the effective lift capacity is 2¼ in.

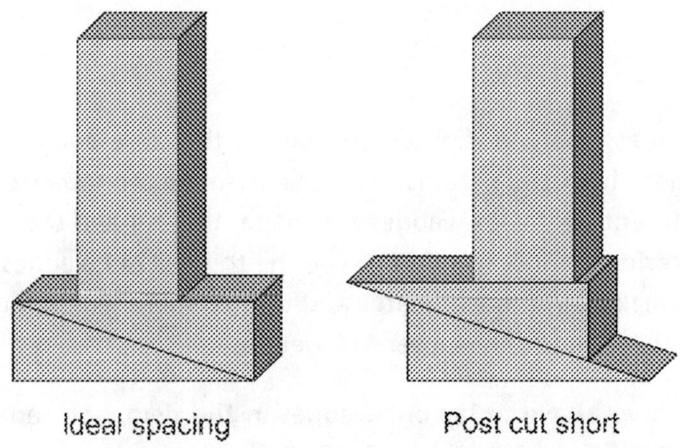

Fig. 3–24 Make sure you have ideal spacing for wedges.

There is a right and a wrong way to install wedges—believe it or not. For the wedges to slide along a parallel plain, the right angles must be directly opposing or opposite each other. If they are not, the top wedge will slide out of level, generally leaving a space, which is unacceptable. The rule of thumb for safety is to place cut side to cut side, or the square right angles of each wedge should be directly opposite each other.

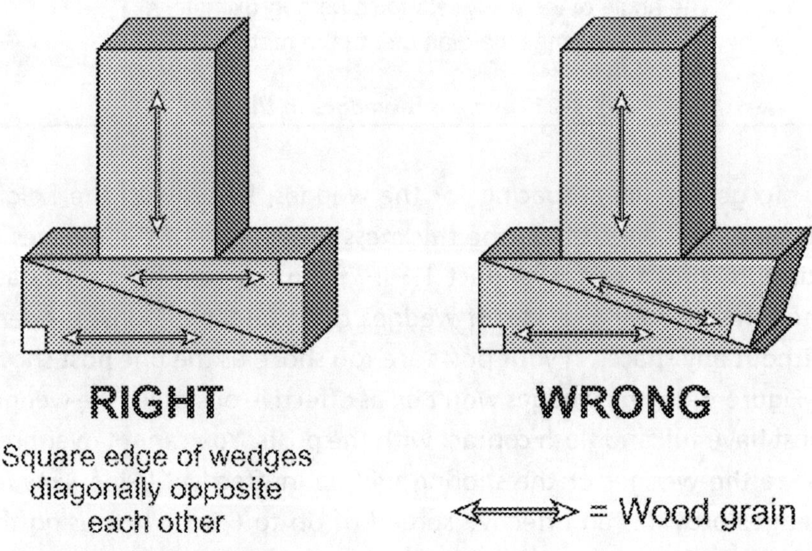

Fig. 3–25 There is a right way and a wrong way to install wedges.

Prefabricated Headers and Posts

There are many instances when you run out of a specific size lumber, or when, for some reason, you need a larger piece than you have on hand. No problem. Improvise and adapt. As you shore up specific openings that are longer that 4 ft wide, you have to have headers larger than a 4x4. The following are some of the options available to your team.

Prefabricated headers

This option for increasing the header to 6 in. in depth uses two sections of 2x6 with a ¾-in. plywood spacer and is the most efficient use of material. This can actually be more efficient than using a 4x6 that is not top grade lumber. Place 5½x12-in. gusset plates on each end. Do not overlap the gusset over the edges of the 2x6s. Nail the three pieces together with 16d nails, using the 5-nail pattern. You should install additional gussets spaced roughly every 2 ft or so. The header is ready for use. Install the header with the 3¾-in. side on top of the posts.

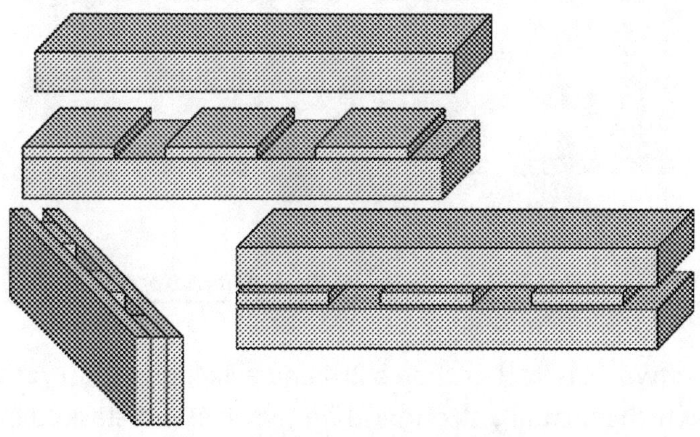

Fig. 3–26 2x6 header construction.

The configuration shown in Figure 3–27 can be used in place of a 4x6. It will be approximately 80% capacity of a 4x6. Place a 2x4 on top of a 4x4 and join them with 16d nails to anchor the 2x4 to the 4x4.

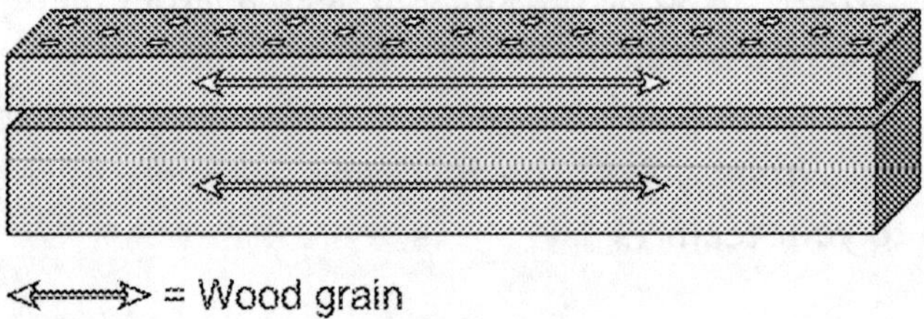

Fig. 3–27 Alternative to 2x6 header construction; 2x4 nailed to 4x4.

Figure 3–28 shows a close-up of the 5-nail pattern on the 2x4 with a space pattern of 6 in., using 16d nails.

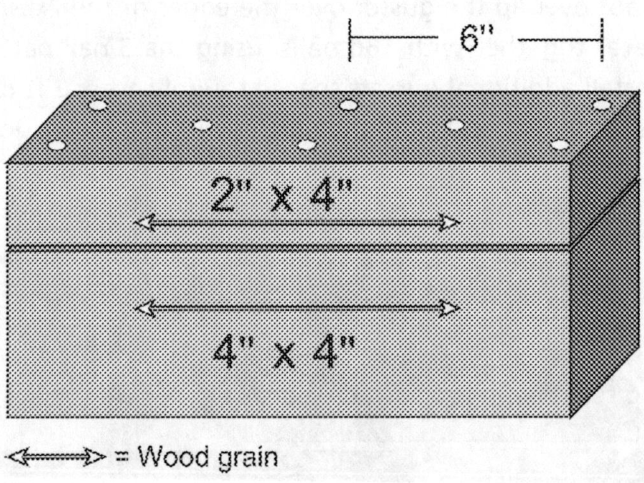

Fig. 3–28 Close-up of the 5-nail pattern.

Using two 4x4s, rather than a 2x4 and a 4x4, gives you yet another option, which is actually stronger than two 4x4s just stacked on top of each other. As long as you nail the gussets on both sides of the 4x4s with the 5-nail pattern as described, spacing the nails 6 in. apart, you will have a header that acts like a 7-in. beam. In this way, the two 4x4s cannot move separately, making them more efficient.

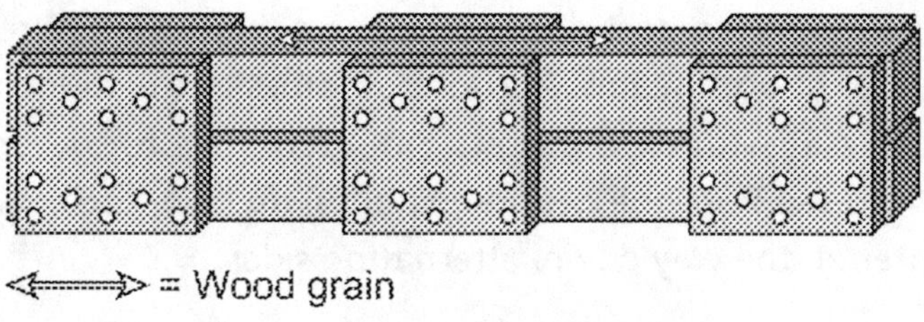

Fig. 3–29 Dual 4x4 header option.

Prefabricated posts

If your team is in need of heavy-duty posts but all you have is 4x4s or 2x6s, you can improvise to create the post you need. There are several methods, but the following are just a few of them.

Anchor four 2x6s together to make a 6x6. To accomplish this, nail one 2x6 into the next, using 16d nails 8 in. on center and stagger the location as (see Fig. 3–30). Try not to nail into other nails by alternating the pattern every piece.

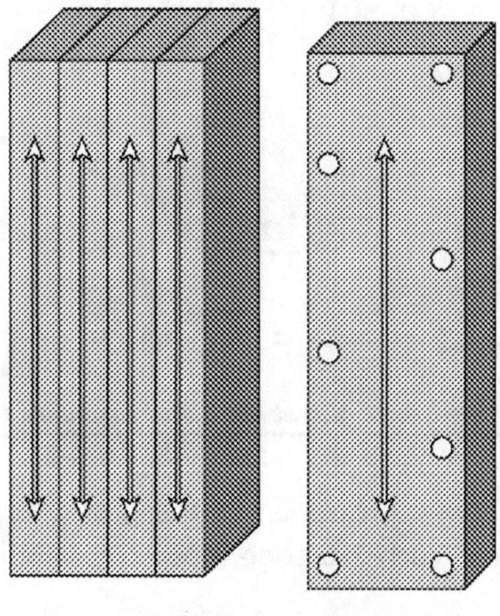

Fig. 3–30 Prefabricated 6x6 post using four 2x6s.

To make sure the post does not separate under pressure, anchor gusset plates to both seams. Anchor both sides of the four connected posts every 4 ft with 6x12 gussets. An alternative to this anchoring method is to use ½ in.-thick carriage bolts 6 in. from each end, then 3 ft on center all the way down, alternating sides.

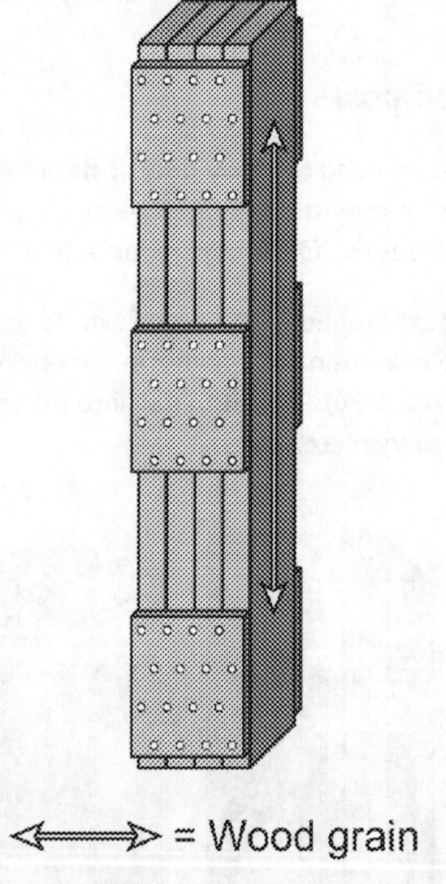

Fig. 3–31 6x12 gusset plate anchors every 4 ft.

Figure 3–32 shows a close-up of the gusset nail pattern, using 8d nails nailed into each 2x6 edge. Keep the nails roughly 4 in. on center.

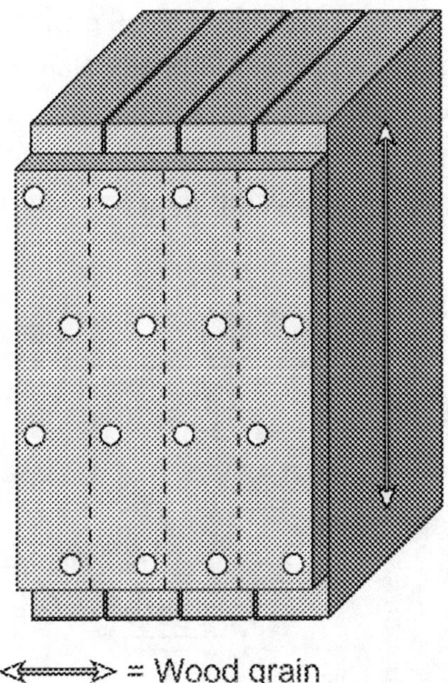

Fig. 3–32 Close-up of gusset plate nail pattern.

To prefab a 4x4 post (should you need to), anchor a piece of ¾-in. plywood in the center of the two 2x4s. Nail with 16d nails every 4 in.

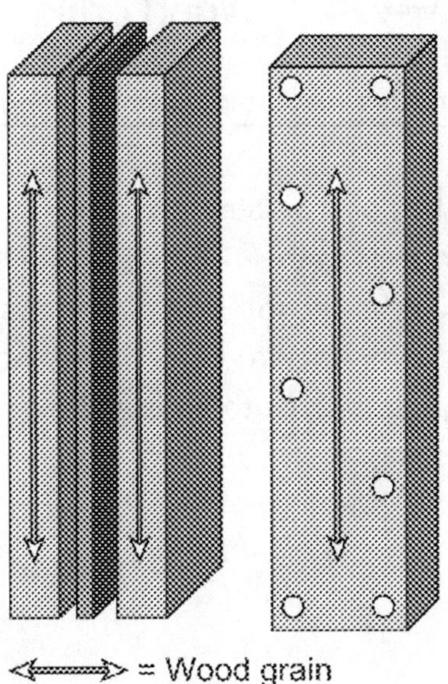

Fig. 3–33 Prefabricated 4x4 post.

Another method of constructing a 6x6 post is to anchor a section of ½–¾-in. plywood to a 2x4 with 8d nails. Nail every 8 in. on center, staggered. Next, nail that piece to a 4x4, using 16d nails, staggering the nailing points every 8 in. on center. Now, anchor a 2x6 to the sides of both the 2x4 and 4x4, using 16d nails 8 in. on center.

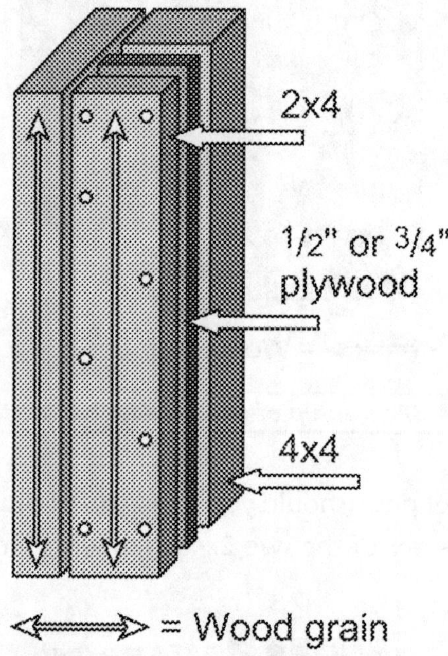

Fig. 3–34 Another method of prefabricating a 6x6 post.

The last step to this post system is to place 6x12-in. gussets where the 2x4 plywood connects to the 4x4 section of the post. The reason to do this is to sandwich the section so it does not separate under pressure. Use the same 5-nail pattern 6 in. apart, anchor the gusset with 8d nails. Place the gussets on both ends and every 4 ft up the post.

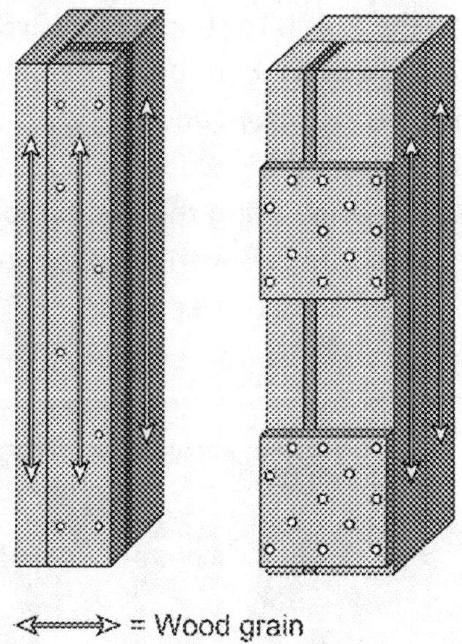

Fig. 3–35 6x12-in. gusset plate anchors every 4 ft.

On the rare occasions when you may need an 8x8 post, you can utilize sections of 4x4s sandwiched together, giving you a 7x7 post, which is generally close enough. Anchor 7x12-in. gussets to all sides at top, bottom, and 3 ft on center throughout the rest of the post, using the 5-nail pattern.

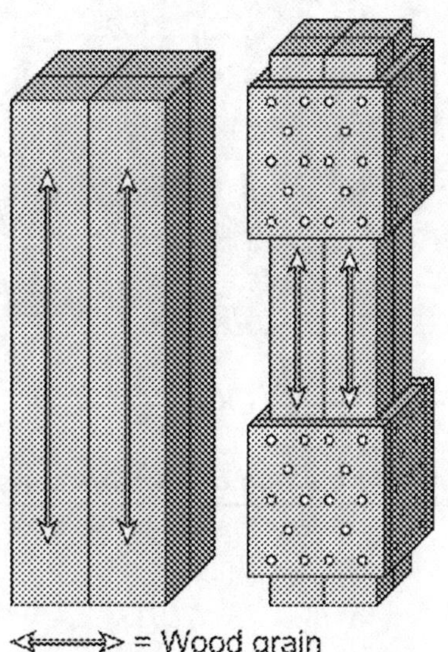

Fig. 3–36 Prefabricated 7x7 post with 7x12-in. gusset plate.

An alternative to the 4x4s is to connect five 2x8s together. Anchor each piece with 16d nails every 8 in. on center. Install ½-in. carriage bolts 6 in. from the top and bottom and every 3 ft on center throughout the whole beam section. Connecting the 2x8s with carriage bolts is the preferred method when using this size lumber. However, if you don't have bolts, connect the 2x8 with 6x12-in. gussets on both 6 in. faces, using five 8d nails 3 ft on center.

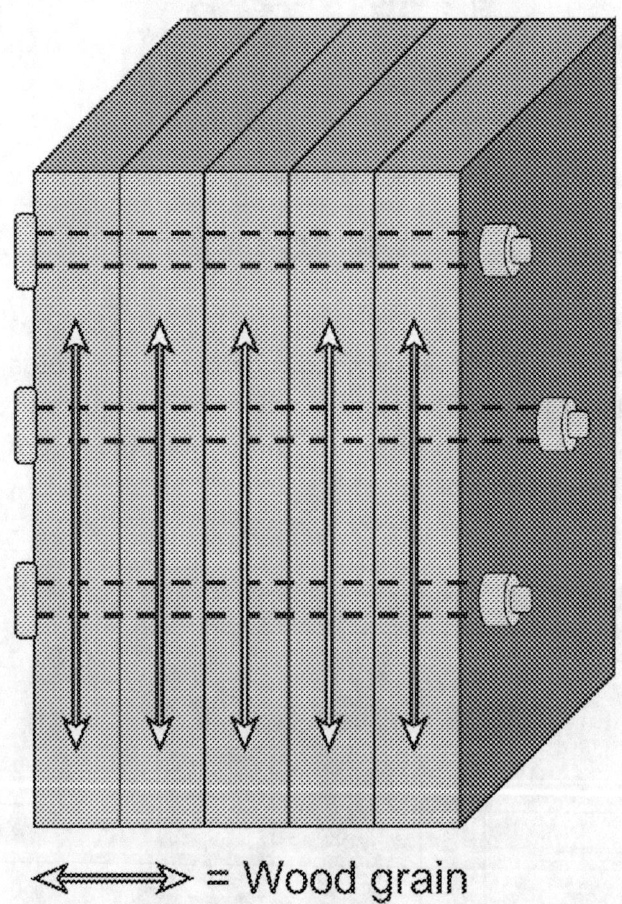

Fig. 3–37 Another method of prefabricating a 7x7 post.

Diagonal Braces
and Raker Shores Angles

Determining the angles for diagonal braces

Although for the most part, you do not have to cut angles into the ends of most diagonal braces, you can do so to make a better fit and to have more surface for nailing. (Also, cutting angles into diagonal braces lets you actually look like you know what you are doing to the rest of the rescue workers.)

To determine the angle to cut, use a tried-and-true method that has been around for many years in the building trades. It takes a little practice to do right, but it is not difficult after you have done it several times. Known in the building trade as the 12-step method, it is how carpenters determine length and angles for rafters. Because the same basic principles apply to rafters and to braces, use the rafter-framing square to determine the length and angles.

The simplest and least confusing way to determine the angles to cut is to make a sketch of the brace and the angles you want. Doing this makes it easy to see the cuts you have to make and the direction of the angles, leaving less chance for error.

The 12-step method starts with determining the height and length of the shore or the wall area to be braced. To determine the angles, take the foot measurement from the shore and transfer it to inch increments on the framing square. For example, if you have a brace that is 8 ft high and 8 ft long, use 8 in. on the tongue of the square and 8 in. on the blade of the square. Figure 3–38 shows a diagonal brace with the angles for a wall section 8 ft high and 8 ft long. When determining the overall length of the brace, take both measurements from the same inside corner.

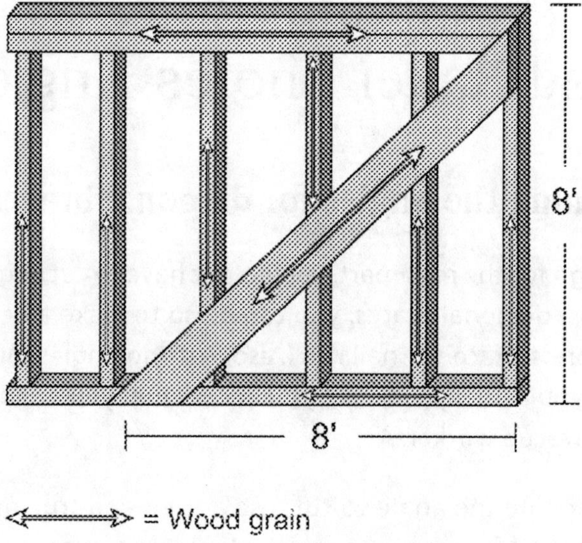

Fig. 3–38 Diagonal brace for 8x8 wall section.

Figure 3–39 shows how to lay out the square on the brace material to determine the angles and the proper length. First, place the square at the 8 in. mark on the blade and the 8 in. mark on the tongue right on the face of the 2x6; this gives you the proper angles for the cut. Keep the same numbers on the face of the lumber at all times, 8 and 8. To do this, make a mark with a pencil at every step, then move the square down to the mark and repeat.

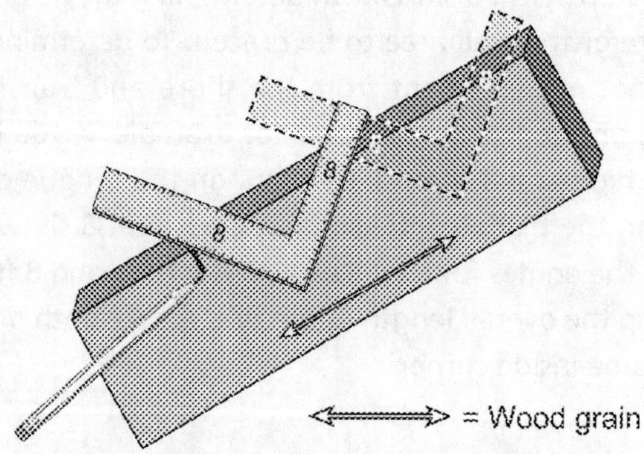

Fig. 3–39 Determining the angles and proper length for the brace material.

The next step is to move the square a total of twelve times down the edge of the lumber. As shown in Figure 3–40, move the square, holding the numbers at each edge. After the twelfth move, make a mark on the bottom of the square across the face of the brace. This is the exact angle you need to cut. The brace will fit perfectly—if you did it right.

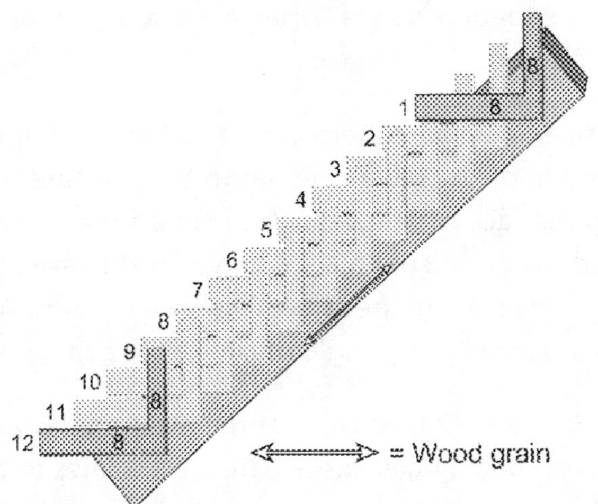

Fig. 3–40 Moving the square 12 times down the edge of the lumber.

Figure 3–41 shows the brace with the proper angles and exact length.

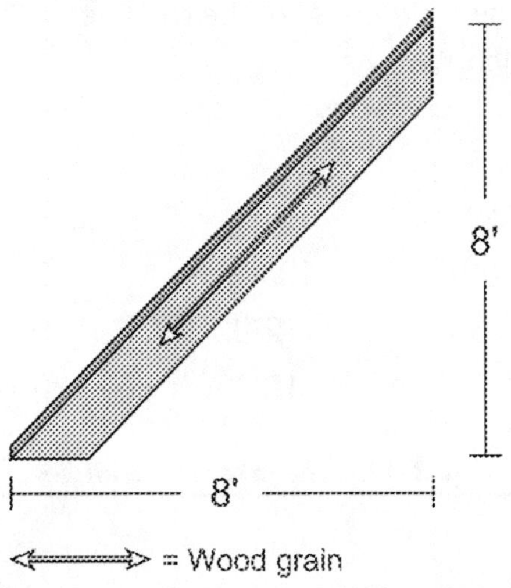

Fig. 3–41 Diagonal brace with the proper angles and length.

Determining the angle of raker shores

Generally speaking, any angle 30–60° works effectively; however, the greater the pitch of an angle above 30°, the less lateral force is applied to the raker. As a result, the raker comes under less compression and is subject to more upward lifting force. The higher the angle, the more this situation occurs. After a 60° angle, shoring basically becomes ineffective.

When utilizing a raker shore with a 45° angle, the pressure applied to the face of the raker is equal to the pressure trying to ride the raker up the face of the wall. This makes for a balanced system, which is good. When you use the 45° angle for the raker, both the top and bottom angle cuts are the same. This makes fabricating the shore much simpler. Normally, the angle of choice for a raker shore is 45°.

Refer to Figure 3–42 for a chart of the three angles you may want to use when erecting a raker shore. The pitch relates to the numbers on a carpenter's framing square. Pitch is the slope, or the angle, of the raker. Without getting too complex, pitch has to do with the rise and run. For example, a 45° angle has a pitch of 12 on 12 (12/12). What this means is that for every foot of height, the base of the raker will be out from the face of the wall 1 ft. This is a constant; it doesn't matter if the height is 6 ft or 60 ft. The length of the distance between those two 12-in. points, or the hypotenuse of the triangle, is 16.97 in.; therefore, use 17 in. as the length.

Degree	Pitch	Length
45°	12/12	17
54°	12/9	15
60°	12/7	14

Fig. 3–42 Three raker shore angles.

The next angle you may use is roughly 54° although this angle really isn't used too often. The pitch on the square is 12/9, and the hypotenuse is 15 in. Many carpenters use this pitch to square things up. It's the old 3, 4, 5 method; 9, 12, 15 is just an extension of these numbers. Some personnel use this angle because of its ease of measuring.

The steepest angle most firefighters work with is the 60° angle and it is used when the length of lumber comes up a bit short for the insertion point. It is also the recommended angle for the split-sole raker shore. The pitch for this one is 12/7, and the length between those two points (hypotenuse) is 13⅞ in. Most personnel round the length up to 14 in. The pitch numbers are the ones to use on the framing square to determine the different angles.

Raker foot lengths per foot of rise

It's okay; I did badly in math too. Therefore, this section of the chapter breaks the subject down to explain where all the numbers come from and why they are used.

To determine the length of a raker, you must know two things: the angle you want and the height up the wall you want the face of the raker to be. The foot height up the wall where the raker intersects the floor is known as the insertion point. Once you know this, all you need to do is determine the angle to use. The angle defines the length per foot of the raker.

The formula determining the length of the raker is pretty simple. Multiply the wall height of the raker face (insertion point) by the hypotenuse of the angle. The answer is the length of the raker in inches. Let's say you want to place a raker at a 9 ft insertion point and will construct the raker at a 45° angle. At a 45° angle, the hypotenuse is 17 (actually 16.97). The equation for this is 9 x 17 = 153 in., which is the raker length from tip to tip.

Assume in all cases that you are dealing with a right triangle; therefore, at the base of the raker shore the angle is 90°.

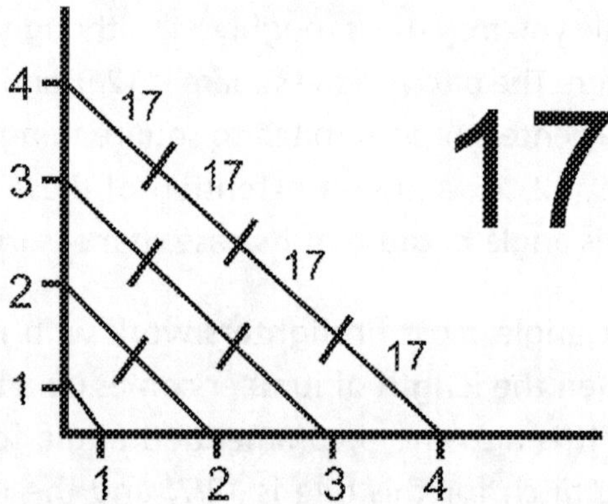

Fig. 3–43 Assume you are dealing with a right triangle.

With a 54° angle for every foot of rise, the length out from the structure will be 9 in., a 12/9 pitch. The hypotenuse for this angle is 15 in. This is an exact measurement. That is why carpenters use 3, 4, 5; 6, 8, 10; and 9, 12, 15 to square items. The formula for the 54° angle is the wall insertion height in feet multiplied by 15. For example, if you have a 9-ft insertion point, the formula is 9 x 15 = 135 in. This means that the length of the raker is 135 in. long tip to tip, and it will sit with the 54° angle 9 ft up the face of the wall. It will extend out from the wall 81 in. This is the proper procedure for determining the 54° angle. Although not often used, it is an option that your team has.

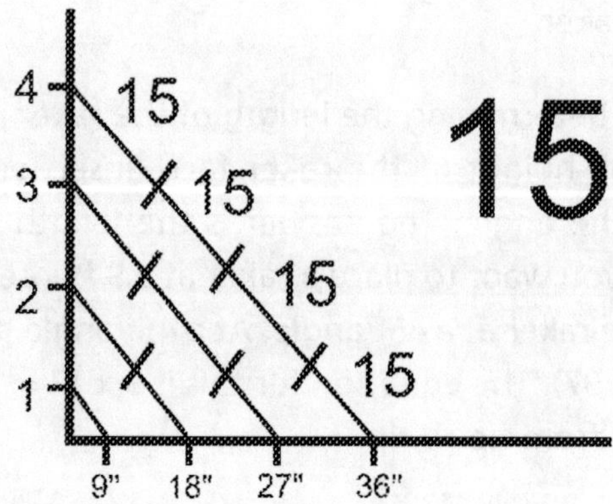

Fig. 3–44 Determining the 54° angle.

With a 60° angle for every foot of rise, your length out from the structure is 7 in., a 12/7 pitch. The hypotenuse for this angle is 14 in. (actually 13⅞ in., or rounded to 14 in. for ease of installation). The formula for the 60° angle is the insertion height in ft multiplied by 14.

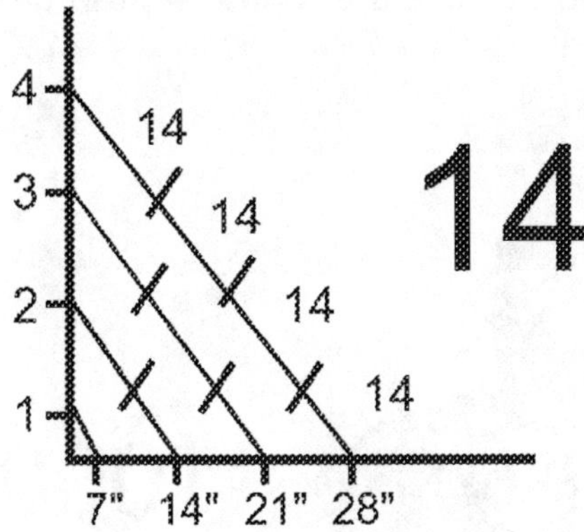

Fig. 3–45 Determining the 60° angle.

For example, for a 9 ft insertion point, the formula is 9 ft x 14 = 126 in. This means the length of the raker is 126 in. long tip to tip. It sits with the 60° angle face 9 ft up the face of the wall and extends out from the wall 63 in. As you can see, this angle is much steeper than the 45° one, and as a result the raker is much shorter. This is a definite advantage when you don't have access to longer lumber.

Angle effect

Whenever a raker shore is placed against a structure, the force from the building pressurizes it. The amount of force affecting the shore depends on the angle of the raker. There are three main directions of force against the shore.

- Down: the direct loading onto the raker itself.

- Horizontal: the force being applied to the raker trying to push it off the wall.

- Up: the force pushing the raker up the face of the wall.

When the raker has a 45° angle, the amount of uplift force is roughly 71% of the diagonal force. The horizontal force trying to push the raker off the wall is also 71% of the diagonal force. At a 45° angle, these forces are equal. Simply, for every 1000 lb of force applied to the raker, 710 lb of force are trying to lift the raker off the ground; and 710 lb of force are trying to push the raker away from the building.

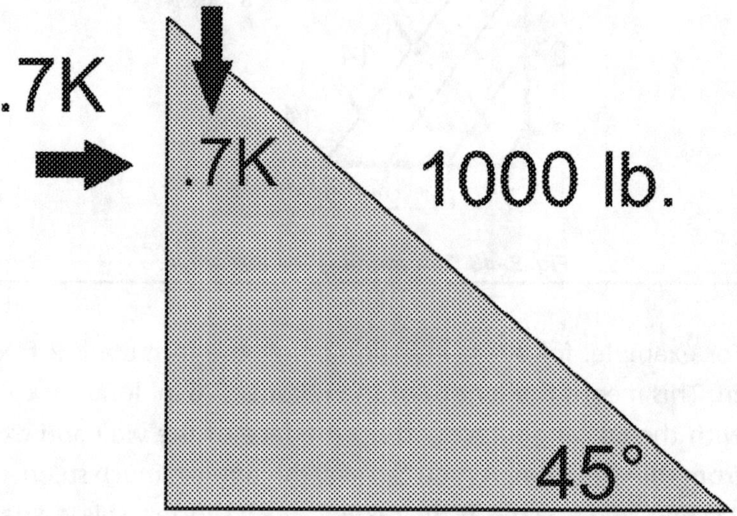

Fig. 3–46 Force affecting a 45° angle.

For a 60° angle, things are little different. Since the angle is much steeper than 45°, the amount of each force is significantly different. At this angle, the horizontal force is 50% of the diagonal force. However—and this can be a problem—the force trying to push the raker up the face of the building is 87% of the diagonal force. This means for every 1000 lb of force, there will be 500 lb of horizontal pressure and 870 lb of uplift force. For this reason, and in order to counteract the extensive uplift forces when installing rakers with angles above 45°, you have to increase the size of the top cleat from a 2-ft cleat to a 3-ft cleat.

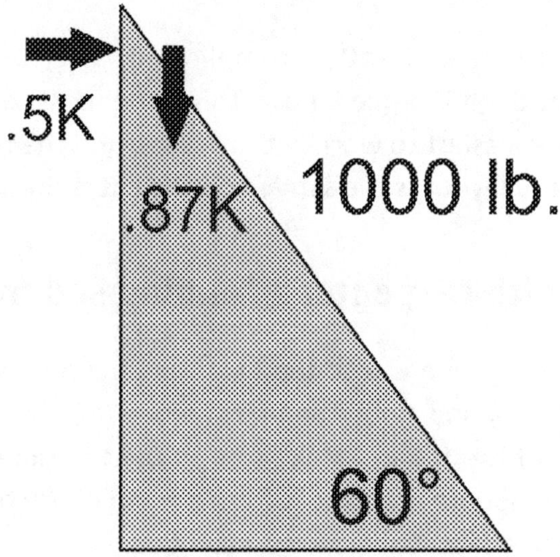

Fig. 3–47 Force affecting a 60° angle.

At the much lower angle of 30°, the effect is just opposite that of the 60° angle. The horizontal force is 87% of diagonal force, and the uplift force is just 50%. If for some reason a raker is installed at this angle, the anchor system has to be twice that used with a 60° angle. You may need as many as (4) 1-in. pins to hold this raker in place.

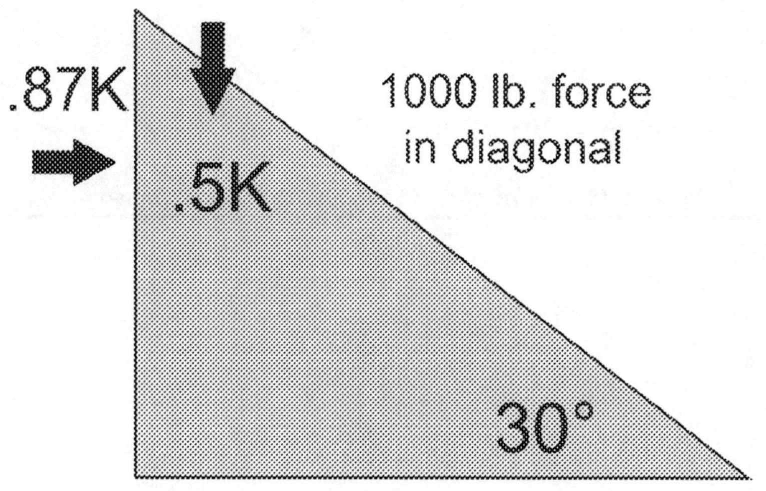

Fig. 3–48 Force affecting a 30° angle.

Laying out the 45° raker

Although it may look rather complicated, after a few practice tries, laying out the 45° angle is easy. This is the most common angle used. It's also the easiest to work with. The top and the bottom angle are the same, the layout is the same, and the cut is the same.

45° angle with carpenter's framing square

Lay the framing square onto the 4x4, using the numbers 12 on the tongue (1½-in. end) and 12 on the body (2-in. end), and scribe a mark along the face of the square. Make sure to use the same number on both ends of the square and take it from the same face of the square, usually the outside face. This action always guarantees a 45° angle.

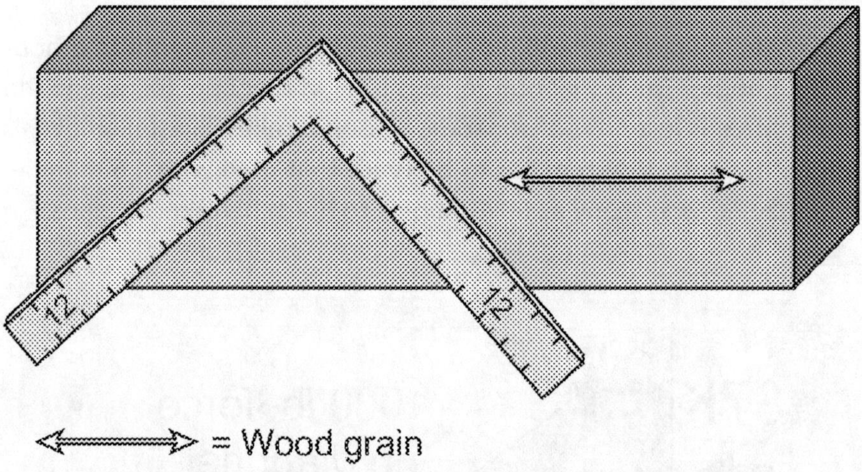

Fig. 3–49 Measuring the 45° angle with the framing square.

Figure 3–50 shows the way the mark should look. It is 3½ in. from the end face of the lumber. Note that with a 45° mark, the distance from the outside face is the same as the thickness of the material.

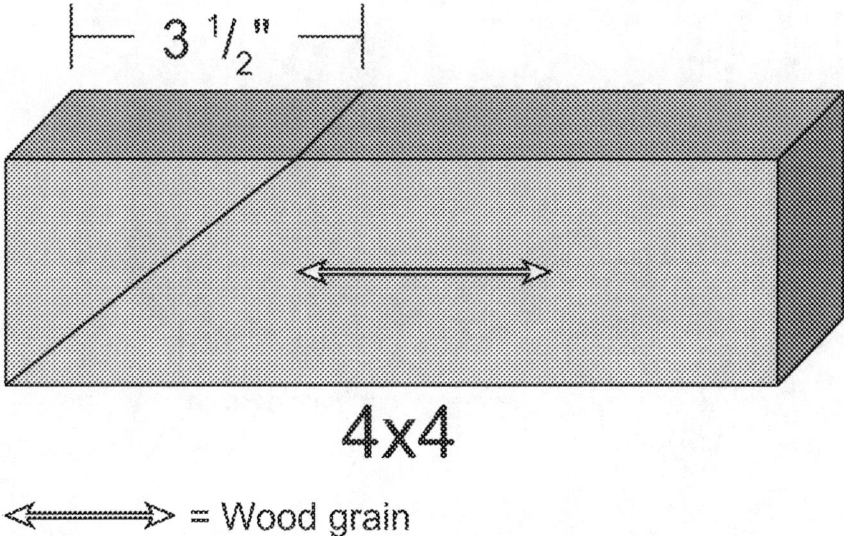

Fig. 3–50 What 45° angle markings should look like.

In Figure 3–51, the angle has been cut. You can use a chain saw or a 10¼-in. circular saw to get this cut. Both tools are acceptable. A power miter box can also be utilized for perfect cuts.

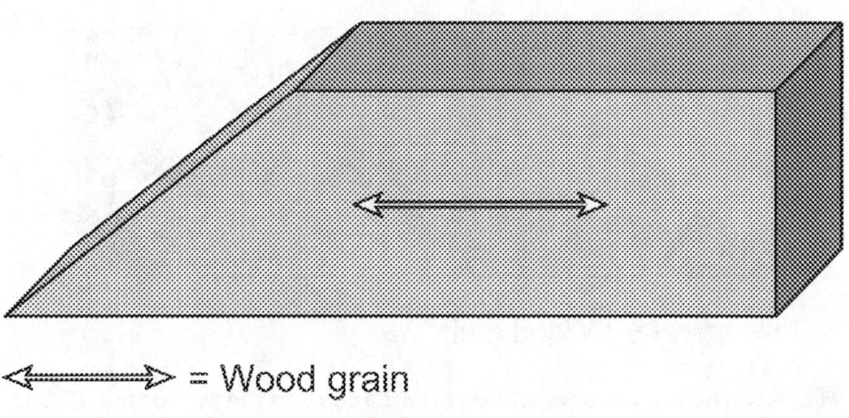

Fig. 3–51 Completed 45° angle cut.

Now place the inside face of the square against the angle you just cut. From the inside corner of the square, slide the square down the face until the space on the tongue is 1½ in. from the face of the cut to the edge of the lumber. Mark this line; this is the return cut.

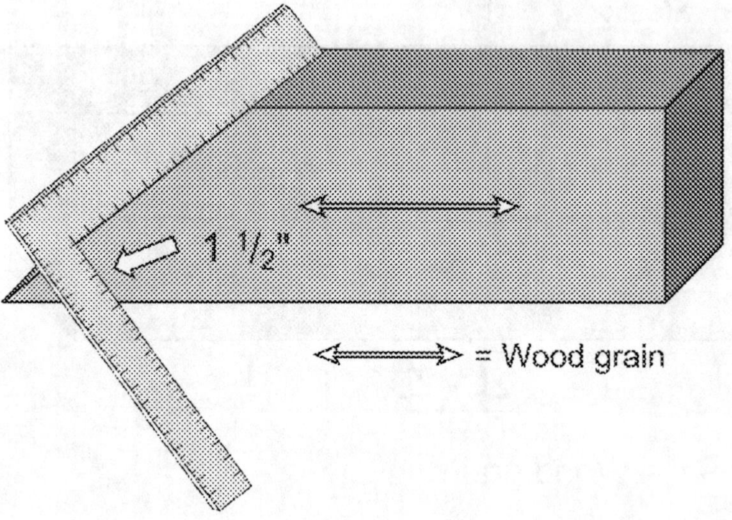

Fig. 3–52 Measuring the return cut on the 45° angle with the framing square.

The 1½-in. return cut must be at right angles to the face of the 45° cut in order for the return cut to work properly. The cut is 1½ in. deep to accommodate the cleat, which will be a 2x4.

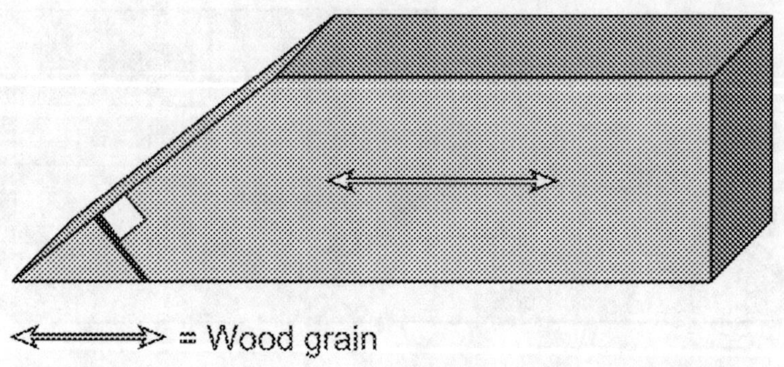

Fig. 3–53 The return cut must be at right angles to the face of the 45° cut.

Figure 3–54 shows how the end of the 4x4 raker looks after both cuts have been made.

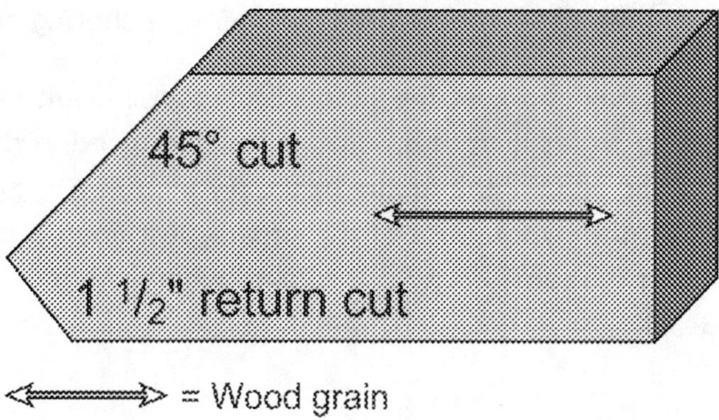

Fig. 3–54 The completed 45° 4x4 raker end using the framing square.

Utilizing the speed square to mark angles

The speed square is a little different from the framing square. Instead of utilizing two legs like the framing square, the speed square utilizes only one leg. Make sure you take your mark on the same leg of the square all the time. It is important that all the marks you make are on the side of the speed square where the pivot point is. If you set up your angle and scribe off the hypotenuse of the square, the angle will be wrong. Also, when marking the opposite side of an angle using a speed square, you must flip the square over to get the angles opposing each other. It you don't flip the square over, the angles you mark are parallel with each other, making the shore inaccurate.

The speed square comes in two sizes: 6x6 and 12x12. Notice that the square is in the shape of a triangle. On the 6x6-in. square, both sides of the square are 6 in. long, which makes the angle of the hypotenuse a 45° angle. The triangle shape of the tool is known as an *isosceles* triangle. Isosceles was a Greek mathematician who lived several thousand years before Christ, so this stuff was figured out quite a while ago. Basically, what he was trying to say is that if you have a triangle in which one angle is 90° and the two opposite legs are both congruent (equal), then the angles formed by the other two legs must be 45°. This design of the speed square is based on this geometric principle. It is also the principle used to determine shoring angles.

Instead of having to set the square at two points with this tool, you only need to set it at one point. All of the marking should be done on the top edge of the square by the pivot point. The angles are all marked on the hypotenuse of the square. Just line up the angle number along the same edge of the raker where the pivot point is (see Fig. 3–55).

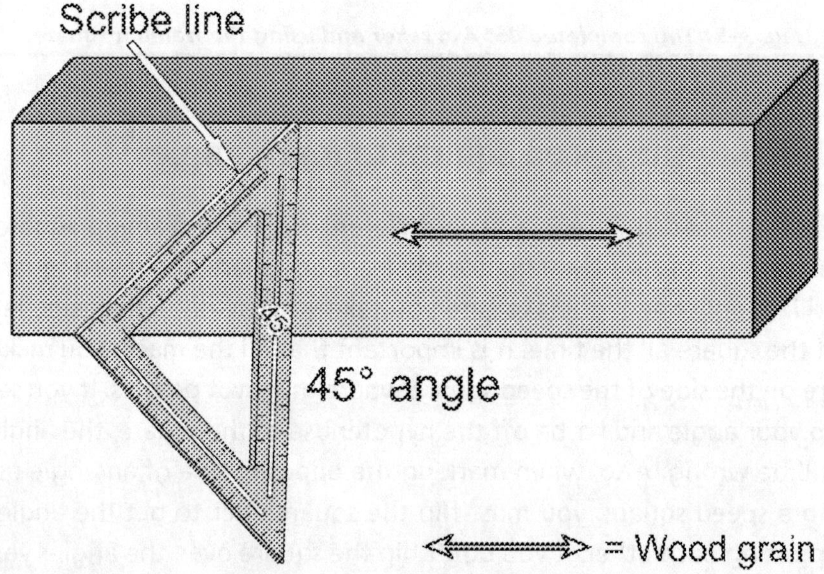

Fig. 3–55 Line up the angle number along the same edge as the pivot point.

As with the framing square, take the speed square and place it along the face of the 45° cut. Flip the square around and place the flange with the right angle against the cut you just made. Slide the square up until you have that 1½-in. return space that you need for the cleat.

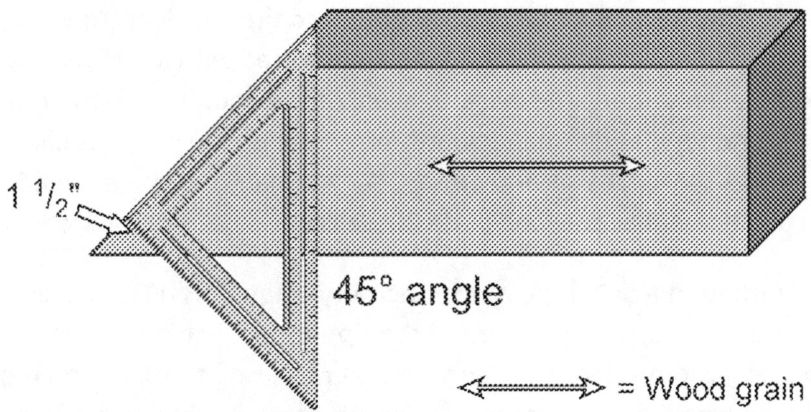

Fig. 3–56 Measuring the return cut on the 45° angle with the framing square.

Figure 3–57 illustrates what the cut looks like if you used the square properly. Either square works fine; your team can decide which one to use.

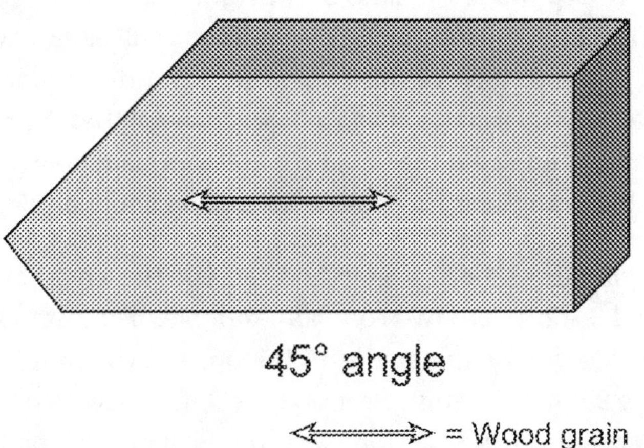

Fig. 3–57 The completed 45° 4x4 raker end using the speed square.

Laying out the 54° raker. Whenever you decide to erect raker shores at angles greater than 45°, additional factors come into play. First, as the angle of the raker gets steeper, additional upward forces apply to the shore. To counteract this additional stress on the raker, you must do one of two things.

The first option is to add another foot to the top cleat. This will result in your top cleat being 3 ft long and secured with additional nails, spaced evenly apart. The extra nails will provide additional strength to counteract the additional upward force that is now being applied to the raker shore. The second option, not normally recommended, is to place a notch into the wall plate at the raker's insertion point.

Although a 54° angle is not used very often, it is relatively simple to figure out and can be one of the options available to your team. You can lay out a 54° angle for your raker by using the 3, 4, 5 method, which immediately gives you the length of your raker with any multiple of the numbers above. This method can be especially useful if you must install the raker shore higher than usual, and you are limited on the lumber lengths available.

The 3, 4, 5 method is based on the Pythagorean Theorem, a mathematical theorem for determining the hypotenuse of right triangles—the length of the rake on the various types of raker shores. The numbers correspond to the sides of the right triangle. A right triangle with one leg 3 ft long and another leg 4 ft long will have a hypotenuse 5 ft long. In the context of this book, the legs of the triangle are the raker's wall height and the distance from the wall that the raker rests. The two angles opposite the 90° angle are approximately 54° and 36°. Either of these angles is acceptable for a raker shore angle.

Any multiples of the 3, 4, 5 method give you the length of the raker; 6, 8, 10 and 9, 12, 15 are two common examples. For instance, let's say your available shoring material is 16 ft long, and you must assemble a raker shore 9 ft high. In this case, using the 9, 12, 15 sequence gives you a 36° raker shore. The raker intersects the wall at 9 ft; the bottom of the raker lies on the sole plate back 12 ft from the wall plate; the raker length is 15 ft and set at an angle of approximately 36°. This shore is quite substantial and effective at supporting a heavy load.

Now, let's assume you want to shore a wall at a raker insertion height of 12 ft. Place the raker against the wall plate at a height of 12 ft. Make sure the wall plate is at least 3 ft higher than the insertion point. The base of the raker should rest 9 ft back from the wall plate on the sole plate. Once again, your raker will be 15 ft long. The angle of this raker, however, will be approximately 54°, which cannot support as great a load as a raker set at 36° without some more substantial anchoring into the building and larger top cleats. This is due to the greater slope of the raker shore against the structure. This angle is not commonly utilized but is available to your rescue team if you should decide to use it.

To lay out a 54° angle, place the framing square at one end (see Fig. 3–58). Using the *9* on the tongue and *12* on the body, lay it down similar to way it is shown in the graphic. Place a mark on the left end face of the square at the number *9* to indicate the cut line for the base of the raker. Cut this angle and then place a return cut into it also, to accept the 2x4 cleat.

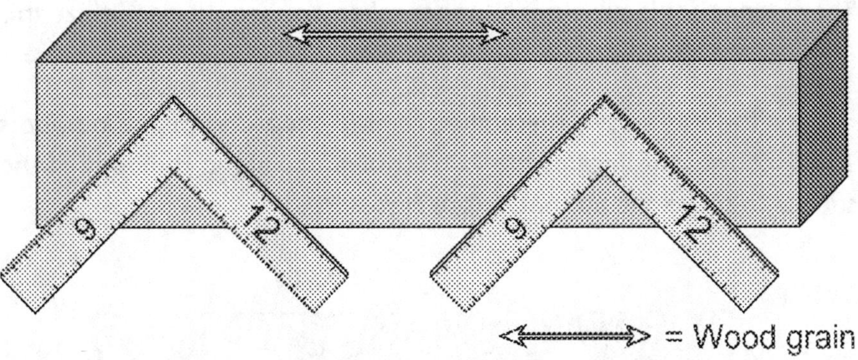

Fig. 3–58 Measuring the 54° angle using the framing square.

For the second angle cut, with the square facing the same way as in the previous paragraph and on the same side of the lumber as the first cut, walk down to the desired length and place the square as shown in Figure 3–58. Scribe a line along the 12-edge side of the body of the square. This angle is the opposite of the one for the first cut. Don't forget to mark your return cut also.

Figure 3–59 shows how the raker looks when cut properly. You will note that the angle cuts are different. The steeper cut on the raker is the face that will be applied to the wall.

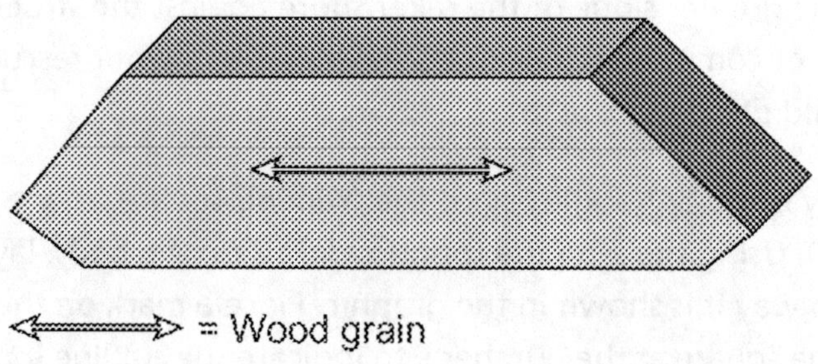

Fig. 3–59 The completed 54° 4x4 raker.

Laying out the 60° raker. This angle of raker is used when the available material is not long enough to reach the wall insertion point. The steeper angle provides a longer raker. You would also utilize this angle when installing the split sole raker.

For the 60° angle, place the framing square with the numbers 7 and 12 onto the face of the 4x4. Scribe a line along the body of the square to make the 60° angle (see Fig. 3–60).

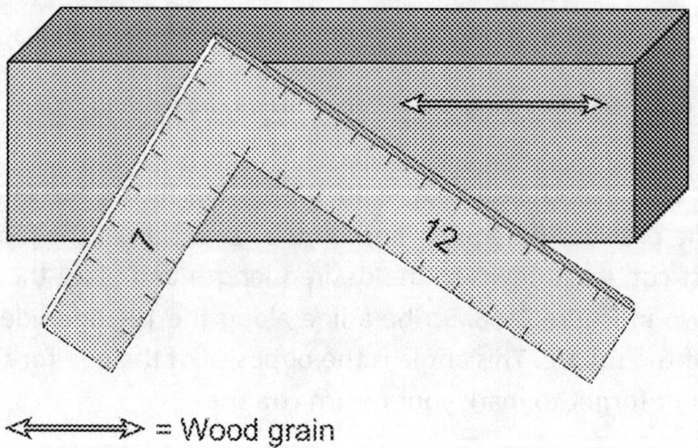

Fig. 3–60 Measuring the 60° angle using the framing square.

Figure 3–61 shows what the first angle cut for 60° would look like.

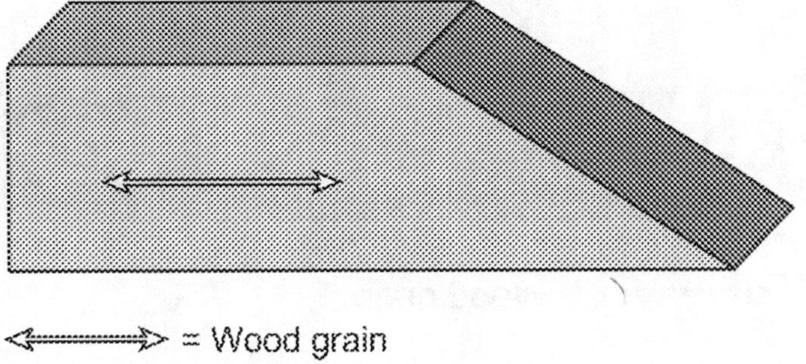

Fig. 3–61 The completed 60° angle cut.

Place the square along the cut face of the raker, slide the square along the face until you have a 1½-in. space then scribe the line.

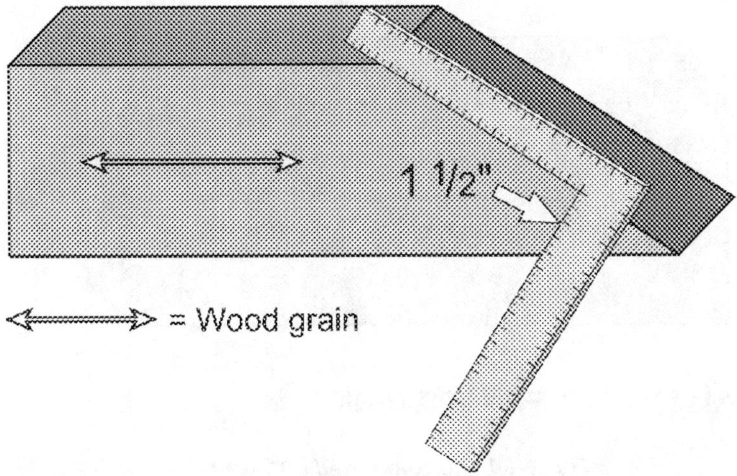

Fig. 3–62 Measuring the return cut on the 60° angle with the framing square.

As with the 45° cut, you must place a return cut to accept the cleat. This is 1½ in. deep and at right angles to the face of the cut. This is very important. Notice that the return cut is in a much different location than the 45° angle return cut (see Fig. 3–58).

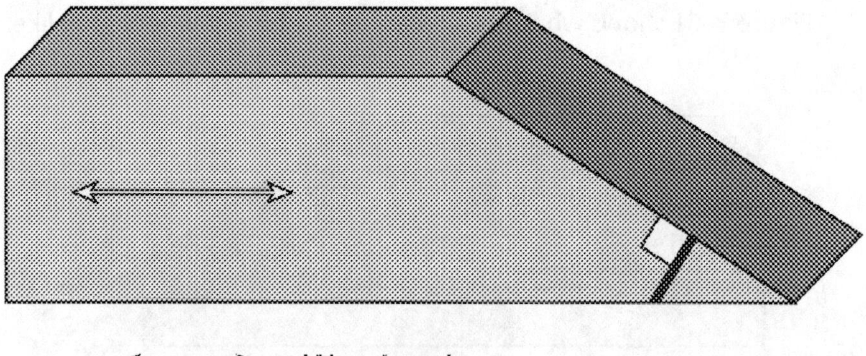

Fig. 3–63 Place the return cut to accept the cleat—1½ in.-deep and at right angles to the face of the cut.

Figure 3–64 shows the 60° angle cut with the 1½-in. return cut finished. This part of the raker is ready to be installed.

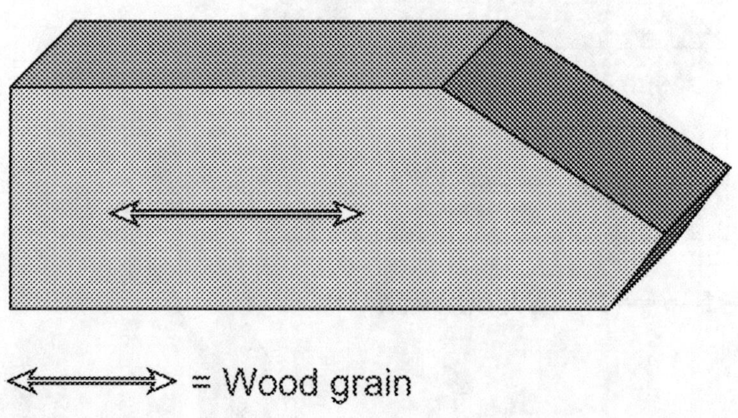

Fig. 3–64 The completed 60° 4x4 raker.

Cutting the 60° angle with the speed square. At the bottom of the 4x4 (the face closest to you), place the speed square and slide it along the pivot point until you reach the 60° mark. Scribe your line along the top and extend it to the end of the raker.

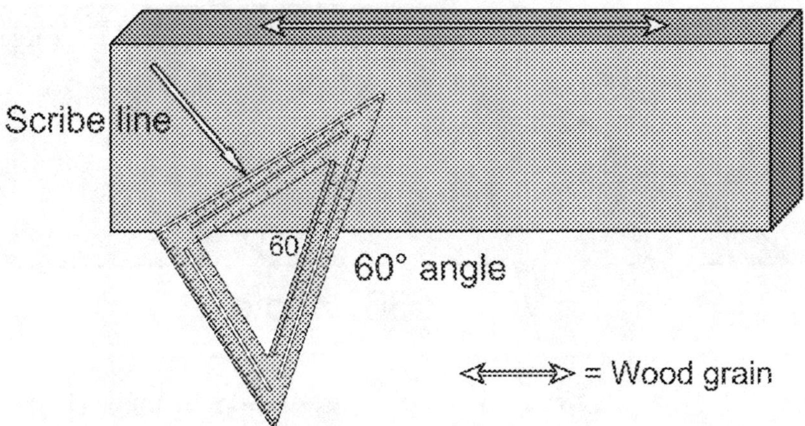

Fig. 3–65 Measuring the 60° angle using the speed square.

Figure 3–66 shows the angle properly cut.

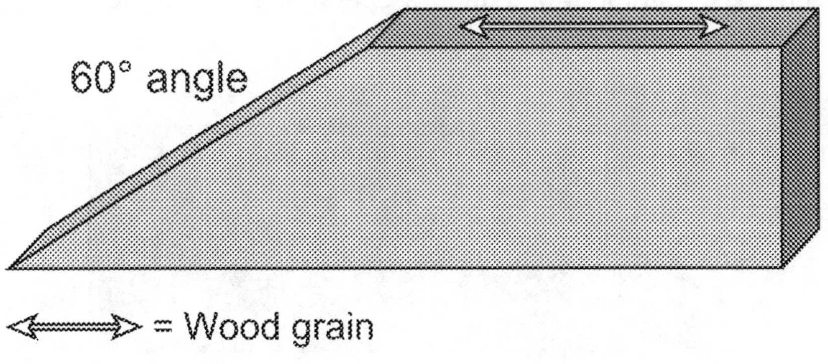

Fig. 3–66 The completed 60° angle cut.

Next, make your return cut. Place the square on the face of the cut and slide the square up until you have a 1½-in. space.

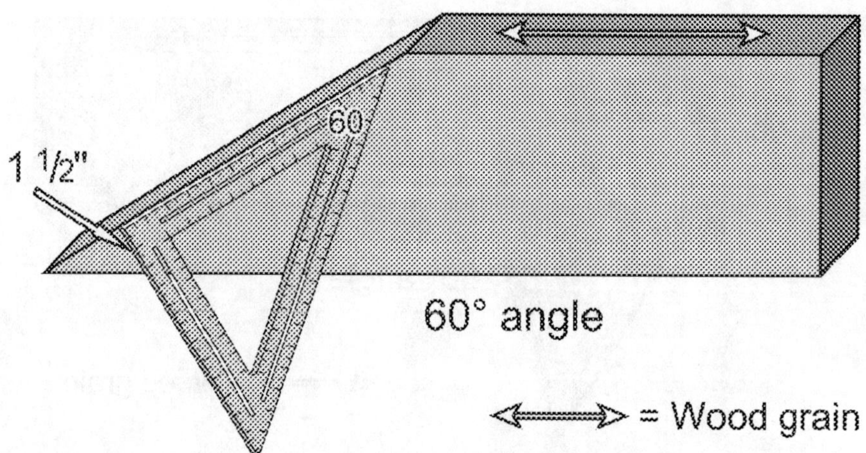

Fig. 3–67 Measuring the return cut on the 60° angle with the speed square.

Figure 3–68 shows the way the cut looks when all the marks have been laid out properly and the saw operator cut along them.

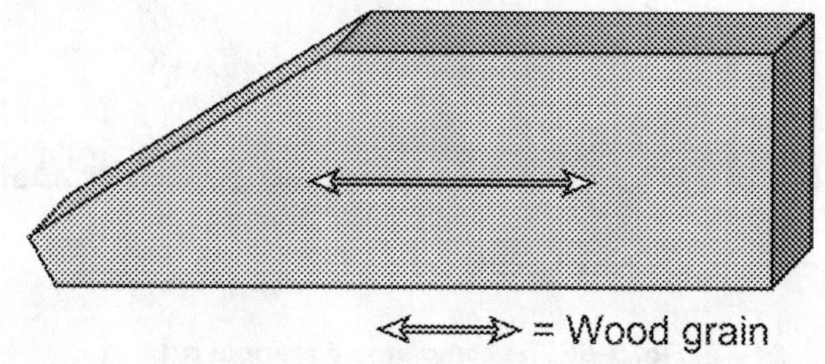

Fig. 3–68 The completed 60° 4x4 raker.

Laying out the 30° cut. For the proper angles, utilize the framing square with the numbers 7 and 12.

For the 30° cut, use the tongue side of the square showing the number 7. Place a mark along the outside face of the square.

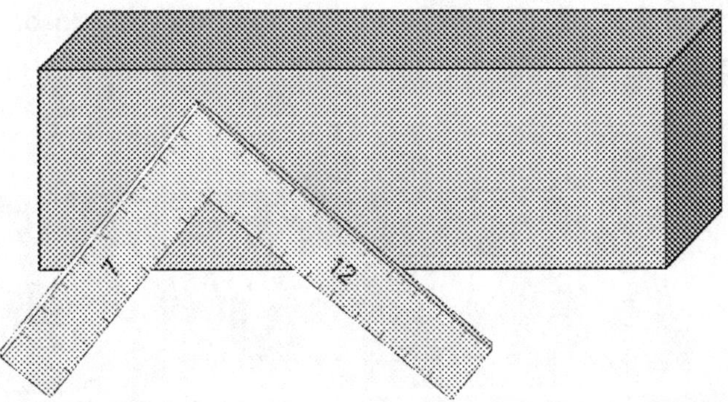

Fig. 3–69 Measuring the 30° angle using the framing square.

Figure 3–70 shows the angle cut. Notice how much tighter this angle is.

Fig. 3–70 The completed 30° angle cut.

As with all the other angle cuts, you must make a return cut into the raker. Slide the square up along the cut face until you have a 1½-in. space then mark it (see Fig. 3–71).

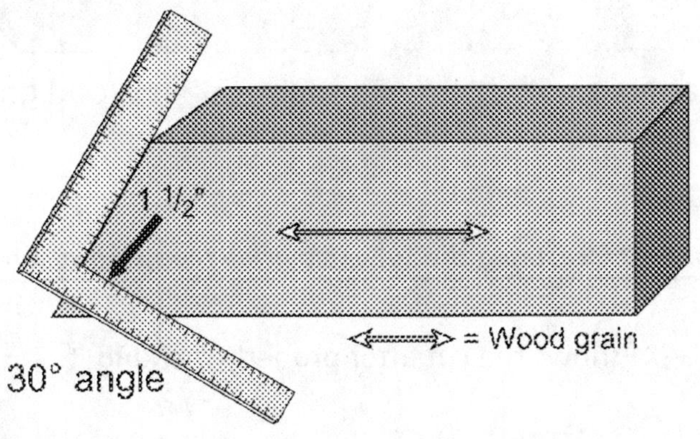

Fig. 3–71 Measuring the return cut on the 30° angle using the framing square.

Now your 30° cut is done, and the raker is ready to be placed in the shore.

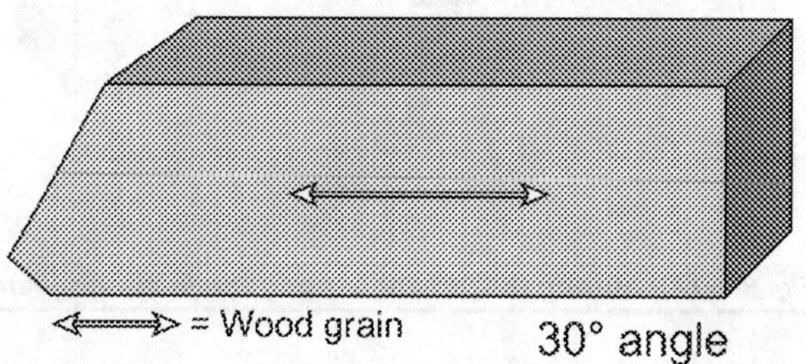

Fig. 3–72 The completed 30° 4x4 raker.

Cutting the 30° angle with the speed square. From the inside corner point, pivot the square around until you have the 30° mark on the hypotenuse. Place that mark along the same face as the pivot point and scribe the line on the top face of the square.

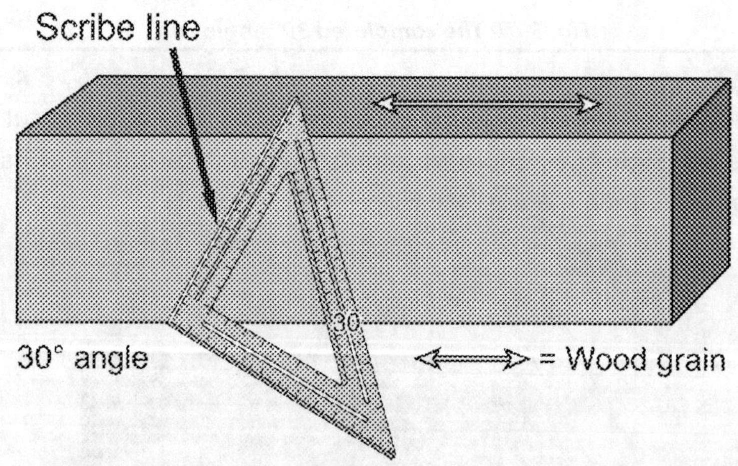

Fig. 3–73 Measuring the 30° angle using the speed square.

Figure 3–74 shows the cut after properly marking it.

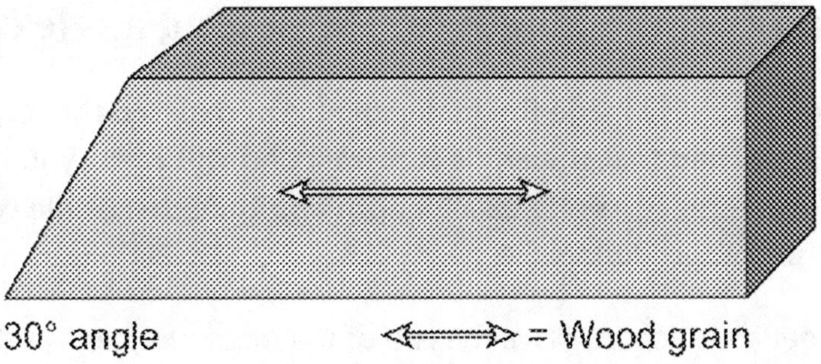

Fig. 3–74 The completed 30° angle cut.

Slide the square along the face of the 30° cut and make sure you have a space of 1½ in. at right angles to the face.

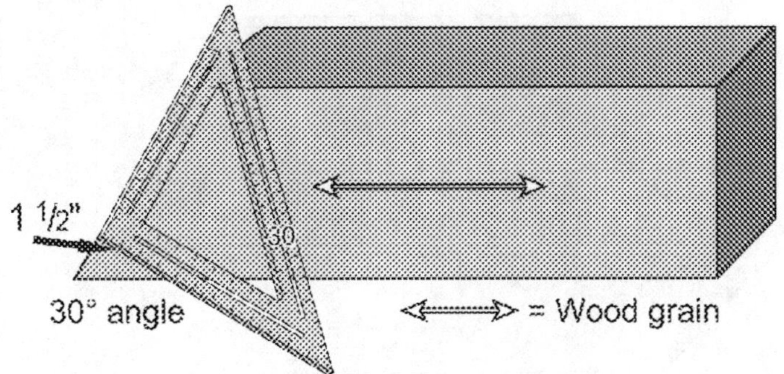

Fig. 3–75 Measuring the return cut on the 30° angle using the speed square.

Figure 3–76 shows the final cut. When both angles have been cut then the raker can be placed in position.

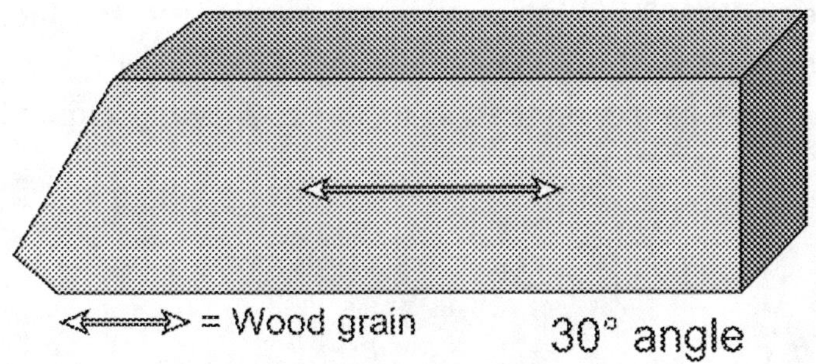

Fig. 3–76 The completed 30° 4x4 raker.

Marking off the length for the second angle cut

After you have made the first angle cut and placed the return cut and are satisfied that the angle is correct and will work, you have to tape out the raker for the exact length and the other angle cut. The correct procedure follows.

To get an exact cut, place the end of the measuring tape at the very tip of the raker. This is always where the return cut meets the angle cut. Notice that this point is up from the bottom 1⅛ in. when using a 45° angle. From this point, using the tape, measure back the proper length.

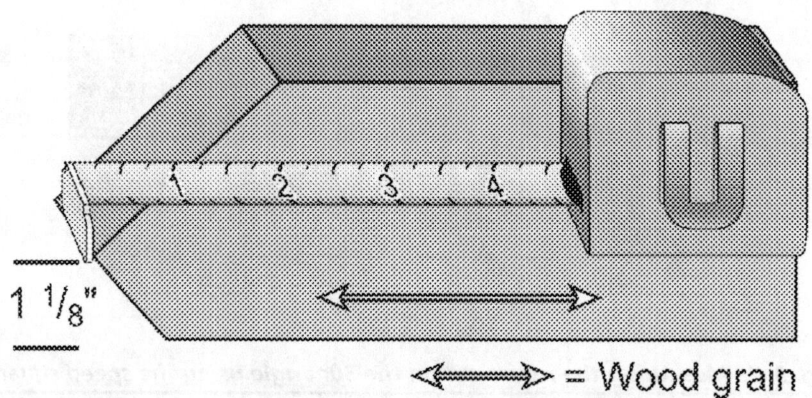

Fig. 3–77 Measuring the raker for the exact length and second angle cut.

For illustration purposes, assume an insertion point at 9 ft. At a 45° angle, the length of the raker is 153 in., tip to tip. At the proper measurement (153 in.), place a square mark along the face of the raker. Since the furthest point of the raker is actually 1⅛ in. in from the face, place a mark on the 153-in. line as shown in Figure 3–78.

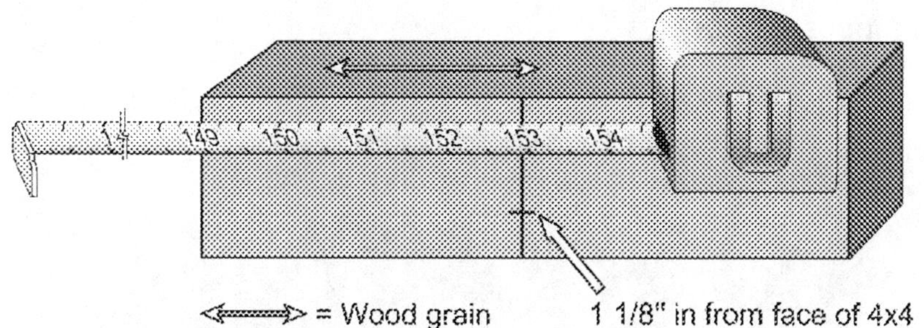

*Fig. 3–78 At the proper measurement, place a square mark
along the face of the raker.*

At the point where the two lines intersect, place the square. Remember, this angle faces the opposite direction from the first angle cut. Scribe the line through the material and cut it. Make your return cut.

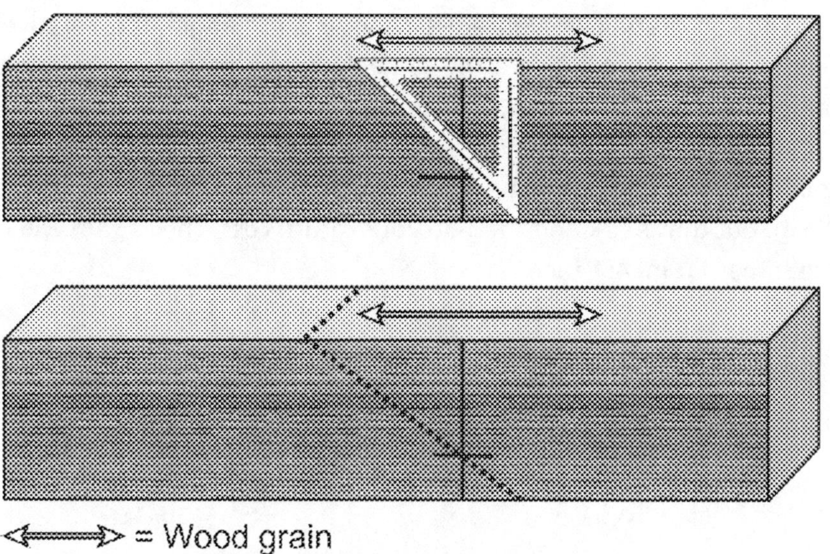

Fig. 3–79 Scribe the line where the two lines intersect.

The shore is now finished and is exactly 153 in. long—just the way it's supposed to be!

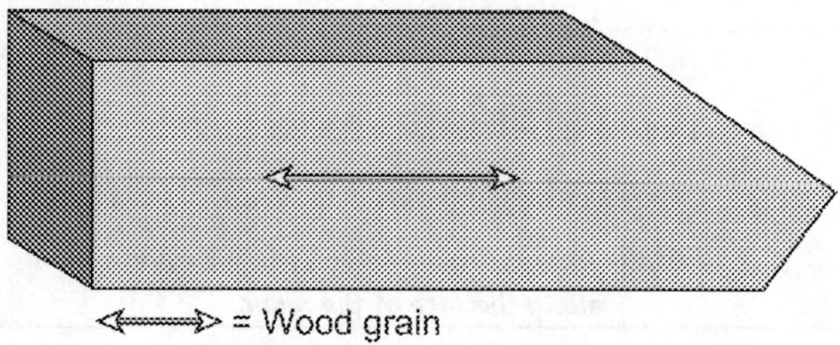

Fig. 3–80 The completed 153-in. shore.

Determining the cuts without the use of a square

The top graphic in Figure 3–81 shows a 45° angle. It is relatively simple. Both of the cuts are the same, but take note that the angles are opposite of each other. They must be cut this way in order to fit properly. For the 45° cuts on a 4x4, you have to come back from a square end and measure back the width of the lumber, in this case 3½ in. Cut that mark and then do your return cuts. They again will be the same: 1½ in. 45° face.

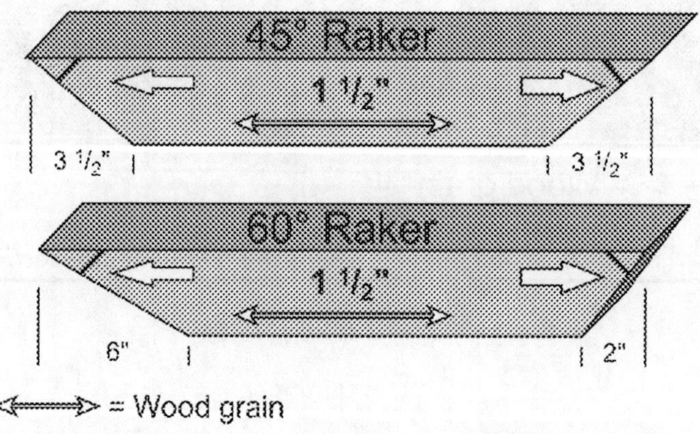

Fig. 3–81 Determining the 45° and 60° angle cuts without the use of a square.

The bottom section shows the cuts for the 60° raker. On the left side is the 60° face. Measure back 6 in. from the square end of the material. Scribe a mark and then cut the angle. Place the return cut into the piece. On the right hand side of the raker, make the bottom cut. This is the 30° cut. Measure back 2 in. from the square end. Cut the line and place the 1½ in. return into the angle. The piece is ready to be installed.

There is a very easy way of laying out the 45° angle without even using a tape measure. Use just a scrap piece of 2x4.

Using a piece of scrap 2x4, follow the square end of the 4x4 and make the 2x4 flush with the end face. Scribe a line on the inside of the 2x4.

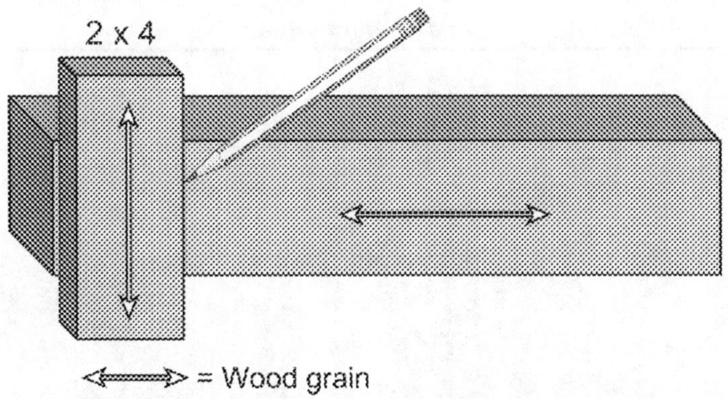

Fig. 3–82 Scribe a line with a scrap piece of 2x4.

Figure 3–83 shows the line scribed. Now you have a box 3½ in. square. A diagonal line joining any two corners gives you a 45° angle.

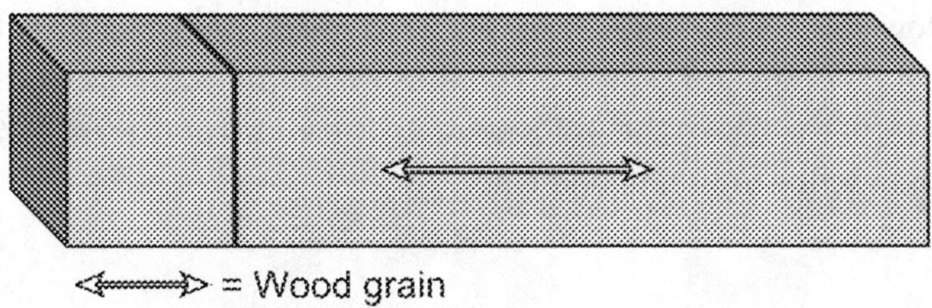

Fig. 3–83 Scribed line creating a box 3½ in. square.

Using the same 2x4 block as a straight edge, scribe a diagonal line from the top of the 4x4 to the bottom corner of your mark.

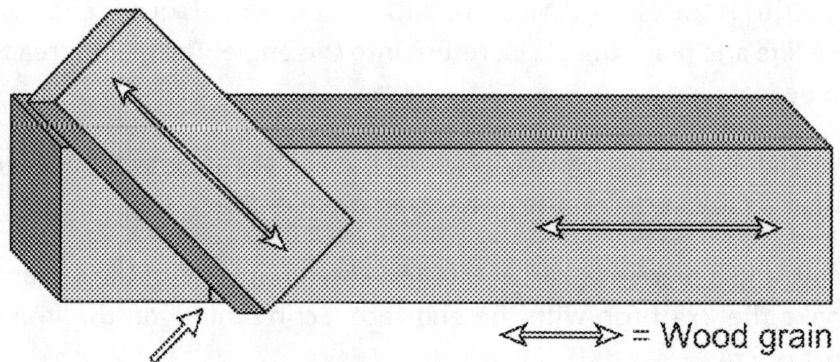

Fig. 3–84 Scribe a diagonal line using the same scrap 2x4 block as a straight edge.

Cut along that mark to get a 45° angle for your shore.

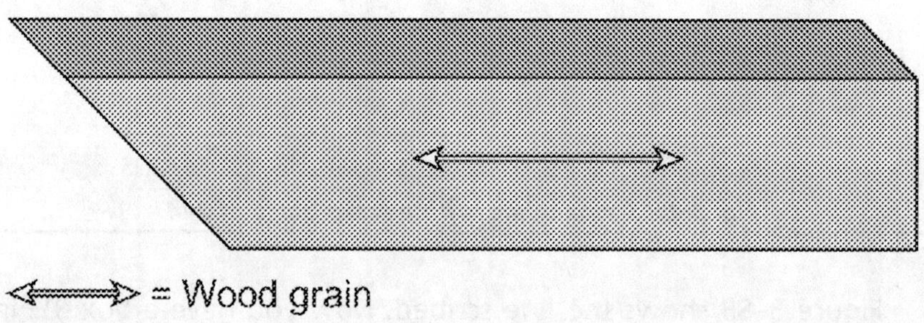

Fig. 3–85 Cutting along the diagonal line produces your 45° angle.

Now you are ready to make the return mark, but that's easy. Place the 2x4 on edge and slide it down, flush with the cut edge, until the space from the top of the 4x4 is equal to the width of the 2x4 (1½ in.), and then scribe your line and cut.

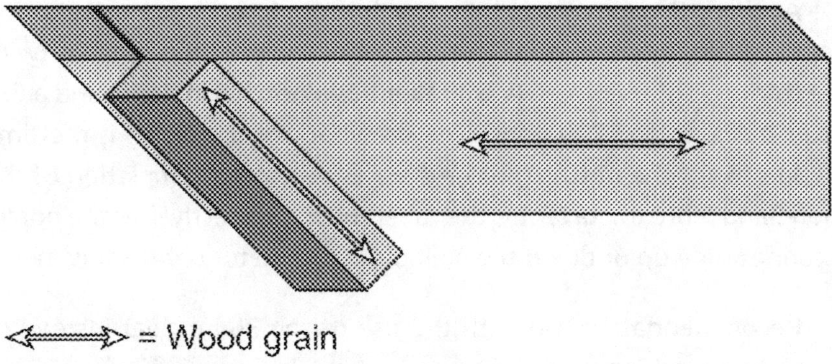

Fig. 3–86 Scribing the return cut using a scrap piece of 2x4.

Figure 3–87 shows the 4x4 cut properly for the 45° cut that was made using just a piece of wood and a pencil.

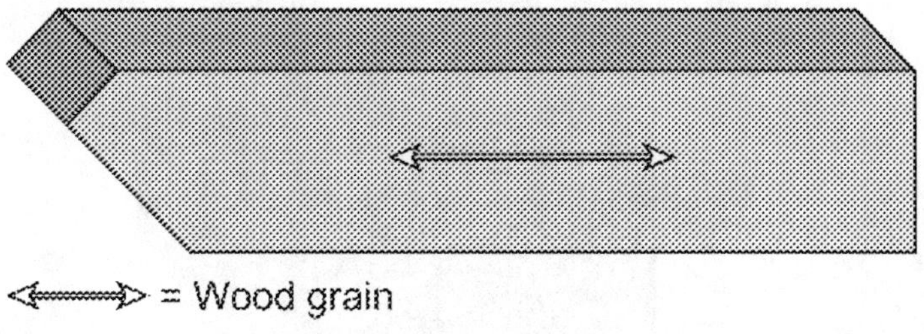

Fig. 3–87 A properly cut 45° 4x4 raker.

Notching

For all raker shores that are erected at an angle greater than 45°, additional strength has to be put in place in order to offset the uplift forces against the raker. There are two ways to accomplish this: notching the wall plate and placing a top cleat in position or increasing the size of the cleat from 2 ft to 3 ft. For a raker installed at an angle of 60°, the top cleat has to resist as much as 20% more force applied to it versus the pressure applied to a raker installed at a 45° angle. As a result, it is necessary to notch or add more length to your cleat.

At times there may not be enough room to add more cleats. The depth of the notch must be 1 in., and the length of the notch is twice the thickness of the raker. A 4x4 raker has a notch 8 in. long, and a 6x6 raker has a notch 12 in. long. The drawback of the notch is the time it takes to cut the notch. That time slows down the completion of the shore, and more importantly, the raker must go directly into the notch. It cannot slide up or down the wall plate to fine tune the adjustment.

Recommendation: Go with the use of the 36-in. cleat instead of doing any notching.

Figure 3–88 shows a view of the notch with the cleat in place. Use a circular saw to cut the notch. Chain saws are very messy, and they don't dress up the notch well.

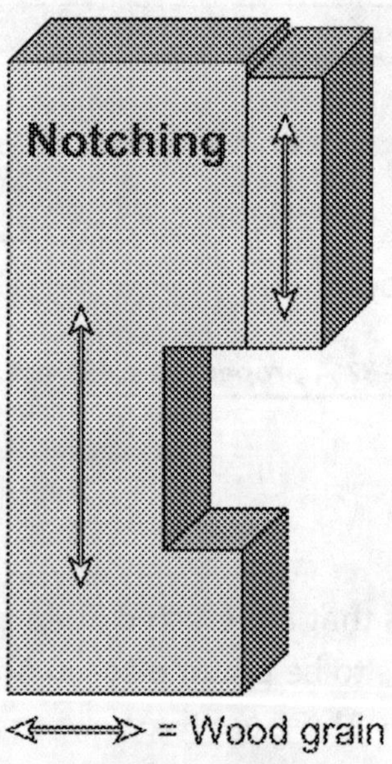

Fig. 3–88 View of the notch with cleat in place.

Lay out the notch and make a line in the wall plate where the face of the raker will go. Cut lines in the notch about every inch.

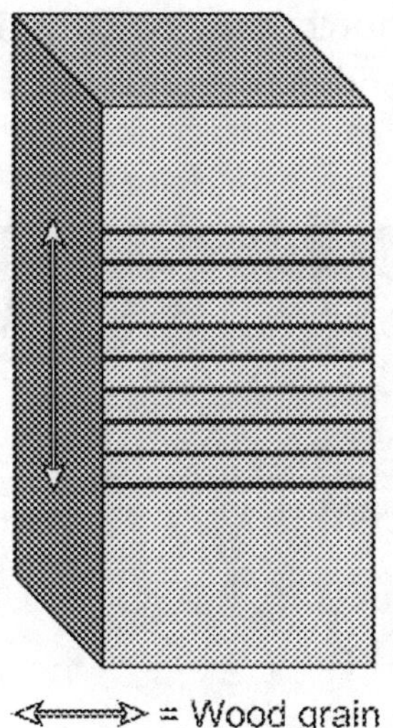

Fig. 3–89 Cut lines in the notch about every inch.

For ease of removing material after you have cut the lines in the plate, cut an *X* in the notch like the one shown in Figure 3–90. Doing this makes it quite a bit easier to remove the cut material from the notch.

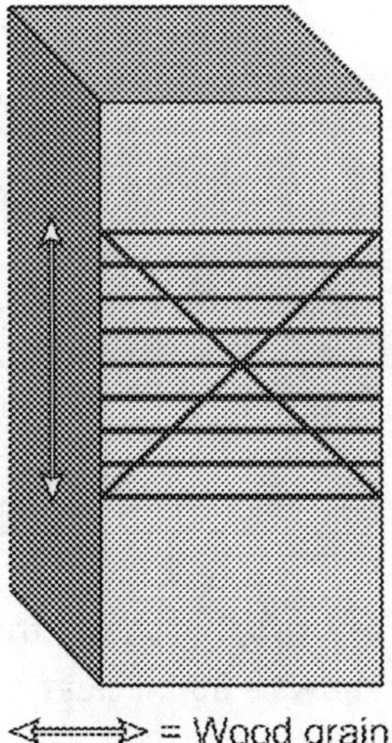

Fig. 3–90 Cut an X in the notch to ease the removal of material.

The depth of the notch must be cut uniformly to fit properly; therefore, set the saw and make consistent cuts 1 in. in depth.

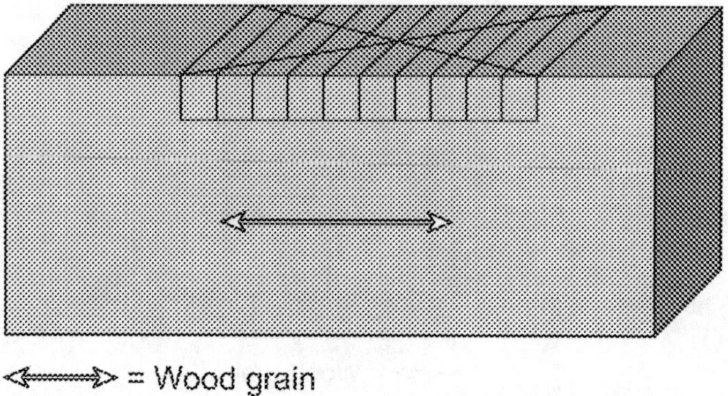

Fig. 3–91 Make your cuts consistent in depth.

With the notch cleaned out, use a hammer to get the remainder of the material out. You may have to use a wood chisel to clean up the rest, especially if there is a knot right in the way.

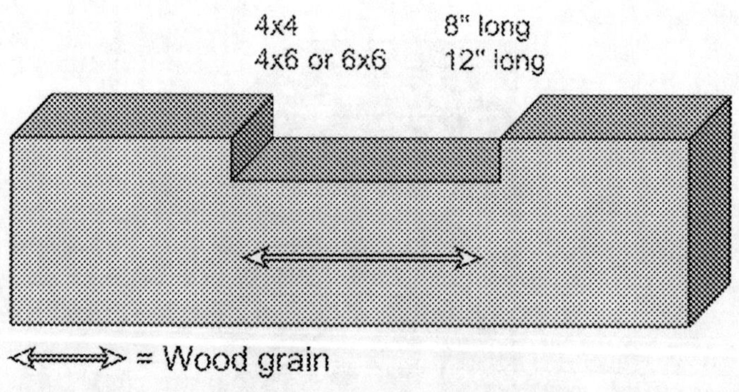

Fig. 3–92 The cleaned-out notch.

In Figure 3–93, the left graphic shows the notch with the 24-in. cleat. The right graphic shows a 36-in. cleat.

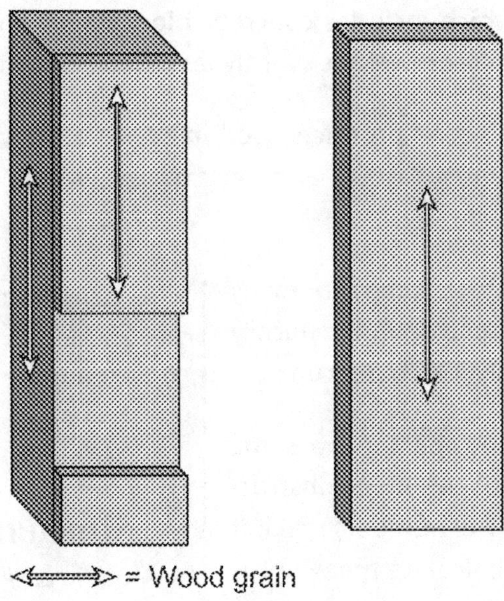

Fig. 3–93 The notch with the 24-in. cleat (l) and the 36-in. cleat (r).

Calculating Load Weights

There are two rules of thumb for determining the weight of specific debris and building materials. Remembering them will aid you in *guesstimating* the loads you may have to shore up in a collapse situation. If the floors are not compromised, then just calculate the weight of the debris for your load. If the floors are structurally compromised, then calculate both the floor weight and the debris load.

The average reinforced concrete typically encountered weighs approximately 145 lb per cu ft, with about 5 lb of steel in it. Therefore, use the ballpark number of 150 lb per cu ft for the weight of reinforced concrete.

Masonry, which includes concrete block, brick and mortar, or a combination thereof, weighs slightly less then reinforced concrete

The typical wood products used in construction, including plywood, studs, beams, rafters, and most trusses, weigh approximately 35 lb per cu ft.

Although steel comes in many shapes and sizes and designs, typically it weighs roughly 490 lb per cu ft.

The following chart shows some of the weights of common construction elements. The numbers make it much easier to calculate the weights of debris and the building's construction materials.

Because poured concrete floors come in all sizes and varieties, their weight is assigned a range of approximately 90–150 lb per cu ft.

Common Weights

- Concrete   150 lb pcf
- Masonry   125 lb pcf
- Wood   35 lb pcf
- Steel   490 lb pcf

Structural Weights

- Poured concrete floors   90–150 lb pcf
- All wood floors   10–25 lb psf
- Contents and interior partitions   20–25 lb psf
- Steel beam/concrete deck   50–70 lb psf
- Masonry rubble   10 lb psf per inch of rubble thickness

Wood floors also vary in size and weight. Combining the floor and joists, the typical weight of a wood floor is in the range of 10–25 lb per sq ft.

The contents of the typical building and the interior partitions for most buildings weigh roughly 20–25 lb per sq ft.

Steel decking with concrete, the typical Q decking, weighs 50–70 lb per sq ft.

Calculating rubble can be a bit tricky, but the aforementioned weight ranges give you a general idea of how to come close to calculating an accurate number. For every square foot of material an inch thick, calculate 10 lb. For example, for an area covered with 12 in. of rubble, the weight of the rubble would be approximately 120 lb per sq ft.

Interior Rescue Shoring Procedures

This chapter covers the most common types of rescue shoring teams erect in collapses of wood-frame, reinforced, or unreinforced masonry structures. The main objective of interior shoring is to re-support, replace, or reinforce damaged structural elements from the inside of the building. It generally is used to stabilize walls, bearing members, windows, doors, and racked or unstable openings.

The interior shoring types covered in this chapter are the following:

- Diagonal brace
- T-shore
- Double T-shore
- Window shore
- Door shore
- Horizontal shore
- Laced-post shore
- Vertical or "dead" shore

The *diagonal brace* is most often erected to laterally stabilize and strengthen leaning walls, preventing further wall movement. The *window shore* is erected to support loose headers or lintels that have shifted or lost their structural stability. The *door shore* is used to reinforce and brace existing or new openings in walls and doorways. The *horizontal shore* is erected to support damaged hallways or access ways, affording safe passage through these areas. The T-shore is used for quick stabilization of an unsafe area. It is a temporary shore. The double T-shore is also for initial stabilization. Altough more stable and stronger thn a single T-shore, it is still an initial safety shore.

The *laced post,* or *shoring tower,* is used to replace columns and support heavy loads. It is a self-supporting shore and can be erected anywhere. It can also be used as a last resort refuge area for personnel if secondary collapses occurs. The *vertical shore*, the most commonly utilized interior shore, is used to stabilize floors or replace or resupport existing damaged beams or girders.

Interior Shoring Size-up

The initial shoring size-up by the shoring officer and firefighters should be a survey of structural damage and victim locations, which are the primary factors used to determine the types of shores to construct and their locations. Size-up should be extensive and ongoing. The safety of rescuers and victims depends on it. Each size-up situation is slightly unique. Even in the same structure, this can be the case. The following are a few of the general points to look for before your team starts shoring operations.

Type of structure

Determining the type of structure is a critical factor in the size-up of a shoring operation. The type of construction and the size of the structural elements are important factors in assessing the size of your shoring and the critical placement of those shores. The weight of the building's elements is an obvious concern and generally dictates the size of the material needed. Whether the structure is framed or unframed will be a major consideration because each has a specific area where your shoring must be placed. An unframed building has the exterior and possibly several interior walls as the main structural supports of the floors. Any damaged or missing sections of these walls must be replaced with some type of rescue shoring. A framed structure doesn't have any bearing walls; its walls are hung on or anchored to a skeletal system, normally consisting of columns, beams, or girders. In the case of a framed structure, damaged or missing columns or girders must be the first items to be looked at and re-shored.

Six-sided approach

Always use a six-sided approach when sizing up any structure for possible shoring operations. It's simple, direct and makes for a quick evaluation. When entering the interior of a collapse incident for shoring purposes, enter from the safest area, which is generally the most stable to begin with. In checking the top above your head, check for any visible structural damage: bulged or cracked walls and floors, missing or damaged structural elements, and any possible unstable debris that may affect your operation. Check all four sides and all interior partitions for damage or instability, whether the sides are load-bearing or not. The condition of the bottom is especially important. Check the floors to determine their condition and whether they can support the shoring and additional loading you will be placing upon them. This is extremely important. Floors must be able to sustain the additional loads shoring distributes to them. If the stability of the floors is in question, additional shoring underneath the floors may be necessary to transfer the building's overload.

Age and condition of the structure

Before you enter the building, try to determine its age and its overall condition. Ask yourself: Is it well maintained or was it in general disrepair? Is it a relatively newer structure or a very old building? It is important to determine both of these situations. The older the structure, the more tired it will be; nature's elements will have taken their toll. There could be some major secondary collapse potential staring you in the face. A newer building generally means two things. One, the structure generally is stable and in relatively good shape—of course you must still determine if that is true. Two, with the lighter-weight building materials being utilized today, there could be a significant possibility of secondary collapse.

Amount of damage

The amount of damage to a structure determines the amount of shoring needed. A simple rule to remember is that the more damage there is, the more need for shoring. Very extensive damage throughout a large building may dictate the use of multi-story shoring systems in order to redirect the unstable loads to a good bearing surface, generally the ground.

Victim location

In many situations, victims are trapped in or around the main debris pile of the structure. This is where a major concentration of weight has been located. As rescuers enter the location to extricate the victims, their weight adds to the load already pressing on the area. Many times one of the primary areas to be stabilized and shored is directly underneath any victim's location. If there is a basement in the structure, it should be an inspection priority and one of the first places in which to erect shoring. In confined areas, box cribbing can be placed, and slope floor shores can support angled floors. Or if there is enough room available, then vertical shores can be erected.

Weight of debris

The key factor in determining what size material to utilize and how close to space the post systems in interior shoring is the amount of weight that the shore must support. Your team's structural specialist should be able to calculate the amount of debris weight that the shore must support. Generally, calculate the weight of the floor in question as well as the weight of the debris on it. This gives a safety factor in the determination of shoring size. The majority of unreinforced masonry debris weighs roughly 125 lb per cu ft. As a rule of thumb, the number 125 works fine. The weight calculated determines the size of shoring lumber and the spacing of the shore's posts. If the shore is not supporting a large amount of weight and if the chances of a secondary collapse occurring in the area are remote, do not go crazy over shoring. You could be wasting precious time and material.

Interior structural members

Almost any building may have one or more interior structural supporting elements. These typically include the following:

- Interior bearing walls (in larger structures)

- Columns

- Arches (usually in much older type construction)

- Girders

- Beams

- Trusses

These, for the most part, are the more commonly used items.

One of your first size-up options is to determine if any of these items is either heavily damaged or missing (destroyed). These items must be checked as soon as possible. The structural members they previously supported may be under extreme stress or—worse yet—ready to collapse at any time. Those items may have to be re-secured or replaced with adequate shoring substitutes. Check on the elements in the following order:

- Interior bearing walls

- Arches

- Columns

- Girders

- Beams or trusses

If any one of these supporting elements fails, you will have some sort of structural failure. The type of interior shore to erect depends on the element in danger. For example, a multi-post, vertical shore can replace a damaged or missing section of interior bearing wall or girder. With damaged columns, a laced-post shore may do the trick.

Sagging floors

Generally, any floor (or roof for that matter) that is sagging, bellied, or deformed is in some sort of overloaded condition. This can be due to dozens of reasons: water, stock, materials, or any number of a combination of reasons. Whatever the reason, the bottom line is that the floor's bearing elements—typically floor joists—are overloaded, causing a deformation of those structural elements. The problem centers on the bearing points of the floor beams. Normally they bear on top of the supporting walls a few inches. In unreinforced masonry and wood-framed construction, 3–4 in. is typical. As the floor beams sag and belly, they can start to slip off of their end supports—obviously, this must be avoided. To stabilize this situation, erect vertical shores in the lowest part of the deformation to arrest movement of the beams. The vertical shores transfer the overload condition of the collapsed or partially collapsed floor to the lower floors or to the ground. Just make sure that the lower floors can handle the additional weight.

Bulged walls

Bulged, bellied, or leaning walls are signs that some type of structural instability exists. Walls are designed to accept loads through their center axis, providing they are plumb. If the walls for any reason become eccentrically loaded, there can be drastic results, especially if the walls are bearing walls. Unable to bear the weight on top of them, these walls can quickly fail. One of the safest ways to counteract this potential instability is to erect interior shores to handle the load from the floors above.

Columns out of plumb

Similar to walls out of plumb, the farther out of plumb columns are, the less weight they can hold. Generally speaking, columns support a joint of some type, usually a set of beams or girders. As they move out of plumb, there is less and less bearing on the column from the beam above. At some point, the beam will slip off and collapse. Columns that are visibly out of plumb are a definite issue for your rescue crew. As a possible solution to the unstable column, you could erect a laced post around the column or two small vertical shores on each side of the column. Either way, you are resupporting the beams around the affected area, i.e., you are basically taking the column out of the equation.

Connection points

One of the most important items that must be looked at is all the structural connection points in the building. After a major collapse has occurred, especially in an explosion situation, the connection points of all the building's structural elements should be examined closely. These are typically the weakest parts of the structure and generally are the first places failure occurs. If there is a scenario where some of these points have been compromised, remedies can be brought in quickly to resupport those damaged connections. There are several types of shores that fit the bill for that situation.

Racked opening

If the structure becomes racked (twisted) due to some type of shifting, it may be necessary to arrest that racking by installing shoring. To accomplish this, install diagonal bracing on the walls and shoring the openings. Normally you would shore the door and window openings that are racked. Use 4x4s for this operation. They are generally sufficient.

Shores bearing support

One of the most important items your size-up must determine is on what will your shoring systems bear. In a multi-story structure, each floor must be checked for integrity and ability to support the additional loads that may be placed on the existing floor system. You must also determine if there is too much debris in the way to properly support the load. If this is the case, then the debris must be removed first before the shores are erected. The overload and weight of your shoring must be calculated—ballpark is fine. Then it must be determined if indeed the floor below can handle the additional weight. If any doubt exists, you may need to continue your shoring system onto each floor, terminating it at ground level. This ensures that the proper amount of bearing necessary to sustain the overloads will be there to handle them.

The Diagonal Brace

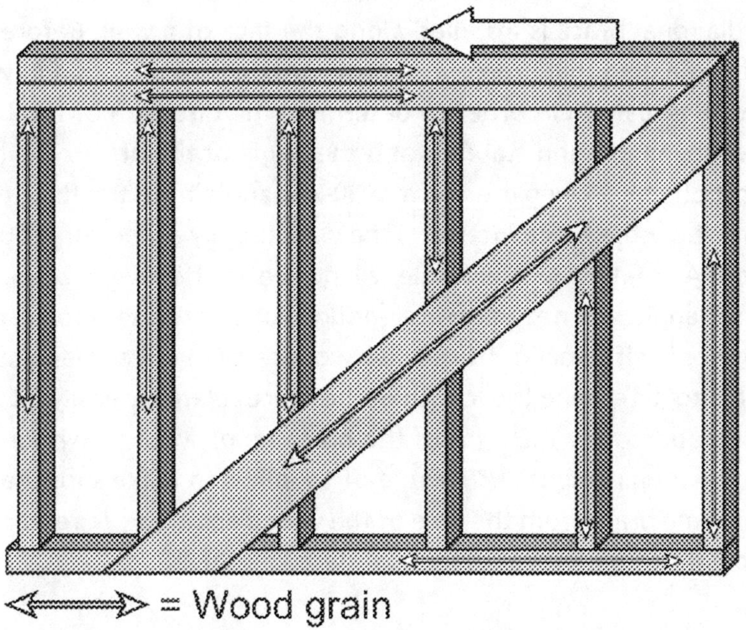

Fig. 4–1 The 2x6 diagonal brace.

The diagonal brace is the simplest and easiest shore to erect and is generally constructed of 2x6 lumber or larger. The main purpose of this shore is to resupport or stabilize damaged or leaning partitions, whether they are bearing or nonbearing. After a structural collapse has occurred, loads can be concentrated almost anywhere. Many times, after a substantial collapse or an explosion, the building may become racked or twisted. Walls can become heavily damaged, destroyed, or start leaning toward the weakest parts of the structure. This may affect the strength of any partition; and many times in that situation, nonbearing partitions become bearing. This additional bracing may help stabilize those walls that have been damaged. When you encounter walls that are leaning in one direction, the diagonal brace anchored to the wall may help it from leaning any further.

In some scenarios, it may be necessary to erect the diagonal braces on both sides of the heavily damaged wall. In this case, as well as any other time this shore is erected, it is important that the diagonal brace be nailed into the wall's studs with (3) 16d nails at each stud. The diagonal brace is installed along the face of a wall. Before constructing a diagonal brace, you must consider the direction in which the wall is leaning in order to determine the direction of the brace, the wall's height and stability, and the angle of the brace. The brace is most effective when erected at a 30–45° angle from the floor to the point where the brace intersects the wall, usually at the corner of the ceiling. A greater or lesser angle will not be as effective. A brace with a lesser angle will not generate enough force into the floor, causing the brace itself to be ineffective. By working with 45° angles, you will be able to determine the length of the brace relatively easily. You will also be able to cut the ends of the brace simply with or without the use of a framing square. The rule of thumb is to make sure that the base of the brace from the face of the wall is equal to or greater than the height of the brace.

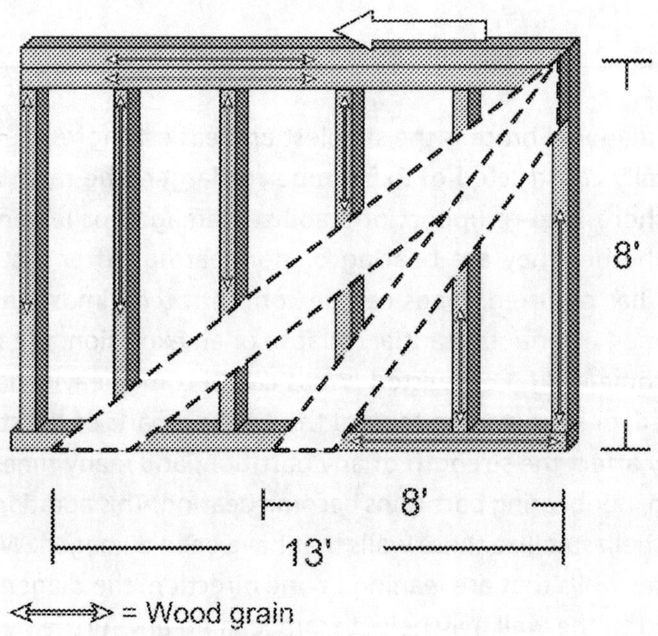

Fig. 4–2 Determining the length of the brace.

For example, if the typical wall height is 8 ft, then the base of the brace generally should be no closer than 8 ft from the wall face and not longer than 13 ft away from the wall face.

The recommended lumber size for a diagonal brace is 2 in. dimensional lumber, preferably Douglas fir. The preferred size lumber would generally be 2x6, 2x8, or two 2x4s used in conjunction one on top of the other; however, it is less effective then the larger dimensional material.

Before constructing the brace, it is important to determine the overall condition and stability of the wall. If the wall is not anchored to the rest of the structure, you will have to re-anchor it to the floor, the ceiling, or another wall. It is also important to identify the direction in which the wall is leaning. If the wall is leaning left, the brace begins at the top, right-hand corner and runs down to the left; that is, the bottom leg of the brace will be on the left as you face the wall. That way, as the wall leans left, the shore comes under compression, stopping further wall movement.

Begin construction of the diagonal brace at the upper corner where the walls intersect. This locks in the corner, increasing the efficiency of the shore by using both walls. The brace must contact the top and bottom plates and bear fully on the floor, otherwise it will move. This places the brace under compression and uses the shore to its full effectiveness. Use a power nailer to anchor each wall stud to the shore with at least three nails. Anchor the top and bottom wall plates to the diagonal brace to increase stability again using three nails each. Erecting braces on both sides of the wall adds further strength and stability, and although not always necessary, doing so can be an insurance policy.

Diagonal brace size-up

Determine the wall to be braced and the direction it is leaning in. Determine the angle and the length of the brace, and cut the angles in both ends of the brace. Place the brace in position, butt the top solid into the corner, and make sure that the brace contacts the floor cleanly at the base. Drive the brace tightly to the wall and nail it to each stud. Use at least three nails per stud. If necessary, place a block behind the base of the brace and nail that into place to help prevent the brace from being dislodged.

Diagonal brace step-by-step procedures

The diagonal brace is installed against a damaged and leaning wall as tightly as possible and as high up the wall as is practical, given the size lumber being used. The high point of the brace should start at the steepest part of the leaning wall and brace into the floor away from that point. In Figure 4–3, the wall is leaning from right to left; the high point of the brace is placed into the angle of the lean. It is important that the brace has full and direct contact into the wall to be stable. You must find the location of the studs in order to properly anchor the brace to the leaning wall. Without this contact, the brace is ineffective and does not do the job. The brace must be nailed to every stud, as well as to the header and the sole plate.

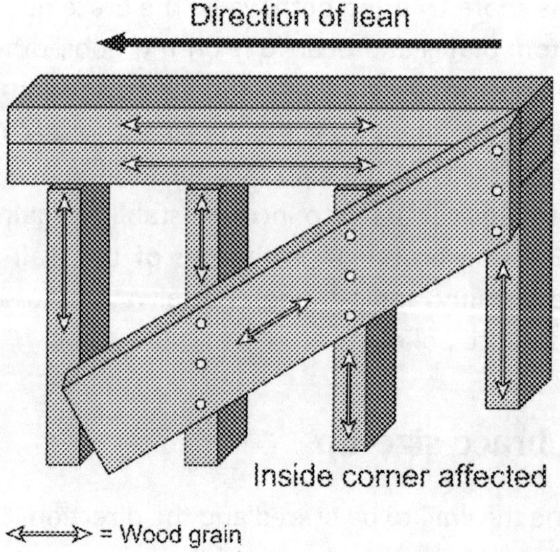

Fig. 4–3 Brace placement for a wall leaning from right to left.

When installing the brace, make sure that the top of the brace contacts the face of an adjoining wall or the outside exterior wall. Cut the brace on the angle, generally a 45° angle. Place it tightly up to the wall, maintaining full contact with the face of the angle cut. Make sure to nail the brace at every stud. This is important to ensure full efficiency of the shore.

On 2x6 and 2x8 size lumber, you must put in at least three 16d nails. Make sure they penetrate the stud the full depth of the nail. Space the nails evenly, starting at the center then keeping the other nails approximately 1 in. in from the ends of the lumber. Use this pattern in order not to split the lumber, rendering the brace ineffective. Figure 4–4 shows a close up of the brace at its contact point to the wall studs.

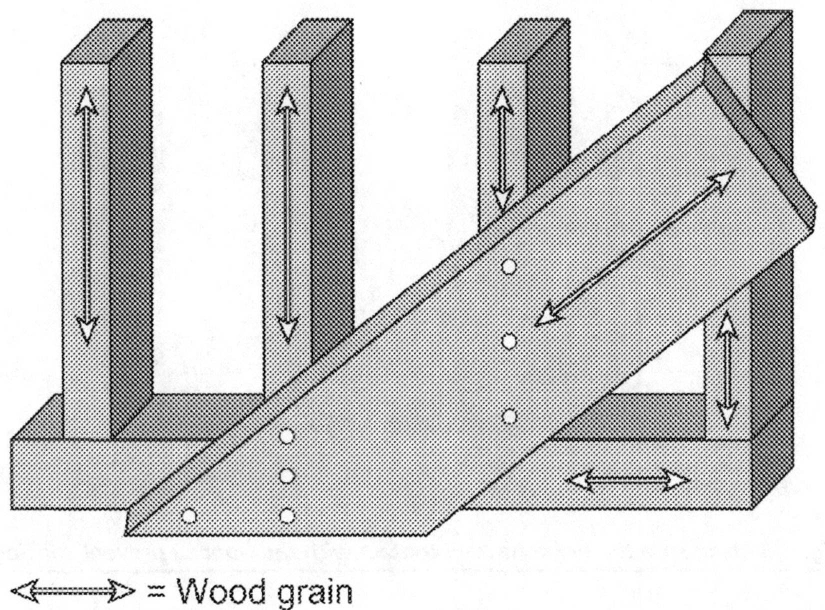

Fig. 4–4 Make sure to nail the brace at every stud, spacing the nails evenly.

At the base of the shore, make sure that the brace contacts the floor. This is very important because without direct contact with the floor, the brace will slide down the wall until it does contact the floor.

Place the brace at the high point and contact the floor. Then gently push it toward the lean of the wall until it is tight. Next, nail the brace in position at all contact points with the top and bottom plates and the wall studs, using three nails at each contact point. Figure 4–4 shows the brace is flush with the floor.

On occasion, the angle cuts on the brace may not fit exactly— not to worry. This can occur when the floors are not level, when the angle cut is off slightly, or when the angle from the wall to the floor is not 45° (see Fig. 4-5). You must, however, make sure you do have full contact with the floor area. This is important for the integrity of the shore. If there is not full contact with the floor, the shore could shift, causing problems when the wall moves.

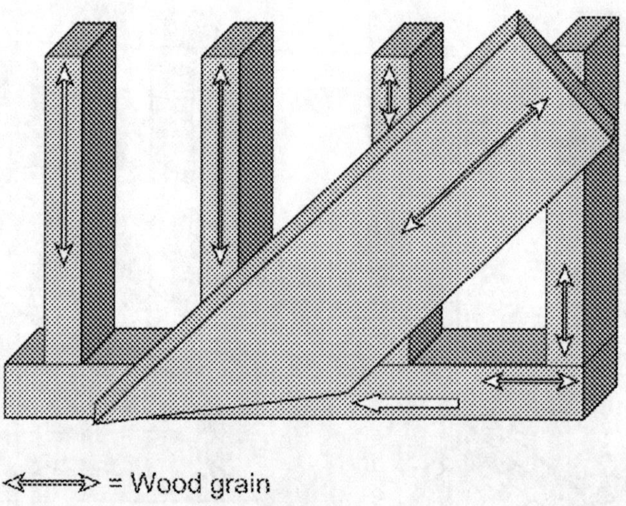

Fig. 4–5 Make sure the brace has full contact with the floor to prevent shifting.

To alleviate the problem of a shifting brace, you generally have three options. The first is to recut the brace. You can measure the gap and cut that length off the front of the brace. Then place a piece of lumber flat on the floor and scribe a line across the face of the brace. This will give you the exact cut necessary to have the brace fit properly. The second option, which is generally quicker, is to slide a shim or a wedge into the gap and fill it in. Regardless of the method your team elects to utilize, you must make sure the brace has full contact with the floor. Both of these solutions are acceptable.

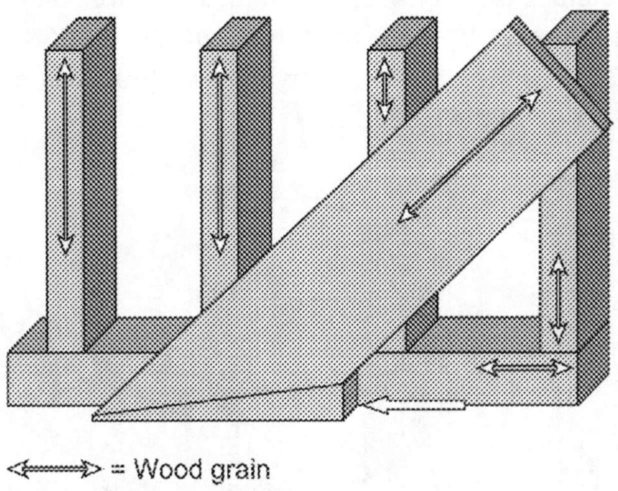

Fig. 4–6 Sliding a shim or wedge into the gap is an acceptable solution.

In order to get the angle you need for the 1½-in. cut back, use a square. Just slide it down until the 1½-in. space is at right angles to the floor. Make your cut; the 2x4 block will fit perfectly.

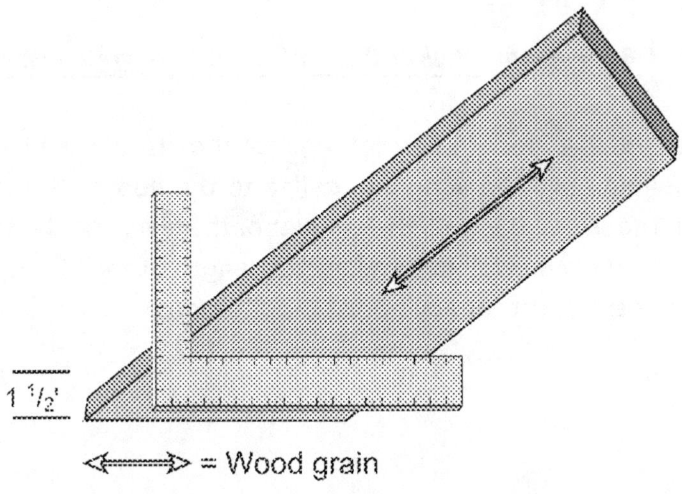

Fig. 4–7 Measuring the angle needed for the return cut.

A third option you have is to anchor a block down to the floor to add strength to the brace. If you are not happy with the contact against the studs, this option is an excellent way to add more support to the brace. Don't forget to put a 1½-in. return cut into the bottom of the brace (see Fig. 4–8). The return cut gives the block much more brace support. Without the return cut, the block is ineffective. Figure 4–8 shows the bottom of the brace with the return cut ready for the insertion of the block.

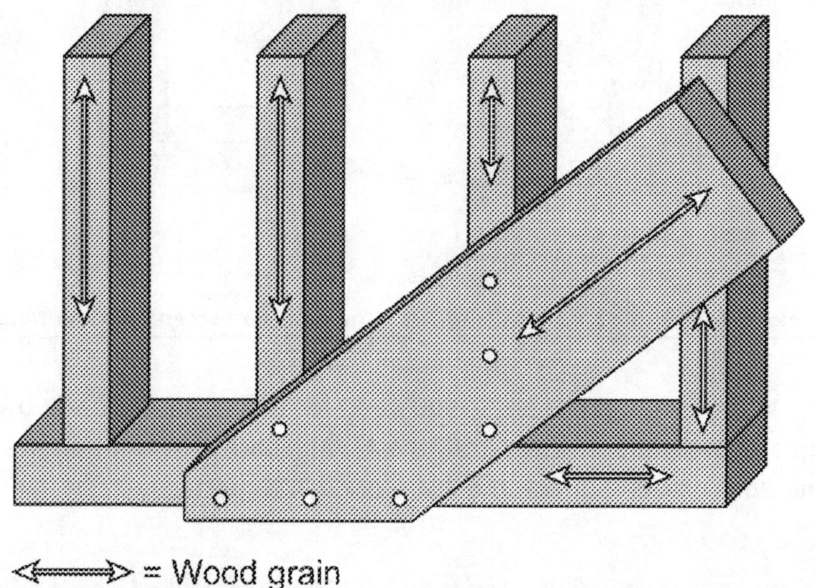

Fig. 4–8 The return cut gives the block much more brace support.

With the 2x4 block in position against the brace, the block must be a minimum of 12 in. long and nailed to the floor with the 5-nail pattern. The size of the wall, the amount of the lean, and the strength and integrity of the wall determines the length of the block. Usually 16–24 in. is sufficient for any normal applications.

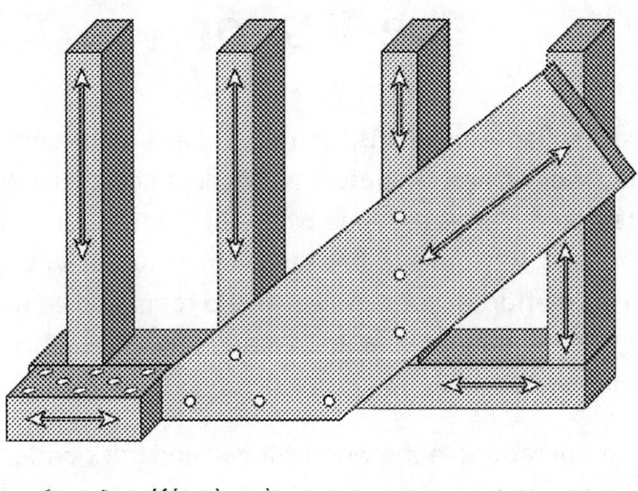

Fig. 4–9 The 2x4 block in position against the brace.

Figure 4–10 shows what the angle of the diagonal brace looks like at a typical 45° angle. This brace is in proper position with a wall that is leaning to the left.

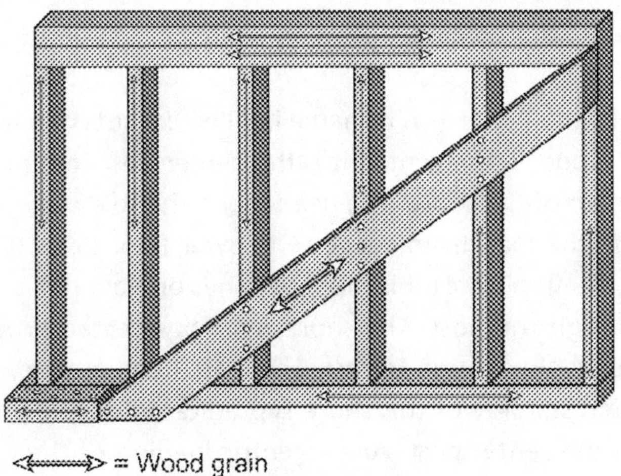

Fig. 4–10 The 45° diagonal brace in proper position against a left-leaning wall.

The T-Shore

For most collapse incidents, there will have to be some type of initial safety shoring erected before a complete and thorough size-up can be attempted. This is the time when the T-shore comes into play. It can be erected very quickly. It is easily moved and can be adjusted to fit with little effort. It takes up very little room and can be put together outside the danger area, then quickly be moved into position, erected, and secured.

For most operations in the wood-framed and brick-and-joist buildings, the T-shore can be constructed of 4x4s with the use of 4x4 wedges. The materials you need to erect one shore are the following:

- Two 12x12¾-in. plywood gusset plates

- One 24-in. 2x4 cleat

- One 36-in. 4x4 header

- One sole plate at least 2 ft long

- One pair of 4x4 wedges 18 in. long

Basically the T-shore, as its name implies, consists of a header supported by a post in the center. It is this center post that provides the main strength of the shore. The header and the sole plate collect and redistribute the load from above to a lower floor or to the ground. Be aware, the stability of this shore depends on how the load is transferred through the post. The shore must be erected properly or it will not be stable enough to carry the load. Bear in mind, the entire strength and stability of this shore is predicated on the proper axial loading of the center post. Any eccentric loading of the post causes the shore to become unstable and, most certainly, fail under load.

The T-shore is marginally stable at best. You must remember this. Do not make this shore too big or you will be asking for trouble. A T-shore should not have a header more than 4 ft long. Any longer and the chance of eccentrically loading the shore is great. If you do erect the shore with a 4 ft header, place a larger gusset plate at the header and post joint. With the use of the longer header, this gusset plate should be 18 in. x 18 in. in order to give the joint a little more stability.

Usually a T-shore is erected as a safety shore before the rescue team erects the more permanent and stronger vertical shoring. Therefore, it is advisable to erect at least two of these initial shores before any other shoring work is attempted in the danger area. Two or more shores are normally enough to protect your shoring team while it erects the vertical shores. Generally vertical shores average 10–12 ft in length. Any longer and they may become much harder to work with and to maneuver and require that an extensive area be cleared without having the area shored.

Figure 4–11 shows the proper terminology for all the parts of a T-shore. The *header* is normally a section of 4x4 3 ft in length; however, it can be erected up to a 4 ft length with additional materials involved.

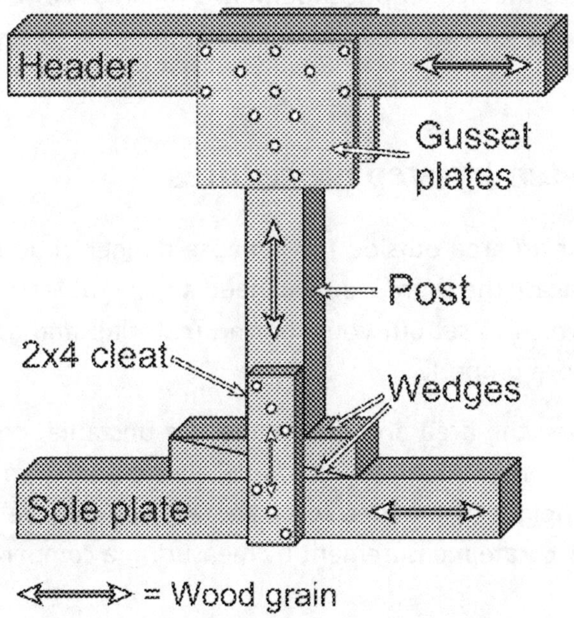

Fig. 4–11 Proper terminology for all parts of a T-shore.

There are always two *gusset plates* installed on every T-shore, regardless of the size of the shore or the length of the header. These gusset plates are constructed of ¾-in. sections of plywood, either 12x12 or 18x18.

The *post* is also constructed of a 4x4. In almost all circumstances, it is placed in the center of the header and takes all the weight of any item being stabilized by the shore. A post can vary in length.

The 2x4 *cleat* usually consists of a length of 2x4 anywhere from 18 to 24 in. in length, depending on the wedges used and the overall length of the shore itself. The cleat is nailed to the post, the sole plate, and the wedges.

Wedges are always placed on top of the sole plate and directly under the center of the post. There are two wedges used at all times for adjustment of the shore. A wedge can be either 2x4 or 4x4 lengths of material.

The *sole plate* normally consists of a length of 4x4. It generally is not as long as the header but should be a minimum of 2 ft long. The sole plate helps spread the load out from the post to several of the floor beams below.

T-shore step-by-step procedures

1. Clear an area outside the collapse danger zone in which to fabricate the shore. You will need an area at least 10 ft x 10 ft in which to set out your shoring materials and assemble the T-shore properly.

2. Survey the area and determine the unstable load displacement and structurally unstable elements. Try to determine the height of the area to be shored. You may be able to get an accurate measurement by measuring a comparable ceiling

in another, undamaged room adjacent to the collapse area. If the room in which the shore will be erected in is not totally damaged, then you can measure the ceiling height near the point at which you entered the area. You may be able to verify the ceiling height by asking the homeowner or, for a building, asking the maintenance personnel assigned to the different areas of the structure. The most reliable way of course is to use the structural drawings for the building, but usually they aren't immediately available. The measurement does not have to be exact. Remember, you will be installing a set of wedges under the post that will allow for several inches of adjustment.

3. Determine the length of the header you need. A 36-in. header should be adequate for operations in most residential and small commercial buildings. However, the decision on the size header to use is made on a case-by-case basis, and the length may vary in the same structure. The length depends on what you intend to hold up and what size structural elements are involved. Depending on the style of framing your team encounters, you can support three or four floor beams with each shore. If the floor joists are 16 in. on center, which is typical, a 3-ft header will catch three beams. When this is the case, always place the post under the center beam to balance the load on the shore.

4. Determine the length of your shore's post. To do this, calculate the ceiling height of the area you are to shore (which you did in step 2). Knowing the overall height of the area gives you the length of the post needed to properly erect the shore.

   If you are using 4x4s, deduct from the overall height of the shore the width of the header, the thickness of the sole plate, and the thickness of the wedges you will place under the post. Deduct 7 in. for the width of the header and the sole plate. The 4x4s are actually 3½ in. wide, so multiplying by two gives you 7 in.

The last item to deduct is the thickness of the wedges. Both of these wedges must be the same size, and you will marry them together, one on top of the other. Remember, the right angle edges of the wedges should face in opposite directions from each other. To deduct the proper amount of space, determine the size wedges you will use. If they are 4x4 wedges, then deduct the thickness of one wedge (3½ in.) from the length of the post. In other words, if you use 4x4s for your T-shore, deduct an overall measurement of 10½ in. from the total height of the shore to determine the proper length of the post. This gives you the exact length of the post you need to properly erect your shore.

5. After the size-up is complete, the first step in erecting the T-shore is to anchor the post to the header. Find the center of the header and place a mark as shown in the graphic. Next determine the center of the 4x4 post. Because the exact measurement of the 4x4 is 3½ in., the center of the 4x4 post can be measured back from the end face of the 4x4 1¾ in. Place a mark as shown in Figure 4–12. Line up both marks with each other and anchor the post to the header with two 16d nails.

   Using a pneumatic or gas-powered nailer, toenail the post into the header. Using two 16d nails, toenail both sides of the post into the header parallel with the grain, using at least one nail on each side of the post. Make sure the post is anchored as tightly as possible and squared (at a right angle) to the header. It is important to make sure to keep the elements of the shore flush with each other. You do not want any of them overlapping or creating a lip. Those conditions would cause problems when securing the gusset plates to the shore.

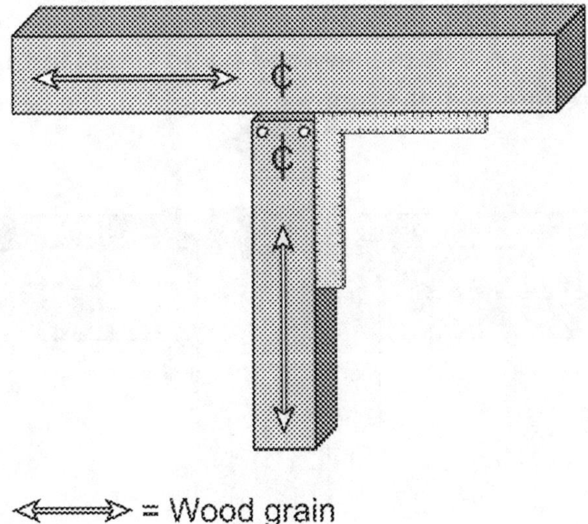

Fig. 4–12 Line up both center marks and anchor the post to the header.

Always try to drive the nails parallel with the grain of the lumber to keep the lumber from splitting. If you attempt to nail the header into the post against the grain, you will have a tendency to split the lumber, something we will try to avoid whenever we are constructing any type of shores.

Another option is to hand nail the top face of the 4x4 post to the header with (2)16d nails and drive them flush with the surface of the material. This way the nails will not interfere with the anchoring of the gusset plate.

After the two items have been nailed together, take a carpenter's framing square and square up the post to the header. This is important! The shore will not work properly if this is not followed. Make sure the post is square to the header. Use a 2-ft square to verify. A 6-in. speed square is not big enough to accomplish this.

6. Anchor the ¾-in. thick plywood gusset plate to the header and post after the post has been properly anchored to the header. Place the gusset plate on top of the shore. Center the plate

over the post, flush with or just slightly below the top of the header so that the plywood doesn't overlap the header and interfere with the transfer of weight to the header.

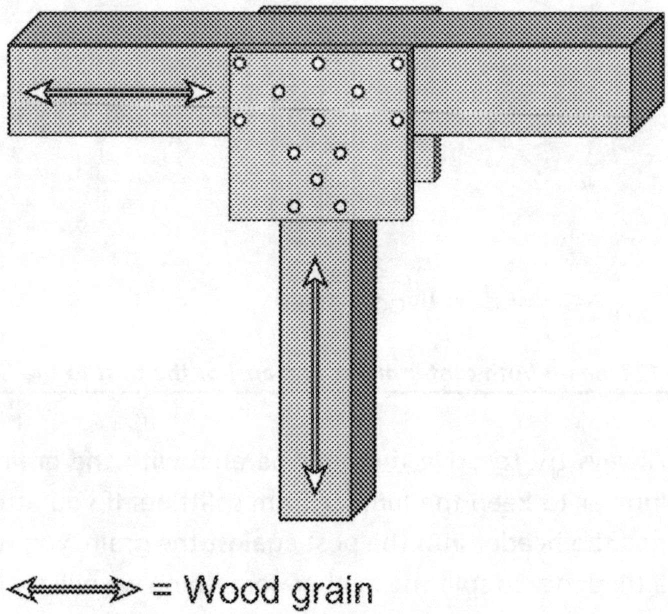

Fig. 4–13 Anchor ¾-in. thick plywood gusset plates
to both sides of the header and post.

The purpose of the gussets is to lock the joint, enabling the shore to more efficiently and safely transfer the weight it supports. For additional safety, strength, and stability, a gusset plate is placed on both sides of the shore. This will effectively stop any twisting action from affecting the shore's stability and effectiveness. Flip the shore over and anchor another gusset plate on the other side of the post and header joint. Nail the plate in position exactly on top of the location of the other plate. Use the same nail pattern you used on the previous gusset plate. Your T-shore is now ready to be installed.

7. Anchor the header with 8 nails and secure the post using the 5-nail pattern. Do this on both sides. Figure 4–14 shows a 12x12-in. gusset anchored properly.

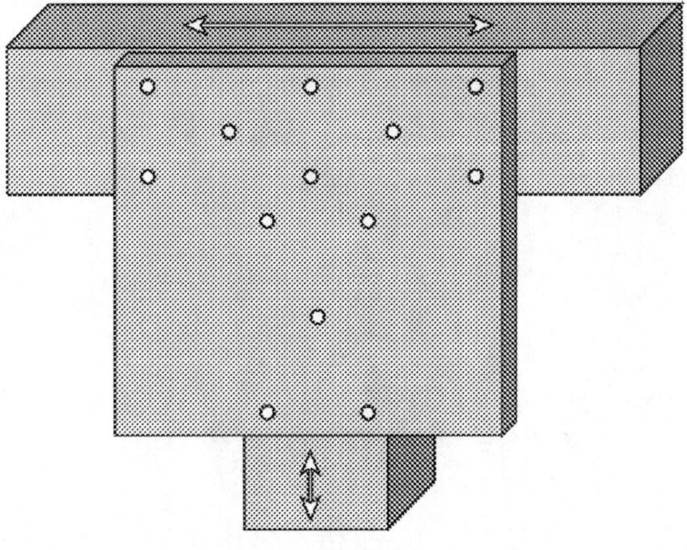

Fig. 4–14 Properly anchored 12x12-in. gusset plate.

8. Clear the floor area underneath the section of the structure you want to support while the T-shore is being brought into position. Remove enough debris to clear an area 2 ft wide and 2 ft longer than the length of the sole plate to provide room for adjustment. The shore must be centered under the load it is to support. It is imperative that the sole plate has good contact with the floor. Loose pieces of wood, plaster, or any other debris should have been moved out of the way. If any debris remains, the shore will be unstable and may be ineffective.

9. Move the shore into position with the sole plate beneath it. If the floor area is clear of debris or the weight to be supported is substantial, it is a good idea to make the sole plate almost as long as the header. One firefighter places the shore in position

and raises it up to the area you wish to stabilize, keeping it as level as possible. At the same time, a second firefighter centers the sole plate under the post, as shown in Figure 4–15.

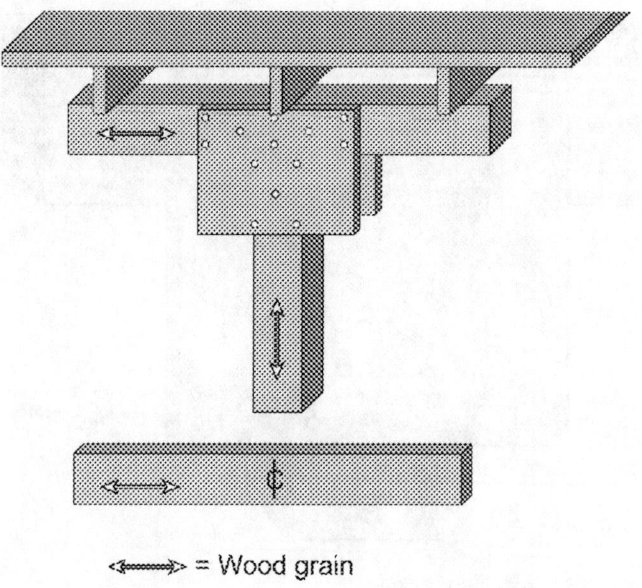

Fig. 4–15 Center the sole plate under the post.

10. Place the wedges in position under the post on the center of the sole plate. When the shore is in position, adjust the wedges. Snug them up tightly enough so the shore cannot move without physically forcing it. At this time, the first fire-fighter ensures that the post is plumb and that the header is at a right angle to the post.

11. Tighten the wedges securely to transfer the load from above. This action safely distributes the overload from above to the

floor below. Toenail the header into the beams to keep it from shifting or accidentally falling. Place one nail in back of each wedge so it will not move with any vibrations that may occur. Your rescue team now should have a safe area in which it can continue to effectively operate.

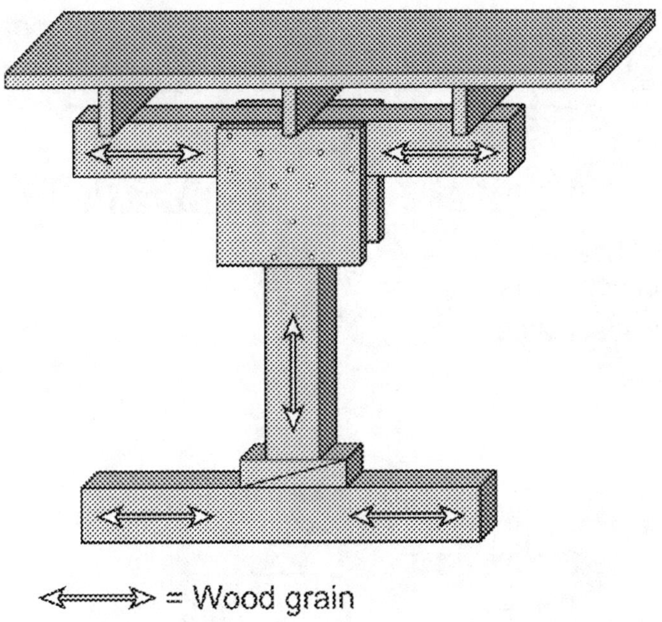

Fig. 4–16 Tighten the wedges securely to transfer the load from above.

12. Place a 2-ft long 2x4 cleat on the post alongside the sole plate, wedges, and post after making sure the wedges are tight and double-checking that the shore is holding securely. Using (3) 16d nails in a diagonal pattern, anchor the cleat into the sole plate. It is necessary to anchor the cleat to one side only. This cleat is to stop the top section of the shore

from being knocked off the sole plate by accident. If your team is erecting shores in an earthquake situation or in an area where heavy vibration will occur, place the cleat on both sides of the shore. This will stop the wedges from walking out from either side.

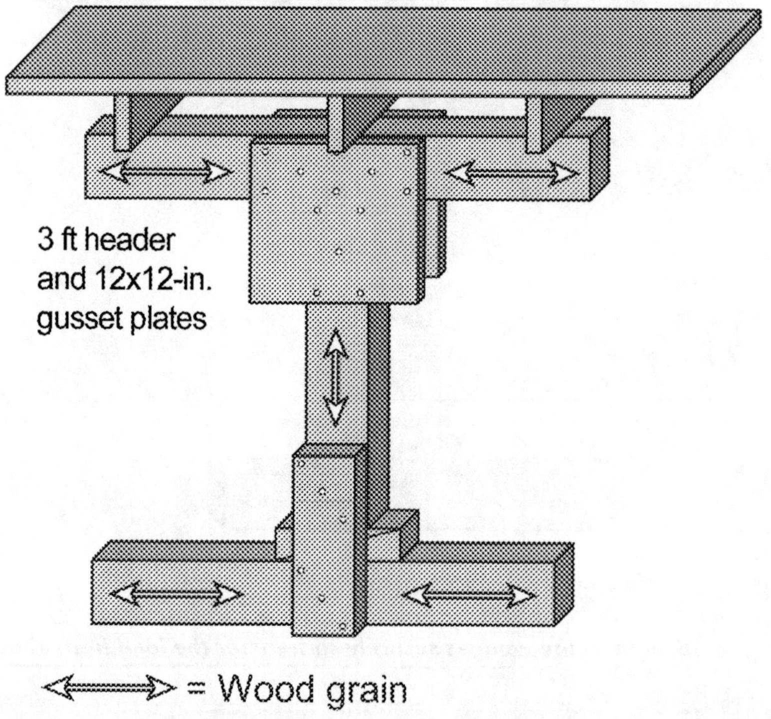

Fig. 4–17 Anchor a 2-ft long 2x4 cleat to one side of the post and sole plate.

Figure 4–18 shows a T-shore with a 4 ft long header. Should you elect to install a longer header, you must employ an 18x18-in. square gusset plate. This size plate helps keep the header and post joint secure and adds additional efficiency to the shore. Note: For a longer header, the nail pattern is continued and the addition of three more nails into the post as well as in the header helps secure the gusset plate to the shore.

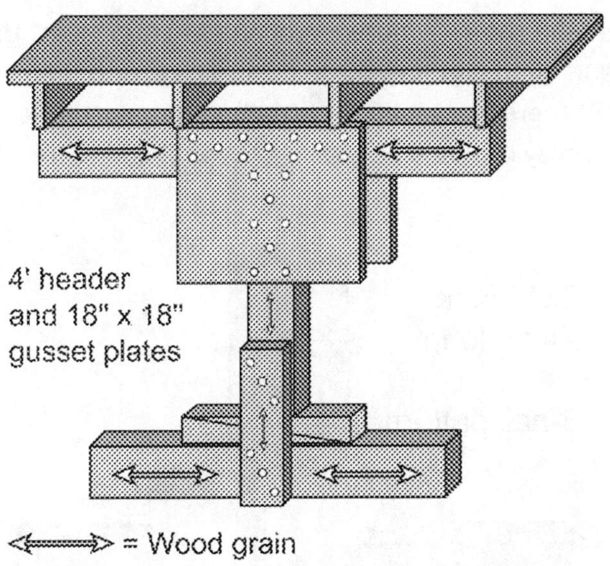

Fig. 4–18 Use 18x18-in. gusset plates with a T-shore with a 4–ft header.

Figure 4–19 shows a close-up of the 18x18-in. gusset plate, secured by the same 5-nail pattern and the same 6-in. spacing. With this method, you wind up using 11 nails on the header and 8 nails on the post.

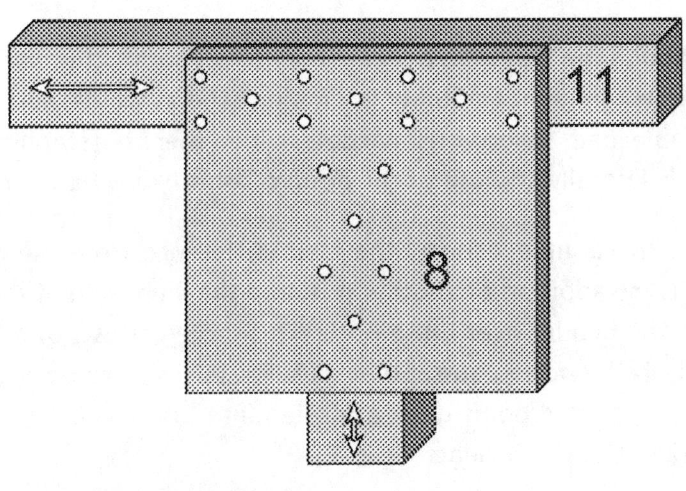

Fig. 4–19 The 18x18-in. gusset plate with 5-nail pattern and 6-in. spacing.

Figure 4–20 shows a close up of the bottom cleat. Using a 2x4 roughly 24 in. long, you can place (3) 16d nails into the post and the sole plate. If there is a problem with vibration, putting one nail into each wedge may not be a bad idea.

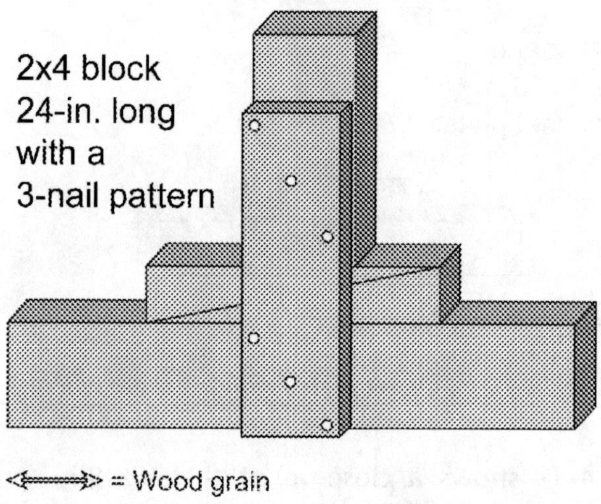

Fig. 4–20 The 24–in. long 2x4 bottom cleat.

Double T-shore options

Although the T-shore is a preliminary safety shore, there are many instances when lateral stability is a major issue as well as supporting an unstable load. When you have determined there is a stability issue, especially laterally, the use of the double T-shore may be warranted.

With the double T-shore, the main difference from the regular T-shore is the addition of another post and the limitation of the over-hang of the header over these posts. This shore consists of a header (typically 4x4), two top gusset plates (various sizes), two posts (gener-ally 4x4s), one mid-point gusset, a sole plate (usually 4x4s), wedges (2x4 x 12 in.), and 2x4 cleats.

There are several varieties of double T-shores. In both of these op-tions, there will only be a 6-in. or a 9-in. overhang of the header over both posts. This is to keep the header tighter and allow less chance of an unbalanced load affecting the stability of the shore. It also helps the shore be more maneuverable with less header length to hang up on anything while shifting and placing the shore in position.

A two-post T-shore with the posts spaced outside to outside, 18 in. apart. This size shore is easy to maneuver in a tight environment. A two-post shore has a header that is 36 in. long and is fabricated with 12x24-in. plywood gusset plates. This option gives you a stable and strong shore without making the dimensions of the shore too much greater than the one-post shore.

The two-post T-shore is prefabricated then carried into position. Assemble the header, gusset plates, and posts together first. Then bring in the preassembled section with two sets of wedges and the sole plate. Assemble all the components in place and pressurize. This shore goes up very quickly, is portable, and is stronger and more stable than the single-post T-shore.

18 in. double T-shore. Figure 4–21 shows the spacing for the 36-in. header and the two 4x4 posts. From outside to outside of the posts, the measurement is 18 in. The overhang on both ends for the header is 9 in.

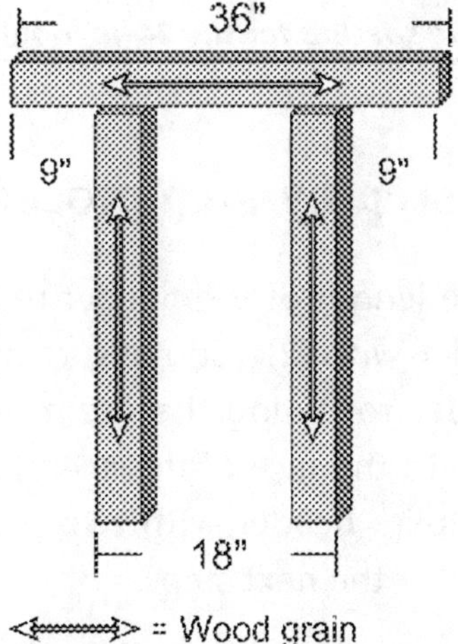

Fig. 4–21 The spacing for the 18-in. double T-shore.

24-in. double T-shore. This shore uses an overall header length of 36 in. Place the two posts 6 in. in from each end on the outside face to get an outside width of the posts of 24 in.

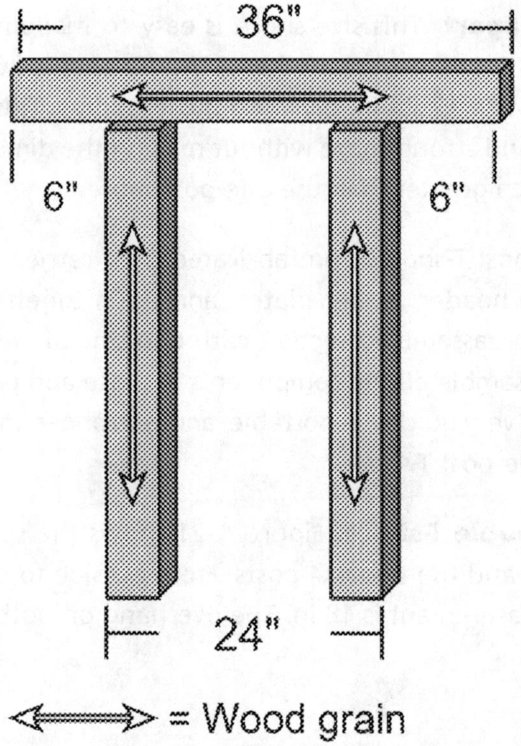

Fig. 4–22 The spacing for the 24–in. double T-shore.

Double T-shore step-by-step procedures

1. Determine the length of the header required. Nail the posts into the header with (2) 16d nails, toenailing from the post into the header following the grain of the material. Drive the nails flush to the face of the 4x4s. Figure 4–23 shows the layout for a 36-in. header with two posts, 18 in. outside to outside ready for the next steps.

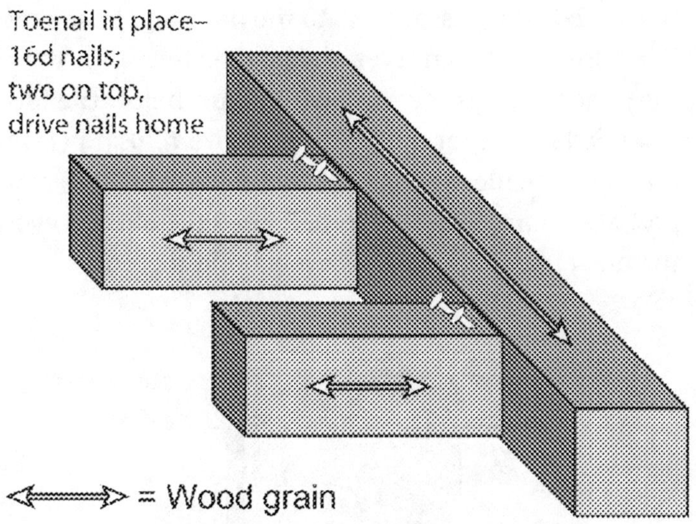

Fig. 4–23 Layout for a 30-in. header with two posts, 18-in. outside to outside.

2. Square up both posts to the header as shown in Figure 4–24. Use a framing square and make sure the toenails are flush with the material. Again make sure the posts are square to the header, this is very important. You do not want to erect the shore out of rack, for if you do your team will have a problem setting it into position and pressurizing it.

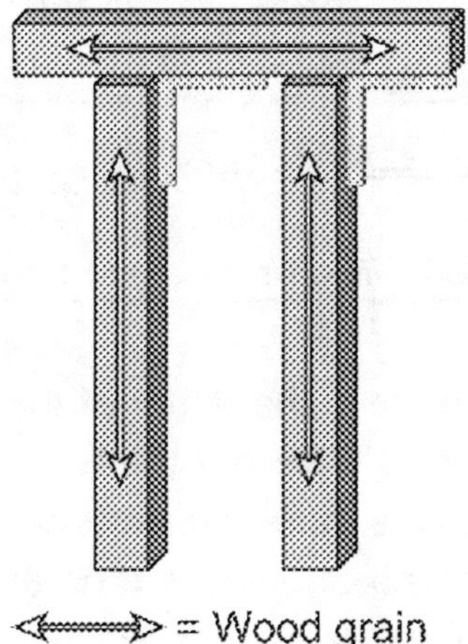

Fig. 4–24 Square up both posts to the header.

3. Nail a 12x24-in. gusset plate to the posts and header as shown in Figure 4–25. Do not let the plate extend past the edge of the header. It must be flush with or just below the face of the 4x4 header. Anchor with a 5-nail pattern, using (17) 8d nails across the header at 5 in. intervals and (5) 8d nails along each post also in the 5-nail pattern. Spread the pattern evenly over the gusset so the spacing is even along the posts.

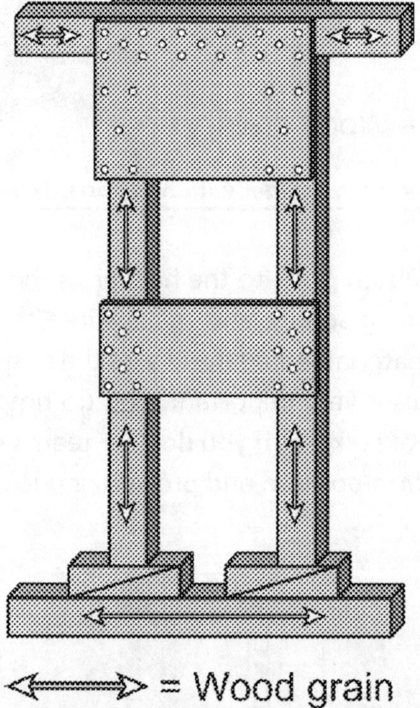

Fig. 4–25 Nail a 12x24-in. gusset plate to the posts and header.

4. Place a gusset plate in the center of the posts. The purpose of this gusset plate is to secure the two posts together so that when you move the shore, the posts do not spread out or slide together. Use a 12x24 and place (8) 8d nails into each post. Make sure the gusset plate is exactly 24 in. long and nail it flush to the outside of both posts.

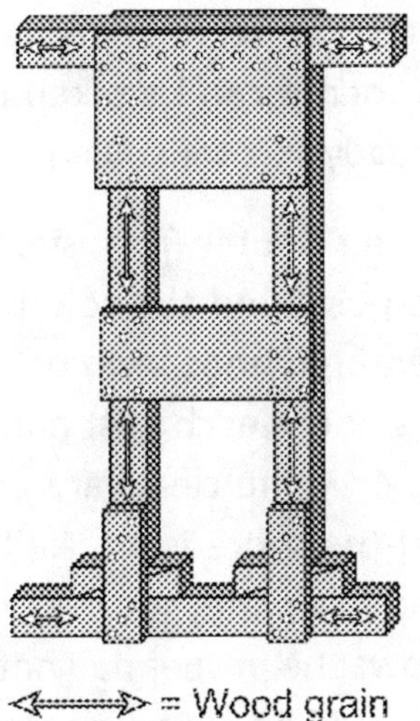

Fig. 4–26 Nail a 12x24-in. gusset plate in the center of the posts.

5. Flip the shore over and anchor another 12x24-in. gusset plate to the header and posts in the same position as the first one. Check to see that all the nails in the first gusset went into the header and posts. If any did not, flip the shore back over and fix it. Figure 4–27 shows the double-*T* section finished: a header, two posts, two 12x24-in. gusset plates, and one 12x24-in. center gusset plate.

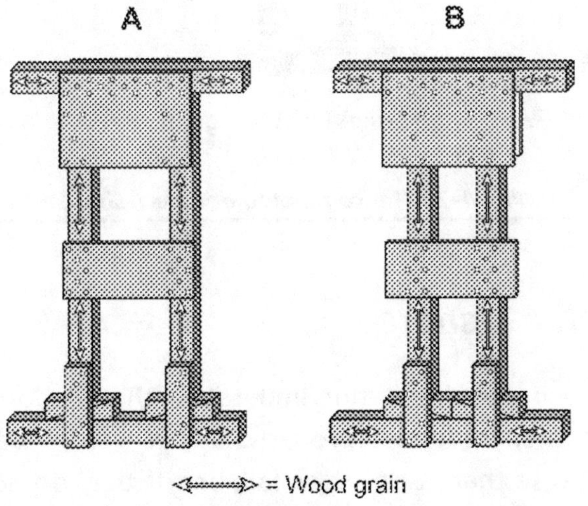

Fig. 4–27 The finished 24" (A) and 18" (B) double-T section.

6. Prefabricate the shore to this configuration outside the collapse area, and then it can be taken inside as one piece. The section is now ready to be installed.

Figure 4–28 shows the double-*T* finished, wedged up, and 2x4 cleats anchored to the posts and the sole plate. This 24-in.-spaced T-shore is placed and centered under two or three floor joists. Place a set of 12 in. 2x4 wedges under each post and tighten. The sole plate should be a 36 in.-long 4x4. The cleats are used to stop the wedges from being dislodged if the shore is hit. Nail the cleats with (3) 16d nails to the posts and the sole plate. You only need these cleats on one side. The figure also shows the proper nail pattern for a shore using a 12x24-in. gusset plate. Use (8) 8d nails with the 5-nail pattern and 6-in. spacing. This configuration delivers 17 nails on the gusset plate.

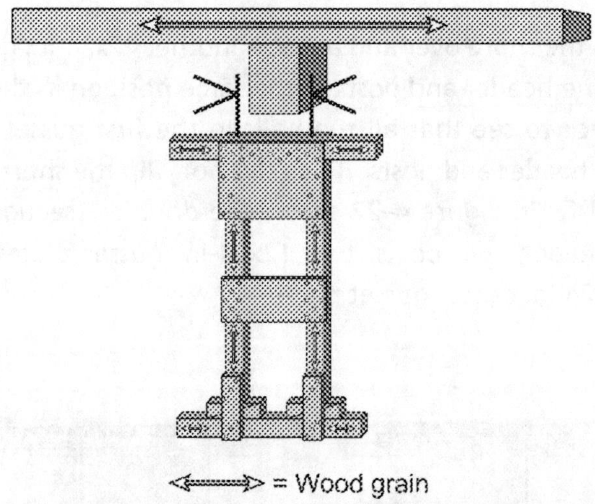

Fig. 4–28 The completed double T-shore.

Loading the T-shore

It is very important if not imperative that T-shores, whether single or double, are loaded properly. Any off-center loading of the shores can cause them to shift or fail; and if they do not fail, their strength would be greatly reduced. In many situations, the single-*T* may be good for only an approximate 4000 lb of support because of its inherent instability issues. Even a double-*T*, which should provide more than 8000 lbs of support, may not come close to supporting that weight if the load is not centered over the two posts. When installing these posts, make sure they are loaded properly or your team may be working under false security.

Always keep the load balanced into the center of the shore for safety reasons. If it is not, the shore will be almost completely ineffective. In fact, the shore is in danger of being kicked over. Instead of going through the center axis of the post, the force of the load is being applied to the post eccentrically. This, in effect, drastically reduces the supporting strength of the post as well as completely compromising its stability. Even the remotest shift of material above may knock the shore to the floor.

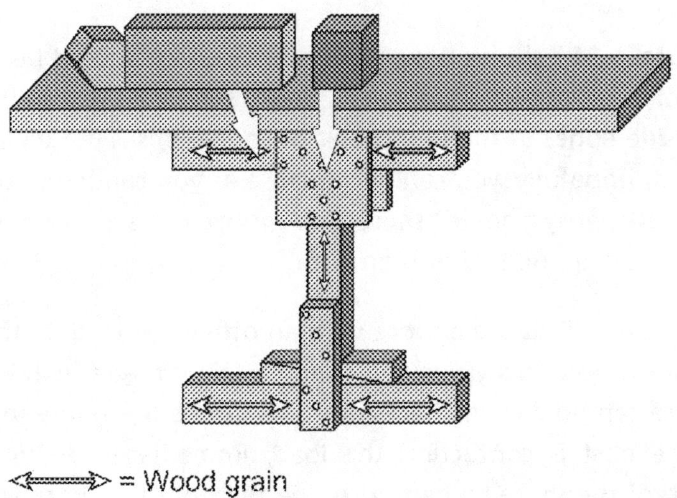

Fig. 4–29 The force of the load is being applied to the post eccentrically reducing support strength and stability.

Figure 4–30 shows a shore that is properly loaded. The material weight from above is transferred through the center of the post. A properly balanced load gives your team the full strength of the post, thereby increasing the effectiveness of the shore.

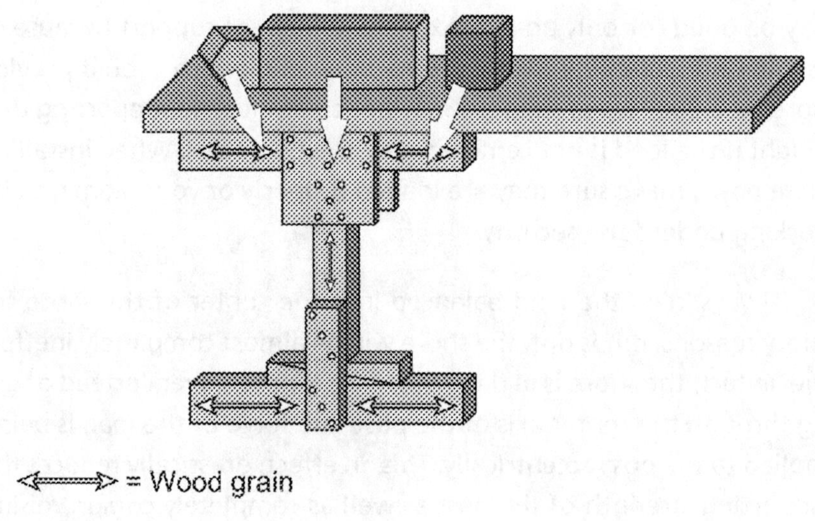

Fig. 4–30 A properly loaded shore.

Figure 4–31A shows how the double T-shore should be loaded. The item or items being supported ideally should be placed between the outside edges of the two posts. Of course this is not always possible; but, hopefully, you can get as close as you can in order to get the best efficiency from the shore. This shore could support more than 8000 lb if it were no higher than 8 ft.

Figure 4–31B shows a shore with an off-center load. In this case, the shore is potentially unstable, and certainly the effective weight the shore can hold is nowhere near 8000 lb. As the figure indicates, only one post is contacting the load, immediately reducing the strength of the shore by half. Also, the fact that the shore is loaded unevenly detracts from the supporting strength, making the real support strength even less than half. Try and make sure that all T-shores are loaded through their centers and are evenly balanced.

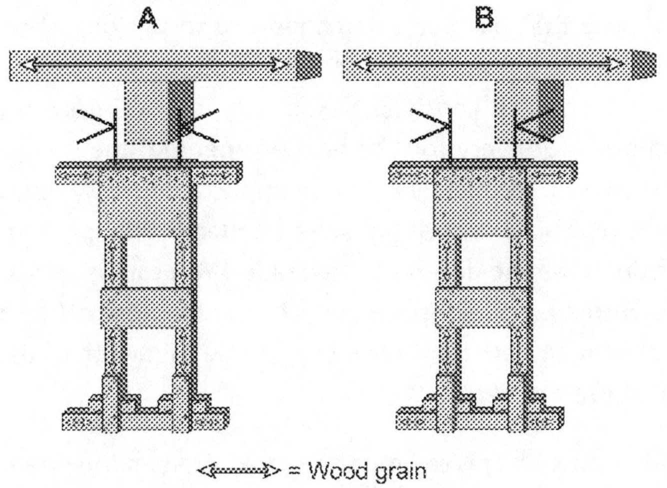

Fig. 4–31 A shore with a centered (A) and off-center (B) load.

The Door Shore

This door shore is basically designed to stabilize any existing door opening or access way, whether damaged or weakened from an incident. It can even be utilized to support an opening created by breaching an existing wall. This shore is especially useful if openings are racked or out of plumb. It is well suited for use in structures constructed of masonry walls.

After a serious collapse, the majority of masonry in the structure is subjected to some type of major stress, both internal and external. While masonry has excellent compressive strength, its ability to withstand tensile stress is minimal. If a long-term operation is being conducted, then it is a good idea to utilize a door shore where your rescue personnel are operating.

In your interior size-up, check all masonry walls in the collapse area for any signs of damage. If your rescue team is going to operate in an area that has been damaged or shows signs of significant stress and cracking, erect the door shore around the access ways. One rule of thumb that is consistent especially in unreinforced masonry construction is for every foot of header opening, the header should have at least 1 in. of thickness. For example, a doorway opening that is 4 ft wide needs a 4x4 minimum as a header. If the opening is larger than 4 ft, increase the depth of the header. You can use a 4x6 or even two 2x6s with a plywood spacer, which may be more efficient than a one-piece header. You don't have to increase the width of the header as you increase the depth.

Another possibility is to limit the space between the openings; this can be done in several ways. Here are two examples. The first is to utilize another 4x4 post in the center of the shore. Generally, this can be done only when the door opening is not being used for building access or egress. The second method is to place two diagonal braces into the middle of the header. Block the ends of the 45° angle diagonal braces with 2x4 blocks to stop them from shifting and to help the braces transfer the weight from the center of the header to the posts and down to the ground.

Door shore step-by-step procedures

The following is a numbered list of the tasks that must be completed to build and place a door shore in a collapsed building. The remainder of this section on door shores provides detailed instructions for constructing and installing the various elements of the shore.

1. Survey the area and determine the load displacement and structurally unstable elements.

2. Clean the area to be shored.

3. Measure for proper lengths of shoring items.

4. Wedge the sole plate in position.

5. Wedge the header in position. Wedges are placed on the opposite side of the bottom wedge set.

6. Set the first post in place under the header at the wedged end; snug the wedges. Wedges under the post are to be on the bottom of the post on top of the sole plate.

7. Set the second post in position and snug the wedges under bottom of post.

8. Check the shore for fit of materials and tightness to the opening and tighten all wedges.

9. Install diagonal braces if the opening is not being utilized for access.

10. Install a gusset plate or cleat posts if necessary.

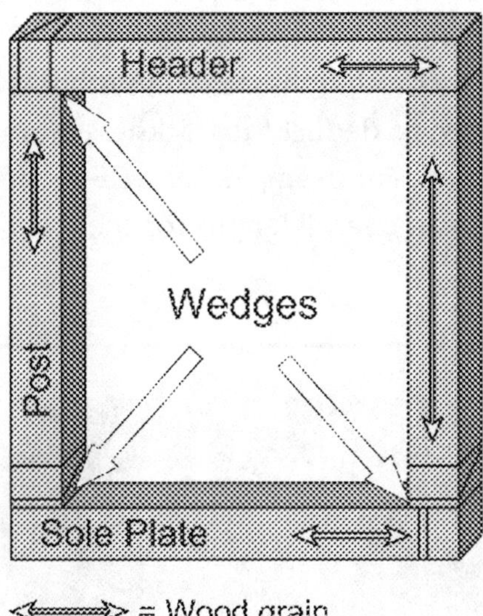

Fig. 4–32 Elements of the door shore.

Typically, a door shore is installed with 4x4s; however, when necessary the header can be two 2x6s or one 4x6 if the opening is larger than 4 ft wide. The header is placed level at the top of the door opening and wedged against the sides of the opening. Posts are also constructed of 4x4s and are placed under the header at both ends of the door opening. They sit on top of the sole plate and wedges. The posts are the items that take the main weight from above and transfer it to the ground. A sole plate is the first item to be installed, placed at the bottom of the opening and wedged tightly. Usually, like all the other door shore pieces, the sole plate is a 4x4. There are four pairs of wedges placed in a door shoring system. The first is between the sole plate and the door opening. The second is between the header and the opening at the opposite side of the wedges placed at the sole plate. The other two sets of wedges are at the bottom of the posts. Anytime you install a post or set of posts in a vertical plane, as in the case of a door shore, the wedges are always at the bottom.

Measure the opening at the bottom of the door and remove the doorstops if possible. Leave a space for the installation of wedges to adjust the tightness of the sole plate. A rule of thumb is to determine the size wedges you need, either 2x4 or 4x4. In interior shoring, for the most part, you would be much better off if using 2x4 wedges because they make for a much tighter shore. With this in mind, for placement of a set of wedges, deduct the thickness of one wedge from the length of the opening. For example, for a 36-in. opening, deduct 1½ in. This calculates to an overall length for the 4x4 of 34½ in.

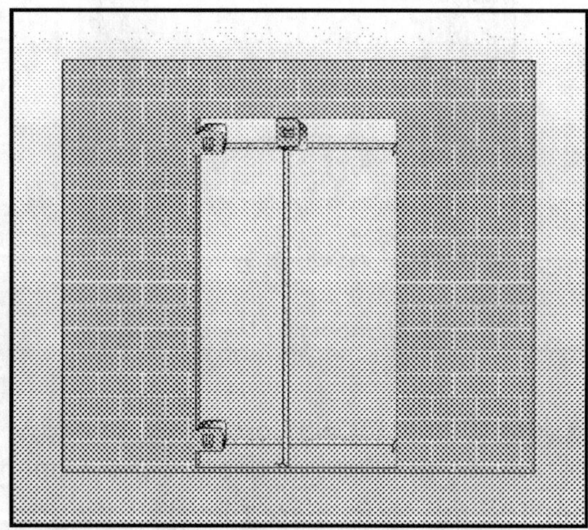

Fig. 4–33 Measure the opening for the header, sole plate, and posts.

With this space, the set of 2x4 wedges fit just right and can be pressurized. Cut the wedges used to keep the plate tight close to the bottom plate to eliminate any tripping hazards. Measure for the length of the header. It typically should be the same size as the sole plate, but to make sure, measure anyway. Deduct for the thickness of the wedges just like you did for the sole plate. Carefully measure for the length of the posts. If the opening is seriously racked, you could have trouble with the accuracy of the measurements. Don't forget, if you do measure for the posts at this time, you must deduct the thickness of the wedges, header, and sole plate from the overall length.

Install the sole plate, clean the area, and keep the plate level. Snug the wedges tightly, keeping the top of the sole plate and the wedges flush with each other. The posts that sit on top fit well because they have full contact with the wood, giving them full bearing on the lumber and making the shore that much more efficient. Pressurize the wedges just enough to tighten the plate to the frame and keep it tight in the opening. Don't tighten the wedges too much. You could cause something to loosen and create problems for yourself.

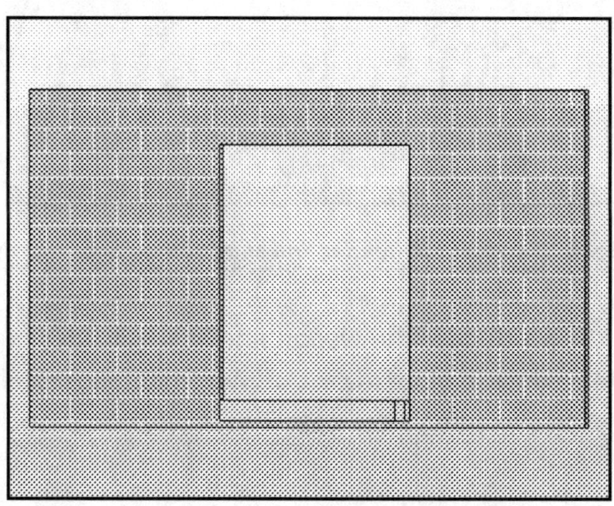

Fig. 4–34 Install the sole plate and snug the wedges tightly.

For a more accurate way of measuring for your posts, especially if the opening is racked, lay the header on top of the sole plate and measure off the top of the header for the posts. Measure both sides for accuracy. If the door header, or lintel, is cracked or bellied, you may want to install the header before you measure for the posts.

Install the header in position, keeping it as level as possible and snugging the wedges. Install the header wedges on the opposite side of the sole plate wedges. Anchor the header to the doorjamb. If there is a belly in the door lintel, keep the shore's header level and shim out the space, especially over the area in which the posts will be assembled. Figure 4–35 shows a view of the two items in position ready for the post installation.

Fig. 4–35 Install the header and snug the wedges tightly.

Install the post under the wedge side of the header first in order to keep the wedges in place in case of movement. Be certain that the wedges are installed under the post. This is very important! With the wedges on the bottom, the entire shore is stable. With more surface contact at the top, the shore is laterally more stable because most of the lateral pressure is at the top of the shore.

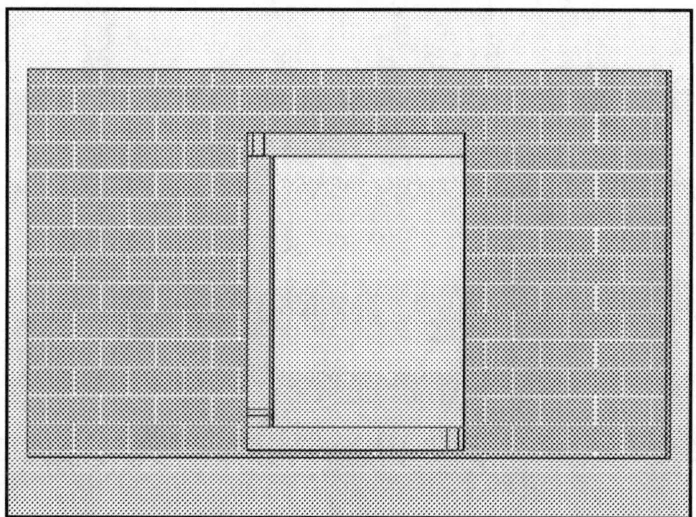

Fig. 4–36 Install the post under the wedge side of the header first, with wedges at the bottom.

Figure 4–37 shows a close-up of the post under the wedge side. If you utilize thin wedges, there will be plenty of bearing on the post. Make sure there is at least 2 in. of minimum bearing contact under the 4x4 header. If you have to pad the post out from the doorjamb, you can use 1x4s or 2x4s, whichever gives you enough room, along the side of the doorway. This padding out of the post has to be done in order to keep the shore's integrity and to enable the post to support a load adequately. The wedge thickness here is 1½ in. Wedges larger than that may require you to pad out the post from the wall. Make sure you have the minimum 2 in. of bearing under the header.

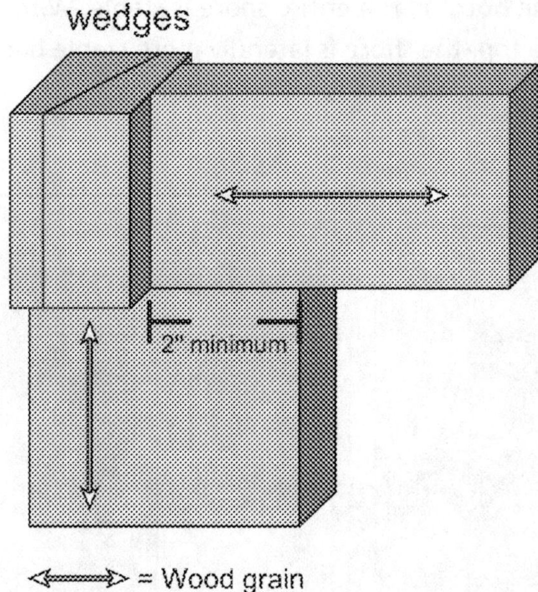

Fig. 4–37 Close-up of the post under the wedge side of the header.

Next, install the post under the opposite side. Anchor both posts to the header and sole plate after all the wedges have been tightened. Anchor the post to the header with at least (2) 16d nails. Toenail from the post into the header, following the grain of the lumber so you don't split the material.

Fig. 4–38 Install the post under the opposite side last.

If the door or opening is not going to be used for team access and egress, you can cross brace the shore to add additional stability to the shore and the wall. Typically, 2x4s are sufficient. Nail them to the posts with three nails at each end. One brace is on one side and one on the other in opposite directions from each other.

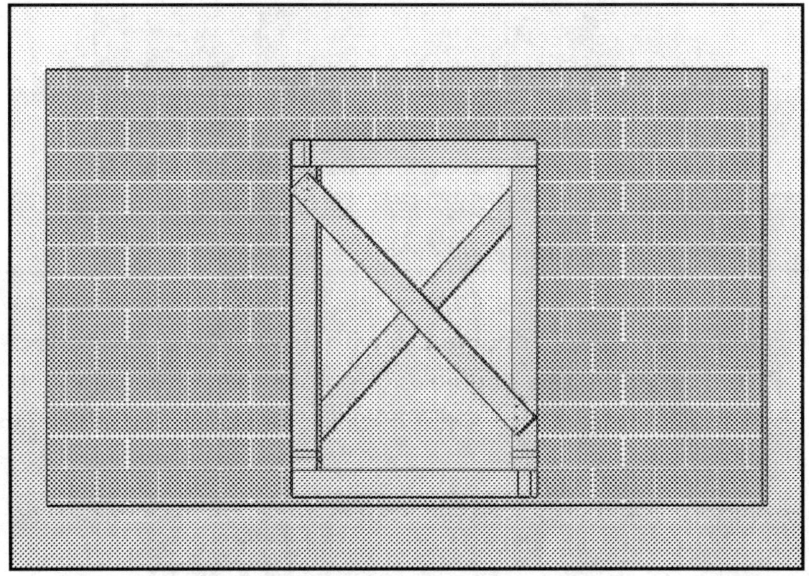

Fig. 4–39 Cross brace the shore if it is not used for team access/egress.

For openings greater than 4 ft, it is usual to increase the header size. There are numerous ways of doing this. One way shown in Figure 4–40 is to decrease the overall length of the header by placing two 45° angle braces into the system. Each brace is the same size and has a 1½ in. return cut on both ends of the braces so that a cleat can be placed in position to butt the braces, stopping them from sliding in either direction. The cleats are sections of 2x4s at least 1 ft long, anchored with a 5-nail pattern.

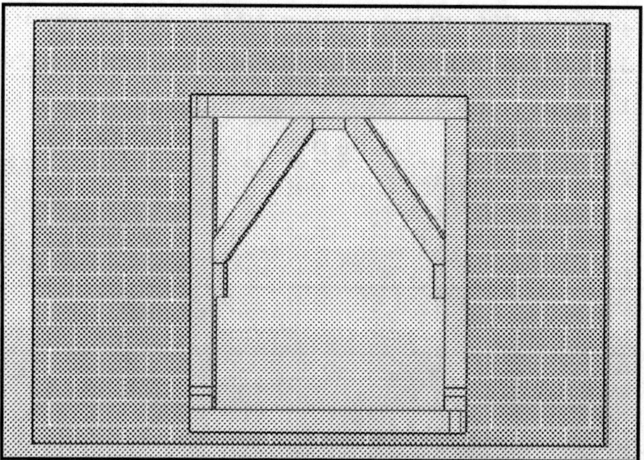

*Fig. 4–40 One way to decrease the overall length of the header—
using two 45° angle braces.*

Figure 4–41 shows a double 2x6 with plywood spacers in the center. Placed on edge, this double 2x6 is more than strong enough to support the loads from above. It is installed and anchored in the same manner as the 4x4s.

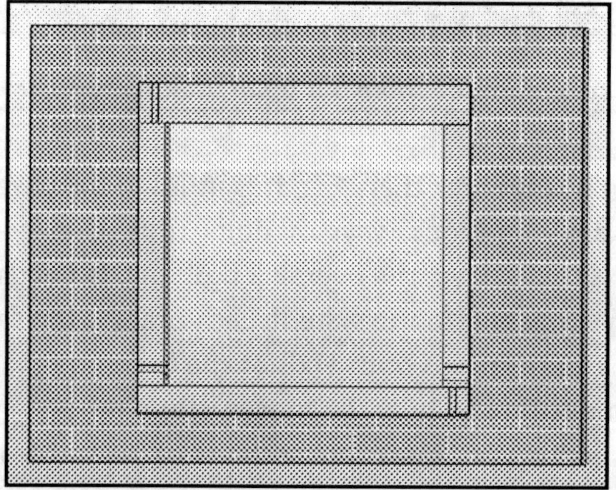

Fig. 4–41 A double 2x6 header with plywood spacers in the center.

If an opening is not being used for access, then an additional post and wedges may be installed, generally in the center. This is usually the easiest method.

Fig. 4–42 An additional post and wedges may be installed in the center if the opening is not used for access/egress.

Figure 4–43 shows another bracing option. In this case, place two 4x4s on top of each other. This bracing technique works well if the 4x4s are tied together. If they free rotate, they provide minimal additional support. However, locking the two pieces together using at least three sections of plywood makes an efficient header. The plywood gussets must be anchored on both sides of the header in order for the connection to be effective. Nail them together using (8) 8d nails in a 5-nail pattern for each 4x4. You can also use 6x12-in. gusset plates in this situation.

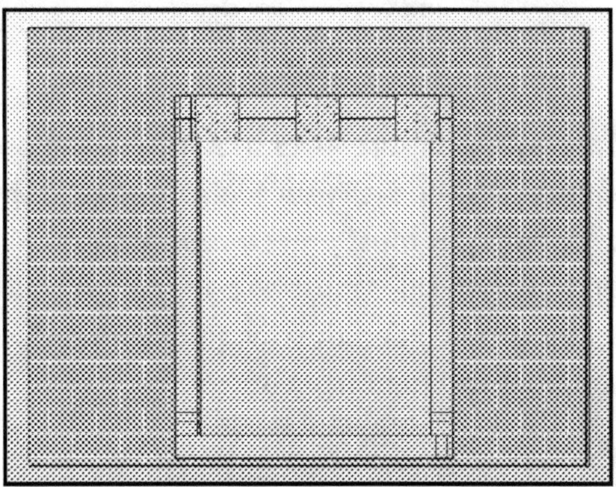

Fig. 4–43 Another bracing option using two 4x4s, ideally using 6x12-in. gusset plates to lock them together.

Figure 4–44 shows a complete door shore in position and fully gusset plated. When the shores are erected in an earthquake situation, the joints must be additionally stabilized. This is done to stop joint shifting. You can place cleats and gussets on the shore in any situation. In Figure 4–44, the shore has one gusset and several 2x4 cleats. Anchoring these cleats on one side only is enough to lock the joints together and stop them from vibrating loose.

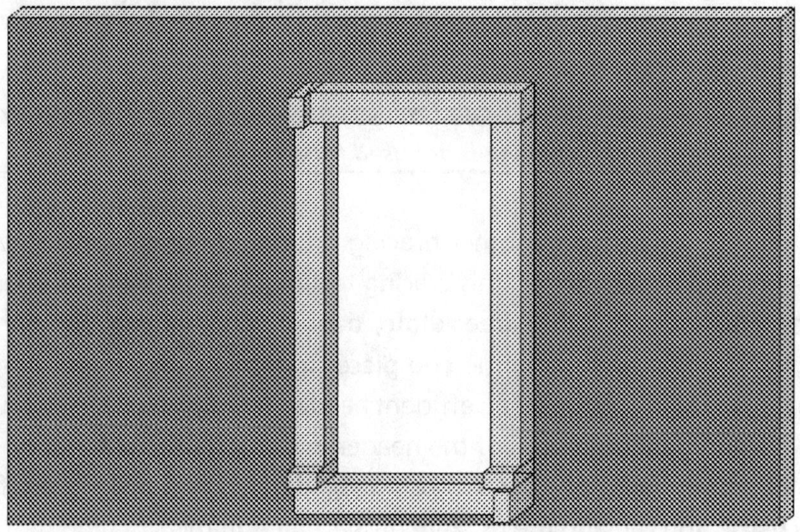

*Fig. 4–44 A complete door shore with one gusset plate
and several 2x4 cleats.*

A racked opening. Figure 4–45 shows how a shore should look in a leaning opening. Whenever your team decides to shore up a leaning, or racked, opening, it must erect a square shore. After the shore is in place, then the open sides must be padded out to the header and the sole plate. This should keep the opening from racking any further as the shore supports the load from above.

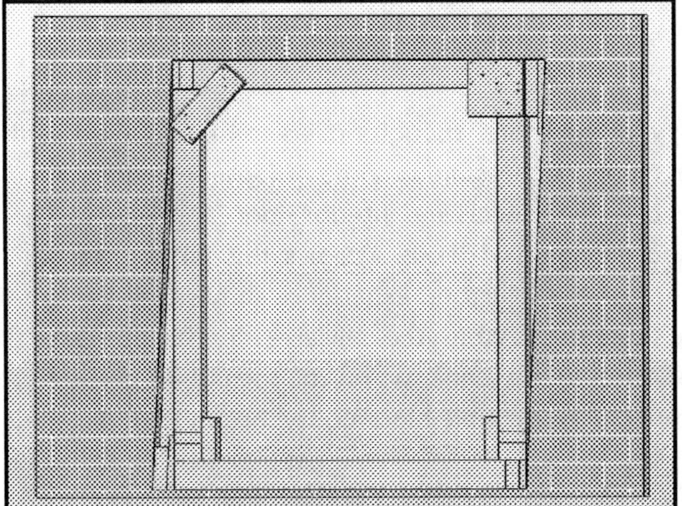

Fig. 4–45 How a shore should look in a leaning opening.

When utilizing pneumatic shoring struts, place them under the header and on top of the sole plate. Tighten them by hand only. Normally do not use air to set these struts.

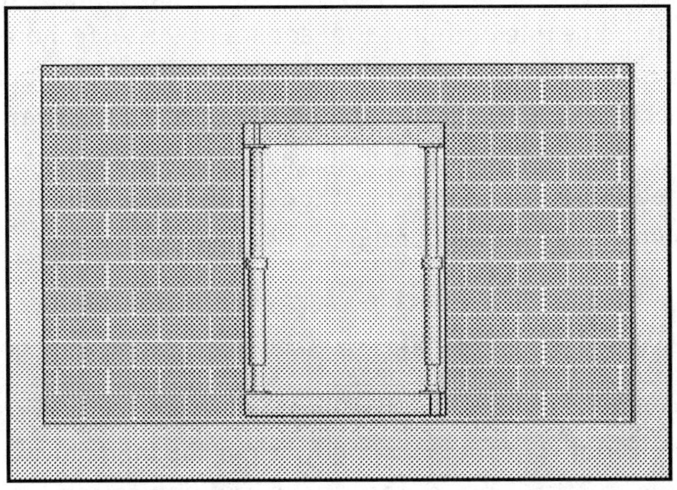

Fig. 4–46 Utilizing pneumatic shoring struts.

The Window Shore

The window shore, though not as commonly used as the other interior shores, is necessary in some collapse situations. An extensive collapse of exterior walls and interior floors and walls can generate a tremendous amount of debris. At times, this debris can block the main egress out of the structure. Experience shows that there have been collapse situations in which the only access to the interior of the structure was through the windows on the second floor due to the large amount of collapse rubble blocking the ground floor entrances.

The main objective is to stabilize, replace, or resupport window openings or damaged window headers. Generally, the window shore is installed to hold up or stabilize loose headers or lintels that have lost their integrity so that rescue personnel have a safe access point. A window shore prevents accidental dislodging of building materials while firefighters use the opening.

The window shore is built the same way as the door shore. Generally, it is shorter than door shores; however, it can easily be the same width. The same rule of thumb that applies to the door shore applies to the window shore. Especially in URM construction, for every foot of header opening, the header should have 1 in. of thickness. A 4 ft-wide doorway needs a header that is a minimum of 4x4. As with any completed shore, firefighters must constantly check it to ensure that no movement of the shore or the opening has occurred.

Window shore size-up

The window shore size-up, while usually brief, still is necessary. If the shore is not constructed properly, it may be ineffective and give the rescuer a false sense of security. In addition to conducting the normal shoring size-up, a specific size-up for window shores adds these two main points: an assessment of the structural stability and an estimate of the amount and direction of the load stress against the opening.

Structural stability. First determine if the area is safe enough for rescue personnel to operate. Then examine the window framework, the opening, and the area around the window for structural stability or, perhaps, for the lack of structural stability. Check the integrity of the framework and the building material around it, making sure it is not damaged beyond safe limits. Determine if shoring the window opening will create a safe access and egress point for rescue operations to continue in that area. This determination is important especially in buildings constructed of unreinforced masonry exterior walls. Some of the items to look for are bulging of the wall, cracks in the masonry joints or bricks, racked or leaning walls, loose headers or lintels, and the overall condition and shape of the wall to begin with before the collapse occurred.

Load stress. A very important consideration concerning load stress is the amount of an additional load stress to be exerted against the opening and the direction in which it is to be exerted. Almost always, the additional stress is applied from above. Gravity constantly tries to pull the building and any debris within to the ground. When the stress is from above, the header or top plate of the window shore must be fully supported by the posts so that the shore takes the largest amount of weight possible from above. If for some reason the load stress is being applied laterally, then additional items may have to be placed in the shore or around it.

Window shore step-by-step procedures

The following is a numbered list of the tasks that must be completed to build and place a window shore in a collapsed building. The remainder of this section on window shores provides detailed instructions for constructing and installing the various elements of the shore.

1. Survey the area and determine the load displacement and structurally unstable elements.

2. Clean the area to be shored.

3. Measure for proper lengths of shoring items.

4. Place a sole plate in position and wedge it in place.

5. Set the header in position and wedge it in place. Place wedges on the opposite side of the bottom set.

6. Set the first post in place under the header at the wedged end and snug the wedges. Wedges under the post are on the bottom of the post on top of the sole plate.

7. Set the second post in position and snug the wedges under bottom of post.

8. Check the shore for fit of materials and tightness to the opening, and then tighten all wedges.

9. Install diagonal braces if the opening is not being used as an access way.

10. Install a gusset plate or cleat posts if necessary.

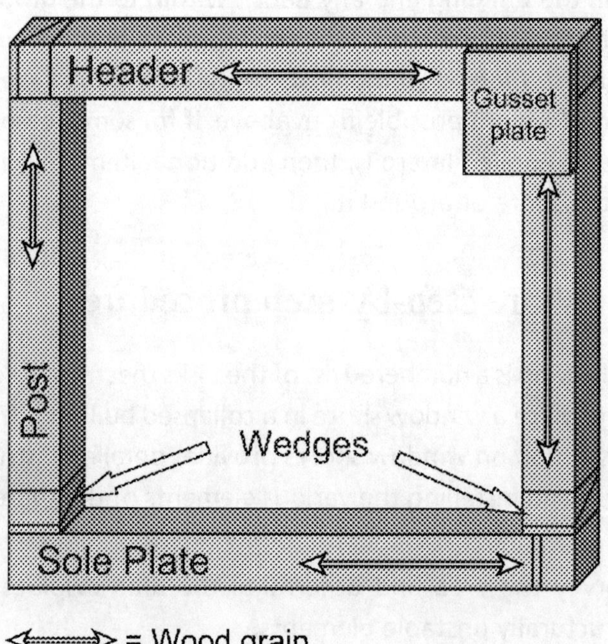

Fig. 4–47 Elements of the window shore.

The window shore is typically installed with 4x4s. Most often the header is also made of 4x4s; however, if necessary, it can be constructed of 2x6s or a 4x6 if the opening is larger than 4 ft wide. The header is placed level at the top of the window opening and wedged against the sides of the opening. The posts also are normally constructed of 4x4s and are placed under the header at both ends of the opening. They also sit on top of the sole plate and wedges and take the main weight from above and transfer it to the ground. The sole plate is the first item to be installed, placed at the bottom of the opening and wedged tight. It also is made from a 4x4. There are four pairs of wedges placed in a window shoring system. The first goes between the sole plate and the window opening. The second goes between the header and the opening at opposite sides of the wedges placed at the sole plate. The other two sets of wedges are placed at the bottom of the posts. Anytime you install a post or set of posts in a vertical plane, such as in a window shore, the wedges are always at the bottom—always.

To determine the length of the header, measure the opening of the window; remove stops or loose and broken framework if possible. You can measure the window opening at the top, the middle, and/or the bottom. Leave a space in your measurement length for the application of wedges to adjust the tightness of the sole plate. When measuring for the length of the header, the measurement should be the same size as the sole plate. Measure anyway to make sure. Deduct for the wedges just like you did for the sole plate. You can also measure for the length of the posts, however, you must be careful with doing that at this time. If the opening is racked or if the header is cracked and deformed, you could have trouble with the accuracy of those measurements. Don't forget, if you do measure for the posts at this time, you will have to deduct the thickness of the wedges, the header, and the sole plate from the overall length.

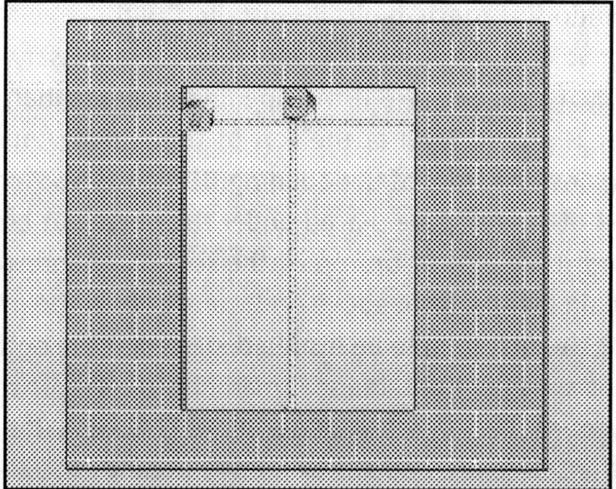

Fig. 4–48 Measure the opening of the window to determine the length of the header, sole plate, and posts.

The most accurate method of measuring for the height of the posts if there is any deformation in the opening or if the header has been compromised is to place the header and sole plate in position, one on top of each other. Then measure at each end where the posts will contact the plates. Measure for both posts, and don't forget to deduct for the thickness of your wedges.

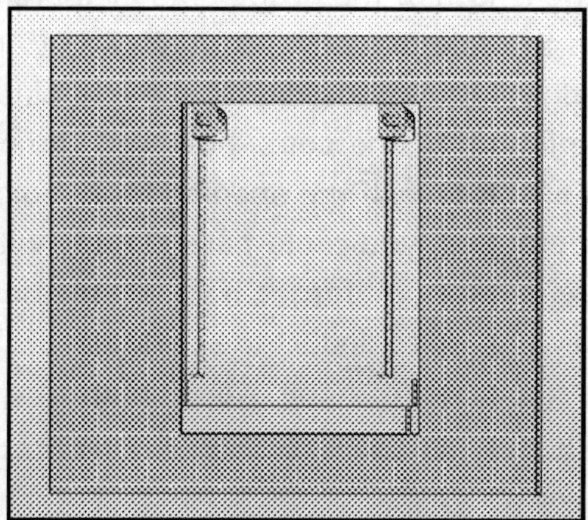

Fig. 4–49 The most accurate method of measuring for the heights of the posts with a deformed opening or compromised header.

Install the sole plate and clear loose debris or framing material away from the bottom of the window opening. Remember to work from a safe area. Don't place your body under the unstable window lintel. Measure the size of the area to be shored and deduct for the width of one of the wedges. Generally in wood-framed and smaller brick and joist construction, 4x4s make good shoring material. Try to keep the sole plate as level as possible during installation. Shim underneath it if necessary. It is especially important to shim underneath the ends on which the posts are to be installed. The ends must have full bearing for the shore to be effective. When the plate has been laid in and shimmed level, snug up the wedges.

Fig. 4–50 Install the sole plate as level as possible.

Install the header and measure the opening at a point just under the lintel, remembering to deduct for the wedges. If the building is large or if the window opening is greater than 4 ft, use larger lumber. Install the header as tightly to the lintel and as level as possible, snug up the wedges, and shim any openings if necessary. By keeping the sole plate and header level, you ensure that the posts are the same size and that the entire shore is square. The elements of the shore will fit better, and as a result, the shore will be stronger and more efficient than if it were not square.

Fig. 4–51 Install the header as level as possible.

Install the first post, take the measurements of both sides, and don't forget to deduct for the wedges. The measurements should be almost identical for both posts; slight differences in length can be compensated for by tightening the wedges. In addition to the added strength factor of a tight, square shore, posts of equal size eliminate any confused about which post goes on which side.

Fig. 4–52 Install the first post under the wedge side of the header.

Install the first post under the wedge side of the header. This holds up the header and wedges in case some accidental or unanticipated movement occurs. Place a set of wedges on top of the sole plate just under the wedge side of the header. Install the post on top of the wedges and as plumb (level in a vertical plane) as possible, and snug up the wedges. Next toenail the post into the header, using at least (2) 16d nails.

Now install the post under the opposite side. Anchor both posts to the header and sole plate after all the wedges have been tightened. Make sure that the wedges stay on the bottom. Keep the posts as plumb as possible and square to the sole plate and header so the joints fit securely. If possible, anchor the posts to the window frame. The shore is now assembled with all the elements square to each other and flush surfaces to nail together. The tighter the fit, the better the shore. The shore should be anchored to the wall in some fashion.

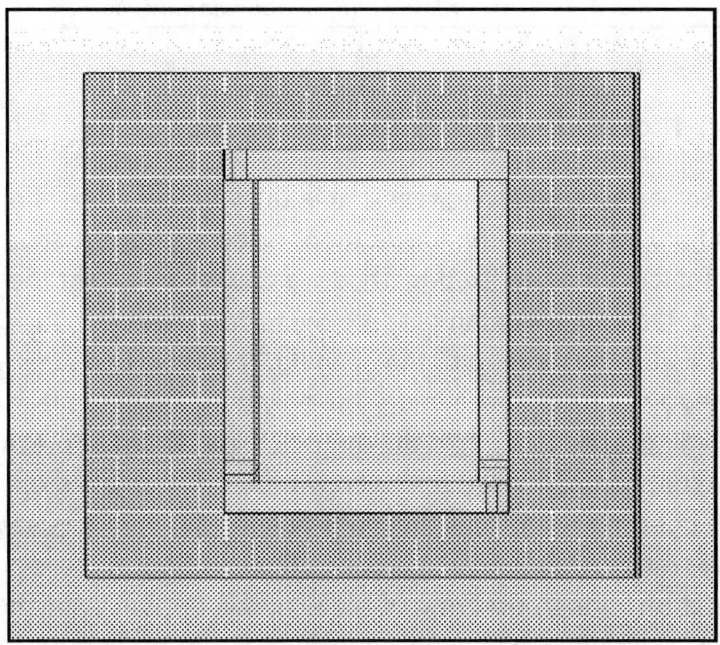

Fig. 4–53 Install the post under the opposite side last.

Figure 4–54 shows the window shore cleated and gusset-plated for installation under earthquake conditions. The purpose of these cleats and gussets is to stop the nailed sections of the shore from coming apart during aftershocks that frequently occur in these incidents.

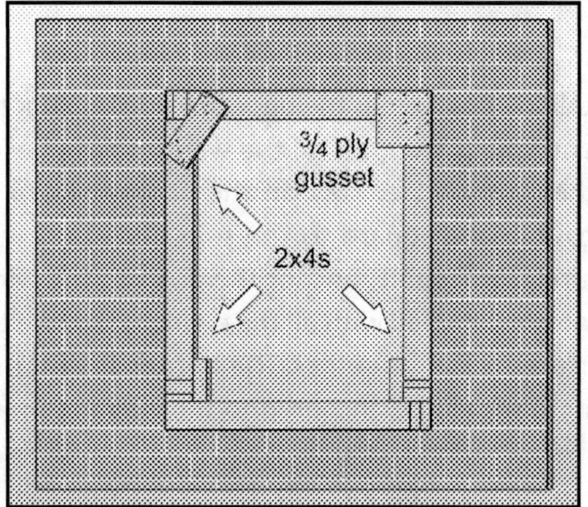

Fig. 4–54 Gusset-plated and cleated window shore.

Fixed-method window shore

In some cases, you can preassemble the window shore and put it in position. You would then have to shim and anchor the shore. The only problem with the fixed-method window shore is that the shore itself contacts the opening on only two sides, if you are lucky. You cannot adjust the shore to fit the opening. The other two sides will be shimmed away from the opening. The only thing holding the shore in the opening is the toenails in the shims or wedges. Therefore, a fixed shore is not recommended for severely racked or damaged openings because getting the proper fit is a problem. It is very important to measure the window opening properly the first time. If there are any bellies, protrusions, or racking of the window, the condition must be taken into account when you measure.

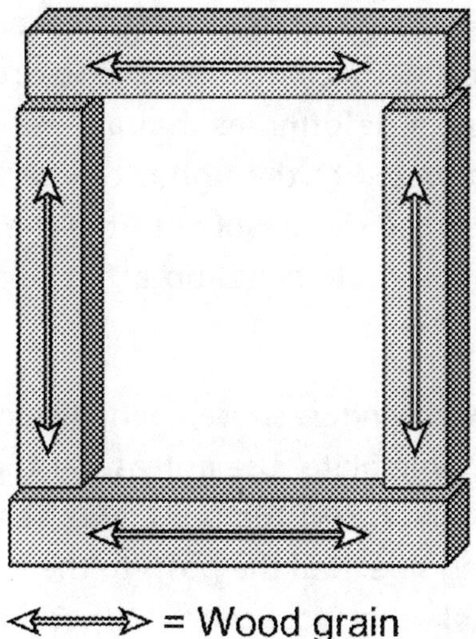

Fig. 4–55 Preassembled window shore.

With the various shore methods available, the first step is to determine the type of shore to utilize. When you have decided which one to use, take the measurements accordingly. If you decide to use the fixed-method window shore, double-check your measurements. The old adage that advises to measure twice and cut once works particularly well in this circumstance. If there are no bellies in the opening, then you can measure the width and deduct the wedge thickness. This will be the overall width of the header and sole plate of the shore.

Once you have taken measurement for the header and sole plate, the next step is to measure for the height of the posts. Measure the center and both ends of the opening. Measure all three in order to determine if there is a belly in the opening. These measurements are very important. Once the shore is constructed, there is no way to change it. If there is a belly in the header, then you must place the shore in the lower part of the opening, and then shim up to the belly. The shore height should be at least one wedge thickness shorter than the opening so that you can properly tighten up the wedges and snug the shore efficiently.

First measure the opening in at least four places, two horizontally and two vertically. This must be done in order to determine if there are any bellies or deformities that are not readily visible. The shore must be constructed to the tightest dimension in each plane. If this does not occur, the shore will not fit in the opening, negating its use. Place the shore's elements on a flat surface and align the pieces in position.

Assemble the fixed window shore, toenailing each end of the post into the header and sole plate. Use at least (2) 16d nails at each end. Preferably, there should be (4) 16d nails, two from one side and two more on an adjoining side. Nail the posts to the header and soleplate at this time. Assemble the shore with the posts in the inside of the header and soleplate; anchor them with 16d nails. If using a pneumatic nailer, place two nails on top of the posts and two nails inside the face of the posts. Do this to all four connection points, making sure that the shore is square when you anchor the lumber. This is important! Use a 2-ft framing square to verify that the shore is square. If you don't square the shore and secure it with a gusset plate, it is a very good possibility that the shore will not fit properly!

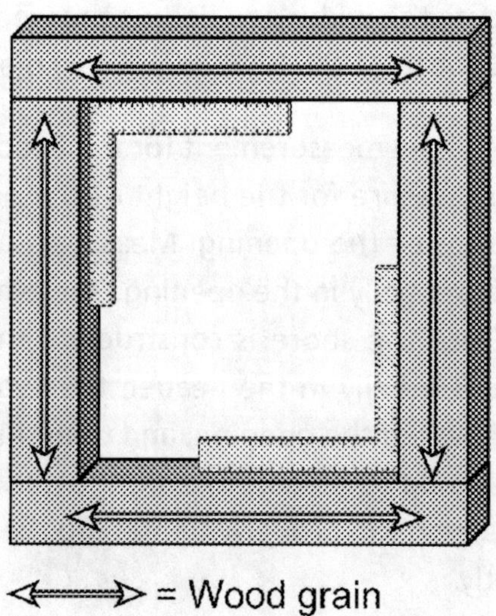

Fig. 4–56 Use a 2 ft framing square to verify that the shore is square.

After you have squared up the shore, you can place gusset plates at all four corners. This is necessary in order for the shore to stay square as well as to stay together as you pick it up and install it in position. Generally, the gussets are ¾-in. plywood, although they can be a 12x12-in. square or a 12x12-in. triangle. They really shouldn't be any smaller than the aforementioned triangle. Make sure the 16d nails on the face of the posts are driven flush with the surface of the lumber. You don't want them to stick up because that would make the gusset ineffective. Secure all four joints with gusset plates to lock the shore together as well as to keep it square during installation. You can use square gussets or triangular gusset plates. Either way, you need 8d nails; eight on one side and five on the other.

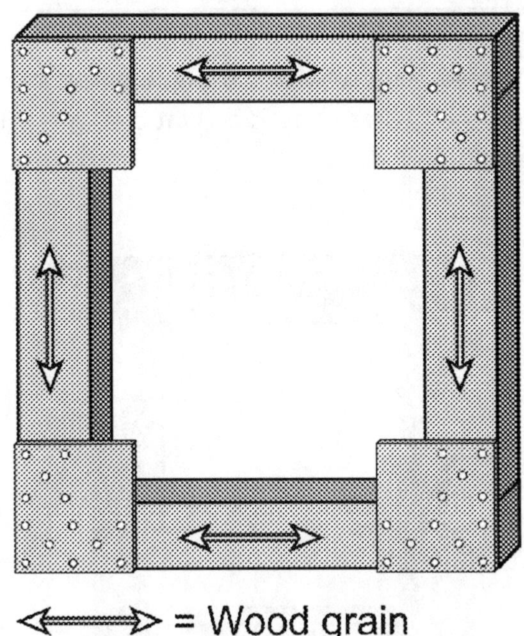

Fig. 4–57 Place gusset plates at all four corners once you have squared up the shore.

Figure 4–58 shows a shore with 6x12-in. gusset plates, which make for easy access if the need arises.

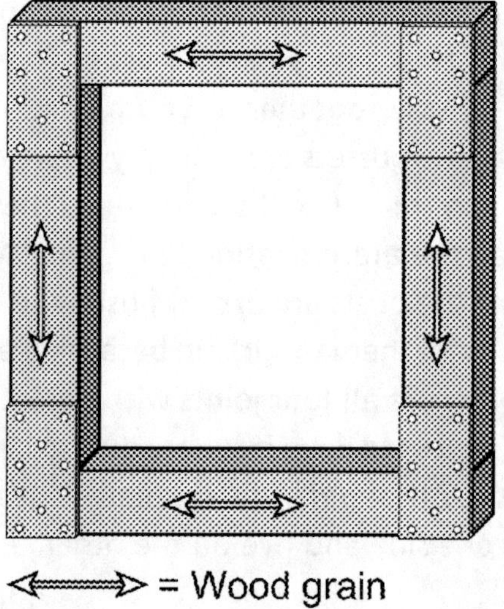

Fig. 4–58 Window shore with 6x12-in. gusset plates.

Figure 4–59 shows a shore with a triangle gusset plates. Each leg is 12 in. long and the gusset plate is a right triangle. This also gives you better access if needed.

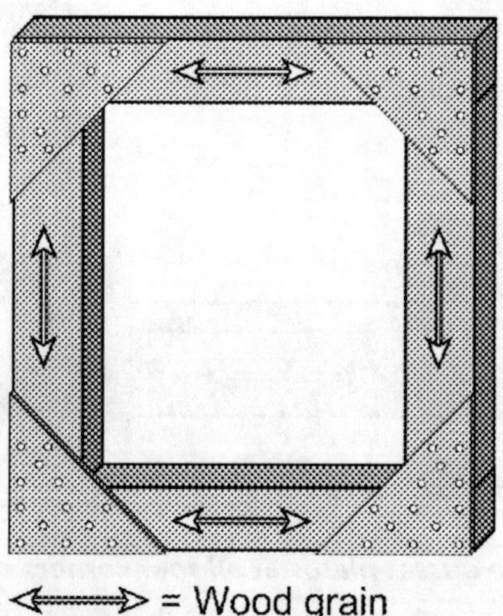

Fig. 4–59 Window shore with 12x12-in. triangle gusset plates.

Install the shore in the opening, keeping it tight to one side, generally the side with the least amount of damage. Next wedge the sole plate. Place a set of wedges in the opening and snug it up tightly. Place the top set of wedges in position and snug up tightly. Place a block or a wedge under this set of wedges and anchor it to prevent the wedges from accidentally dislodging in case of some vibration. Double check both sets of wedges, then install the top set of wedges.

Shim up the lowest point of the header first. Place a set of wedges on top of one post and tighten them. Set the other pair of wedges in place and tighten them too. Double-check all the wedges and toenail them into place. Shim the rest of the shore where necessary. Remember, you must pressurize the shore with the use of wedges at the sole plate, header, and both posts. They must also be nailed into the shore; or they may dislodge, and the opening may collapse.

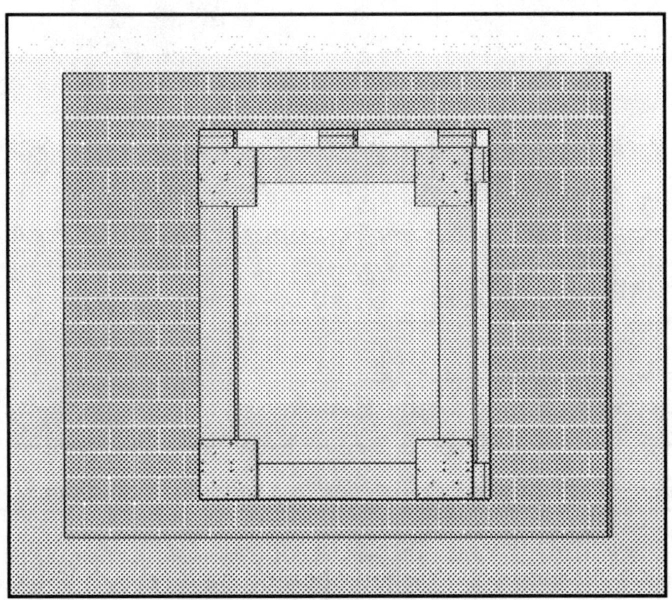

Fig. 4–60 Install the shore in the opening tight to one side, wedging the sole plate and header.

If you encounter a severely racked opening, there are a few things you should do to properly install the shores. Your shores, no matter which one has to be constructed, must be square for them to work properly. This is a *must*. In Figure 4–61, the shore is in a racked opening, erected square, and shimmed, or blocked, to the opening. Note how the sole plate and header are installed. They extend past the posts, which is fine. This way, they secure the top and bottom of the opening. The posts must be installed plumb, as you see in the figure. You can pad out the space from the posts and the racked end of the opening to tighten up the opening and stop any loose material from inadvertently falling out.

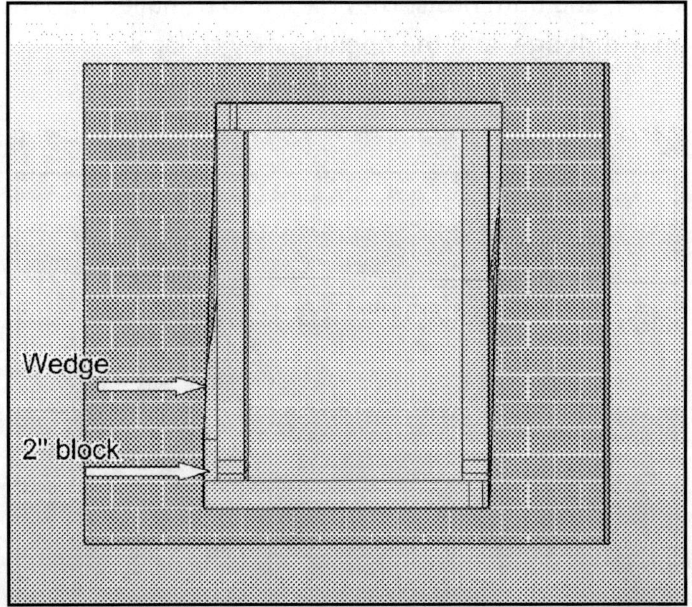

Fig. 4–61 Window shore installed in a severely racked opening.

Figure 4–62 shows the racked opening padded out. You can place additional material inside the window shore to keep it solid against the padding. On the bottom, a 4x4 works fine. For the top, a 2x4 is more than enough. Cut them flush with the inside opening and then tighten up the padding against the building.

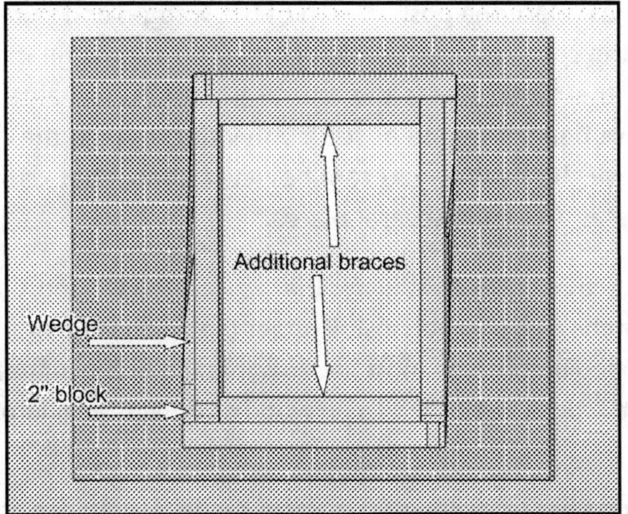

Fig. 4–62 Padded out window shore installed in a severely racked opening.

In this instance you are supporting a damaged masonry header. Your proper measurements will be to the lowest point of the damage. This will be the total height of your shore. Erect the shore in the fashion described earlier.

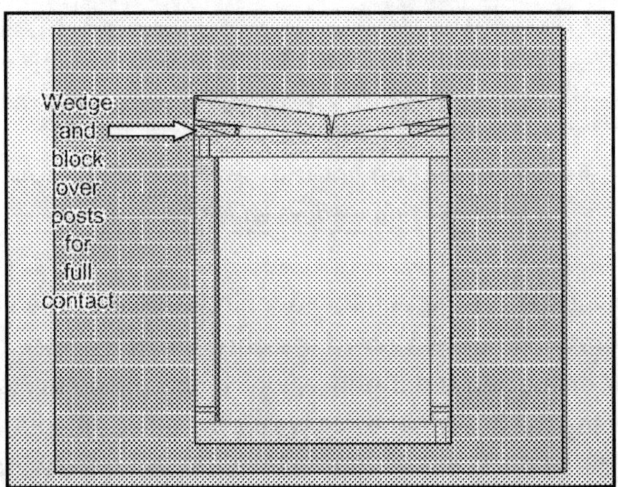

Fig. 4–63 Window shore supporting a damaged masonry header.

Now, just place the posts on top of the wedges and tighten slightly. Fill in the space above the header. You need only fill in the space above by the posts. This is the critical point because all the weight from above bears on these posts. If you want to fill in the rest of the space above the header, you can; however, that is not critical unless the header is broken in several places.

If you are using wedges wider that 1½ in., you will most likely have to pad out the first post you install under the wedge end of the header. As mentioned before, you must have a minimum of 2 in. of header bearing onto the post. Pad out with plywood or dimensional lumber.

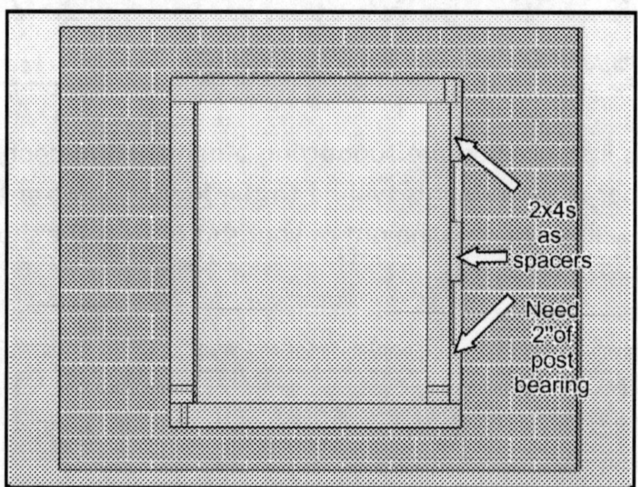

*Fig. 4–64 Pad out the post under the wedge end of the header
if you are using wedges wider than 1½-in.*

In many instances the window opening is greater than 4 ft. For a typical operation in unreinforced masonry, either increase the size of the header or shorten the distance, which is depicted in Figure 4–65. While keeping the header at a 4x4 we just installed an additional post in the middle of the shore. If your team is not using the window as an access point then this can be your simple option.

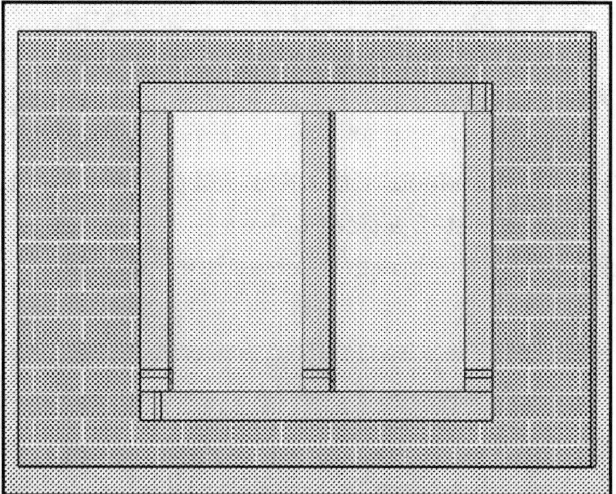

Fig. 4–65 An additional post and wedges may be installed with window openings greater than 4 ft if not used for access/egress.

The preferred method of shoring a window opening when the opening is larger than 4 ft is to use two 2x6s placed on end and put in the center of the 2 sections of plywood as a spacer. This header is very strong as is the one typically used in building construction.

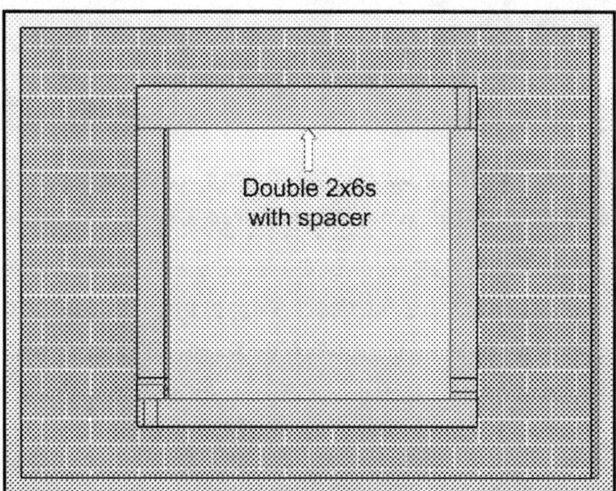

Fig. 4–66 The preferred method of shoring window openings greater than 4 ft is to use two 2x6s with a plywood spacer.

Figure 4–67 shows another fixed window shore option using a longer header. In this case, the assumption is that you can't get any 4x4s in that size. You can place two 45° angle braces into the header and bear them against the posts. Make sure cleats are at least 12 in. long under the braces and nailed with the 5-nail pattern. Place return cuts into the braces and install a cleat between the braces at the bottom of the header where the braces intersect the material.

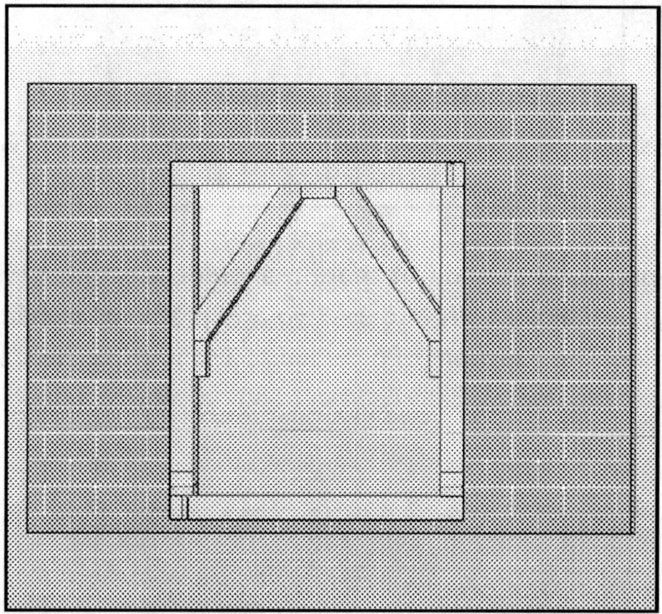

Fig. 4–67 Another shoring option for shoring window openings greater than 4 ft is to use two 45° angle braces and 12-in. cleats.

Figure 4–68 shows one method of shoring up a window with an arch. Here the shore goes in square as usual and another section of shoring material is added to support the main element of the arch, the keystone. If the arch is in good shape, this method works well.

Fig. 4–68 One method of shoring up a window with an arch.

Figure 4–69 shows a situation in which the arch itself is not in good condition. This method supports the keystone as well as the rest of the supporting stones. Here 2-in. dimensional lumber is in place with an additional block installed to keep the spacing tight.

Fig. 4–69 One method of shoring up a window with an arch in bad condition.

The following figure shows the window shore using pneumatic struts. It is assembled the same way as the door shore.

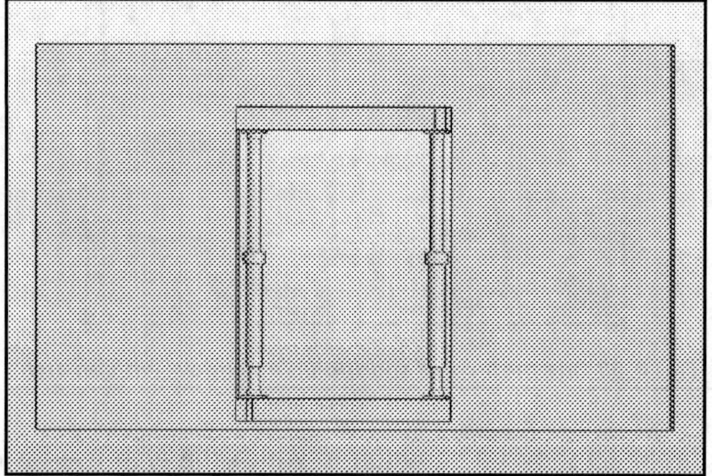

Fig. 4–70 The window shore using pneumatic struts.

The Horizontal Shore

The horizontal shore can be used for interior and exterior shoring. It is used primarily to support damaged or unstable walls in hallways, access ways, shafts, and the like. Generally, it is used in large structures such as office buildings, commercial establishments, or large residential structures of brick and joist or concrete construction. The main purpose of the shoring is to stabilize normal access ways that have been compromised. Horizontal shoring allows for continued collapse rescue operations and provides a relatively safe area for both personnel and rescued victims.

Size-up

In any collapse rescue situation, you must take the time to survey the area and determine the best means of access and egress before heading into an unstable structure. After conducting the survey and deciding the best entrance and exit points, proceed with the size-up for determining the need for horizontal shoring. The size-up inspection should include the following:

- A check of the amount and extent of damage

- Location of possible victims

- Location of leaning walls, damaged structural elements, and bulging walls

Horizontal shore step-by-step procedures

The following is a numbered list of the tasks that must be completed to build and place a horizontal shore in a collapsed building. The remainder of this section on horizontal shores provides detailed instructions for constructing and installing the various elements of the shore.

1. Survey the area and determine the load displacement and structurally unstable elements.

2. Clean the area to be shored.

3. Determine the size and number of struts needed.

4. Measure for proper lengths of wall plates.

5. Set cleats at their proper positions and install wall plates.

6. Install posts; if using three posts, install the center one first. If utilizing two posts, install the top one first. Doing this makes it safer for you to work in the area.

7. Check the shore for fit of materials and tightness to the opening and then tighten all wedges.

8. Install diagonal braces if the opening is not being utilized for access.

9. Install gusset plates or cleat posts if necessary.

Wall plates. Wall plates are the two vertical members of the horizontal shore that are positioned on each wall directly opposite each other and as plumb as possible. For interior operations, use 4x4-in. lumber for the wall plate. In frame buildings, if stability conditions warrant, toenail the wall plates into the top and bottom plates in the existing wall for full effectiveness. For masonry block structures (brick, or concrete walls), install plywood between the wall plate and the wall to provide greater support and to ensure that most of the surface area of the wall plate contacts the wall. Therefore, connections made through the plywood secure the plywood wall plate assembly to the wall. However, when any type of wall (concrete, masonry block, brick, or wood) is so unstable that you must limit vibration, anchor the wall plates by the pressure of the struts alone.

Struts. Struts are supports that keep the wall area open for the rescue team. Depending on the situation, use two or three struts. They should be of the same width or dimension as the wall plates for ease of installation a proper transfer of loads. Determining the number and locations of the struts depends on several factors such as the following:

- Extent of the damage

- Amount and location of debris

- Type of wall construction

- Locations at which the greatest amount of force is being applied to the walls

- Known or possible locations of victims

- Use, or not, of the area for access or egress

Hanger cleats. The hanger cleats support the struts while they are being installed. The cleats also help support the installed struts in case someone inadvertently steps or leans on them. For this reason, always install the cleats as an additional safety factor. Generally, hanger cleats are short pieces of 2x4-in. lumber. They should be approximately 12–14 in. long. If they are any shorter, they tend to split when nailed.

Wedges. Place a set of wedges together, or marry them with each strut, and tighten. This is done to apply pressure securely to the struts against the wall plates that transfer the load from the damaged wall through the shore to good bearing material. Toenail the wedges after the shore has been tightened so they will not become dislodged accidentally.

Gusset plates. These plywood plates, anchored onto the outside face of the connections between the struts and wall plates, help guard against shore members loosening and dislodging. The gusset plates provide additional protection against aftershocks following earthquakes or from vibrations caused by secondary collapse in other sections of the structure. Gusset plates are sufficient on only one side of the shore.

Diagonal braces. The last elements of the horizontal shoring to be installed are diagonal braces, which are used if the area is not an access or egress site. They are installed to lock the entire shore together as one unit and to help provide resistance to eccentric loads that may be applied to the shore. The use of 2x4s or 2x6s is satisfactory. Nail them on both sides of the shore in opposite directions from each other to form an *X* pattern. This technique gives the shore more lateral stability and effectively locks all the shores elements together.

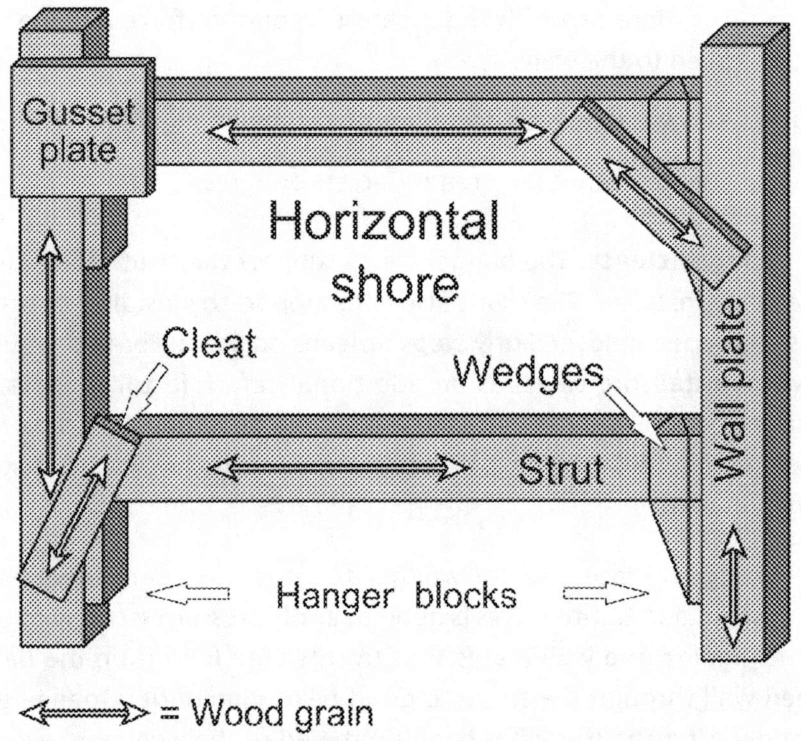

Fig. 4–71 Elements of the horizontal shore.

Considering construction of the shore, the first order of business is to determine the height of the wall plates. They must be cut to the proper size in order to fit properly. Generally, they should be as long as possible to cover the greatest height. Making them a few inches shorter than the ceiling height so they will fit in easily and are not too tight, causing you to force them into position, is something you should not do. Forcing items into position is never a safe way to operate in a building collapse situation.

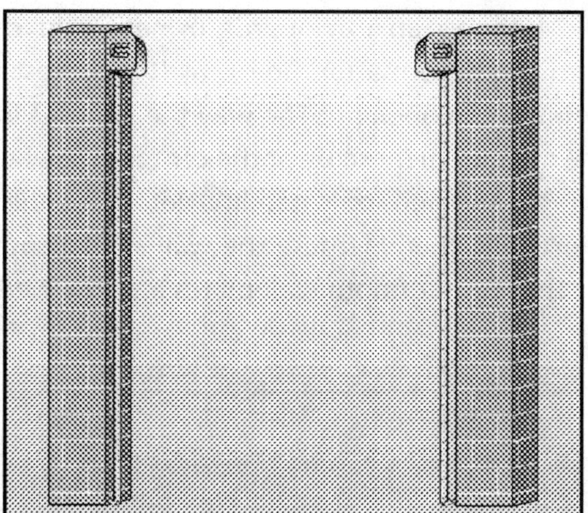

Fig. 4–72 Determine the height of the wall plates.

Measure the opening. Check for cracks and bulges in the wall, damaged structural elements, possible victim locations, large areas of debris, and other signs that shoring is needed. If the walls don't have obvious deformities, measure in two places in order to obtain the overall dimensions. When calculating for the lengths of the struts or posts, always remember to deduct for the width of the wall plates (uprights), and the thickness of the wedges your team is utilizing.

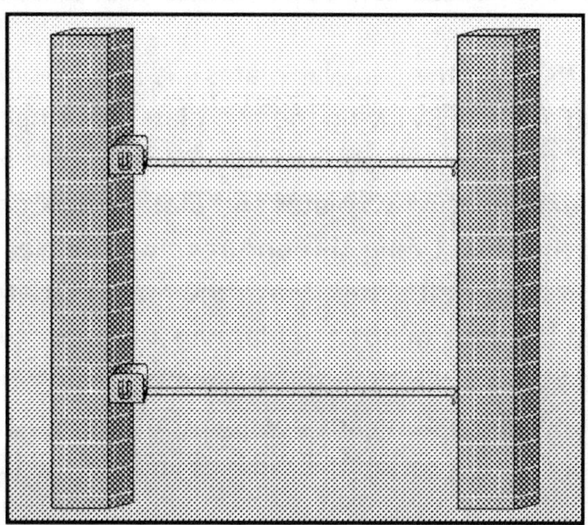

Fig. 4–73 Measure the opening in two places to obtain overall dimensions.

Figure 4–74 shows an obvious deformation in the one wall. Measure up to the deformation point, as it will most likely be the narrowest point in the opening. The way the wall sits in this graphic predicates the measurement to be at the center of the bulge. Examine the wall for cracks and signs of instability, determine the best place to locate the shore, and then measure the wall at that spot. Remember to deduct for the width of both wall plates and the wedges.

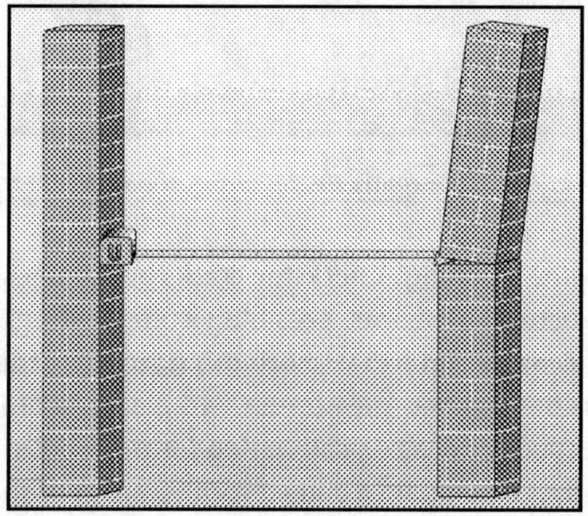

**Fig. 4–74 Measure up to the deformation point,
as it will likely be the narrowest point in the opening.**

Figure 4–75 shows the two wall plates laid side-by-side ready to be measured and marked. By laying the two plates together, you need to measure only one time, thereby saving your team time in assembling the shore. Square up the two plates and place a mark on both plates even with each other. Doing this ensures that the cleats are in the same place on both wall plates. Make the mark first at the bottom, measure from the bottom up then place an *X* to mark the plate position. The *X* is a clear indication to the assembly team where the cleat is to be installed, eliminating any possible confusion. Measure up and mark the upper cleat location the same as you did the bottom cleat.

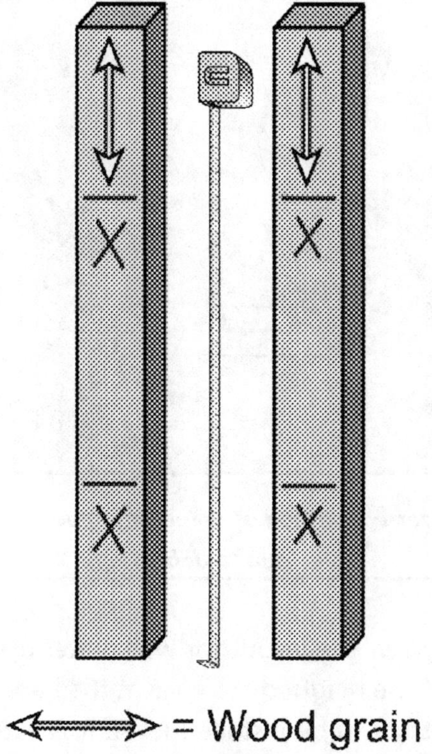

Fig. 4–75 Save your team time in assembling the shore
by measuring only one time.

In a situation in which there is some sort of debris or an object you cannot move out of the way, your method for measuring for the wall plates will be different from that previously described. Determine the heights of both posts. Because the measurements are different, you will subtract the height difference off the post at the bottom. There are several ways to get this measurement. The preferred way is to place a 2x4 and a level on top of the debris and measure from it to the floor as shown in Figure 4–76. Determine the height of the debris or obstruction and deduct its value from the bottom of the wall plate.

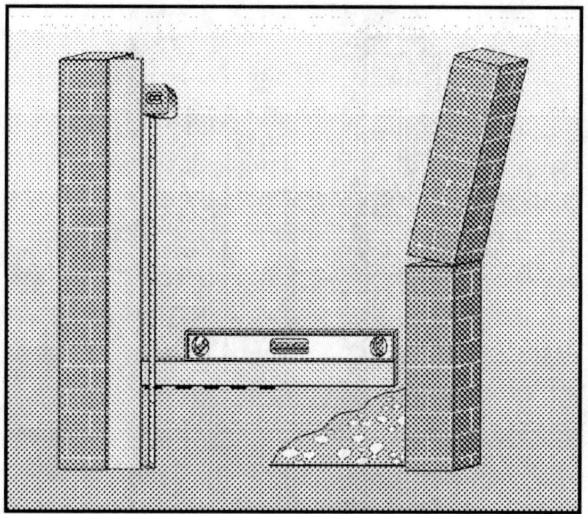

Fig. 4–76 The preferred method of determining posts of different height due to debris.

Figure 4–77 shows the layout for wall plates used when the floor is uneven. Deduct the height difference at the base of one plate. This will enable the struts to be installed level. It is extremely important that they are always installed level.

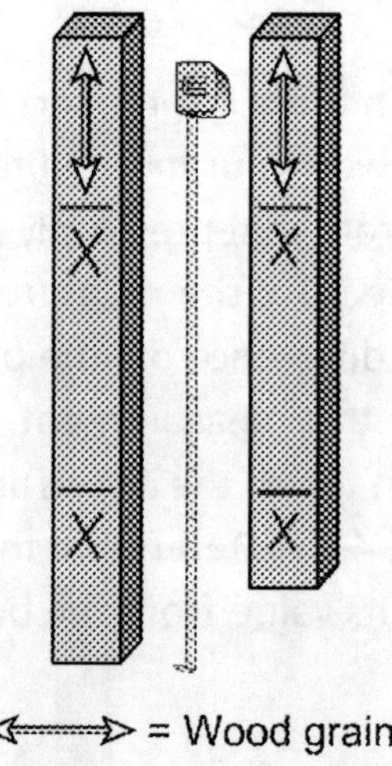

Fig. 4–77 When laying out the posts,
deduct the height difference at the base of one plate.

Figure 4–78 shows several, but not all, cleat options available to the shoring team to use to anchor them to the wall plates. Some options work all the time, and some work only in a specific set-up such as with a wall plate using no wedges.

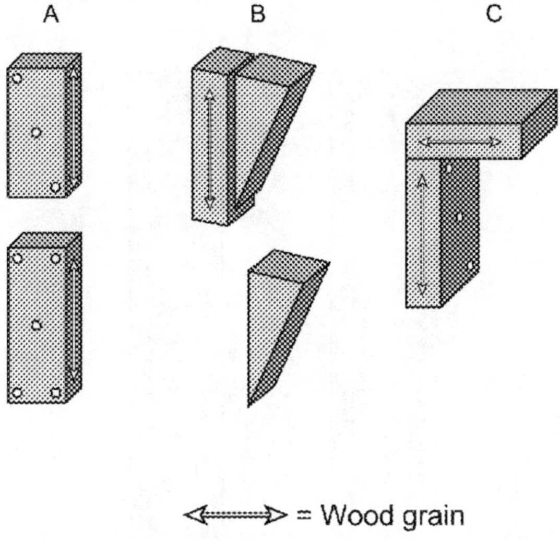

Fig. 4–78 Several of the cleat options available.

A shows the most typically used cleat that is simple and effective. The top cleat is a 12-in. piece of 2x4 with (3) 16d nails anchoring it to the plate. You may use the 5-nail pattern. The bottom cleat is slightly longer, measuring 14 in. long. There is no maximum size; however, with any cleat more than 12 in. long, use the 5-nail pattern.

B shows a typical setup utilized on the wall plate where the wedges will be placed. The top cleat is a 2x4 with a wedge nailed into it. This can be a 2-in. or a 3½-in. wedge. Either one works fine. The bottom cleat is just one wedge, which must be a minimum of 3½ in. for the strut and top wedges to fit properly.

C shows the standard 2x4-in. cleat, but instead of using wedges, it has another section of 2x4 at right angles to and on the top of the cleat. This piece will act as a shelf on which the wedges and strut sit.

Figure 4–79 shows four cleats installed on two wall plates, roughly 12 in. long. These support the struts as they are being installed. Only (3) 16d nails are necessary to secure them. If you want to put in a longer cleat, use the 5-nail pattern. Place the cleats to the line and nailed into the wall plate at the X.

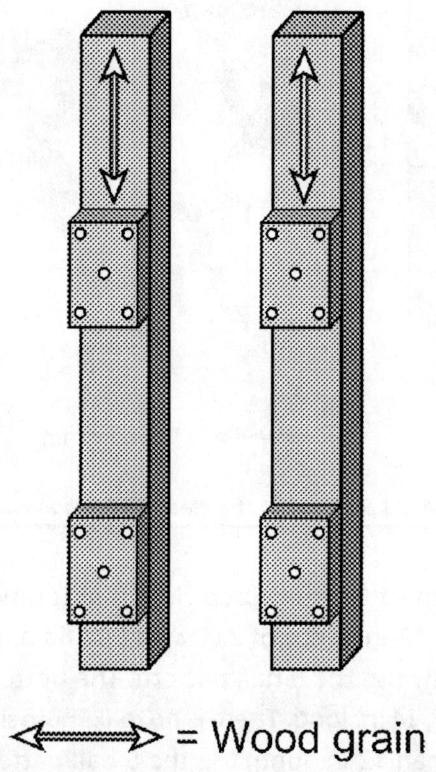

Fig. 4–79 Four 12-in cleats installed on two wall plates.

Figure 4–80 show three wall plates with different cleat options. A shows just two 12-in. long 2x4-in. cleats nailed in place. Usually installed at the bad wall, this plate set-up is fine for the end where the wedges are not placed.

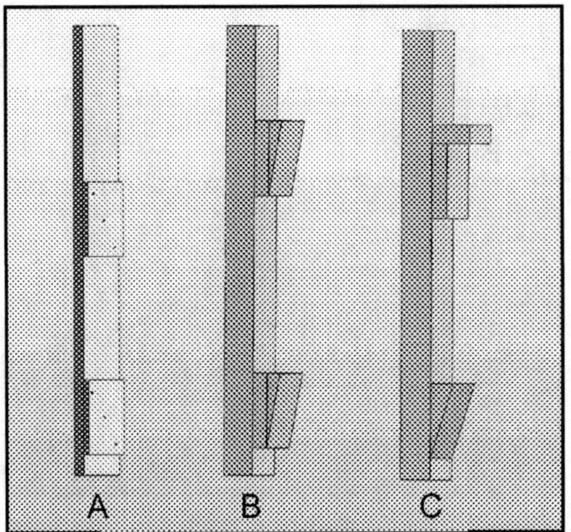

Fig. 4–80 Three wall plates with different cleat options.

B shows a wall plate placed on the side of the wall with the wedges installed. In this case, nail the standard 2x4, but because that will not be enough to catch the struts, anchor wedges on the face of the cleats. This action gives enough room to install the wedges and provides a large enough support base to accommodate a 4x4.

C shows a wall plate for which there are two other cleat options. The bottom wedge is 3½ in. wide. This will hold up the horizontal struts at both ends. At the wedge end, there will be enough space for the 4x4 strut to sit on if your team utilizes 1½ in.-thick wedges, which would be recommended here. The top set-up is just two pieces of 2x4. Place one 12-in. cleat in position and lay the other one at right angles to the first on top of the 12-in. cleat to act as a shelf.

Figure 4–81 shows the wall plates installed on the walls. Erect the plates one on each side and in line with each other, keeping them as plumb as possible. You may have to shim them out to keep them straight. Anchor them if at all possible.

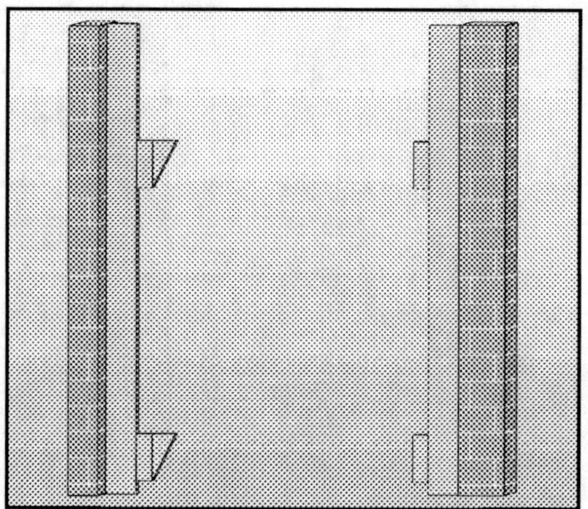

Fig. 4–81 Erect the plates one on each side and in line with each other.

To install a two-strut system, place the wedges on the good wall so that you are not working on the side with the damaged wall. This will keep your body away from the damage and, hopefully, keep you a bit safer. Measure for all struts, remembering to deduct for the thickness of the wedges. Try to keep the same measurement for all the struts to make cutting and installations as simple as possible. Install the struts as level as possible. They will work more efficiently if you do. Tighten the bottom strut first, and then tighten the upper strut.

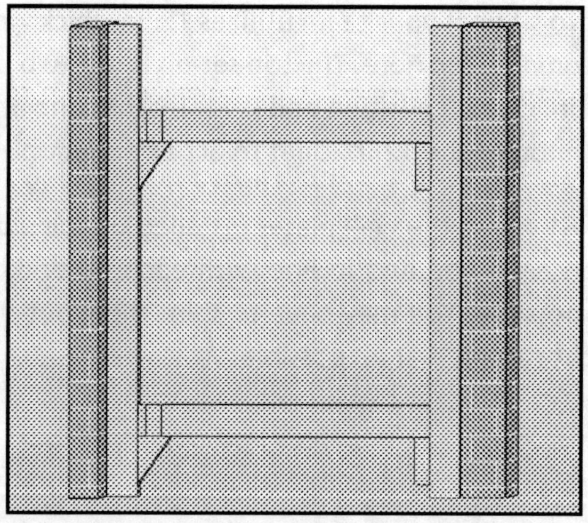

Fig. 4–82 Installing a two-strut system.

Figure 4–83 shows a close-up of the typical wedge placement. Generally use 2x4-in. wedges, always installing them in pairs. Pressurize both at the same time. Place the wedges in position on top of the support blocks after the strut has been laid in place. Snug them up, keeping everything in alignment and evenly spacing the wedges in the center of the strut. Now, pressurize the wedges.

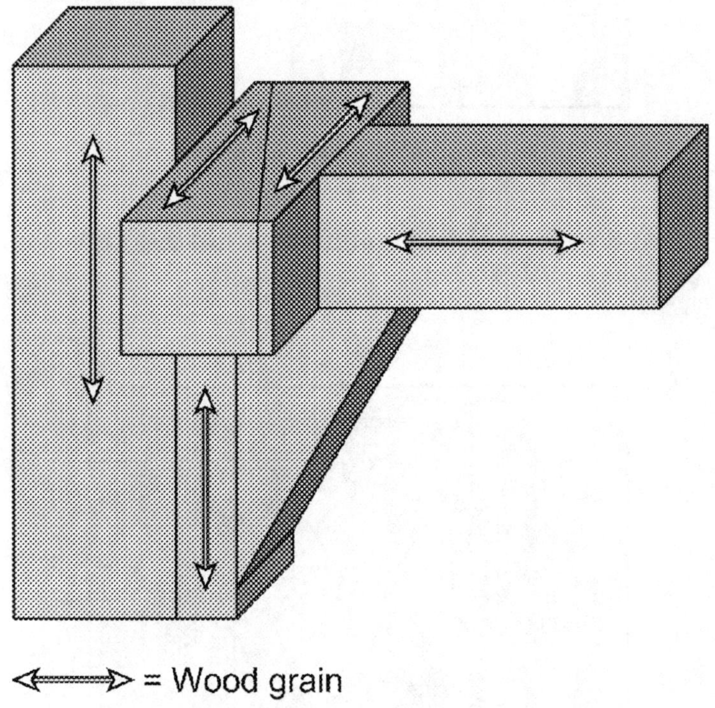

Fig. 4–83 Close-up of a typical wedge placement.

If your system contains three struts, place the center one in position first then pressurize the strut. Next install the bottom strut and pressurize it. Install the top strut last and pressurize. After this is done, review all the struts' fit and retighten if necessary.

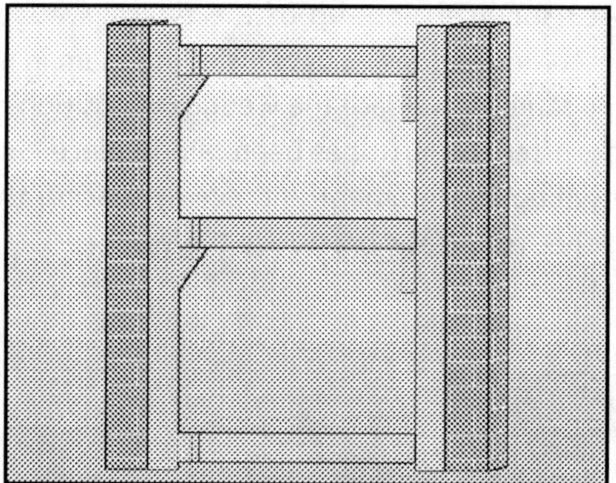

Fig. 4–84 Installing a three-strut system.

Figure 4–85 shows a diagonal brace in place in the horizontal shore section.

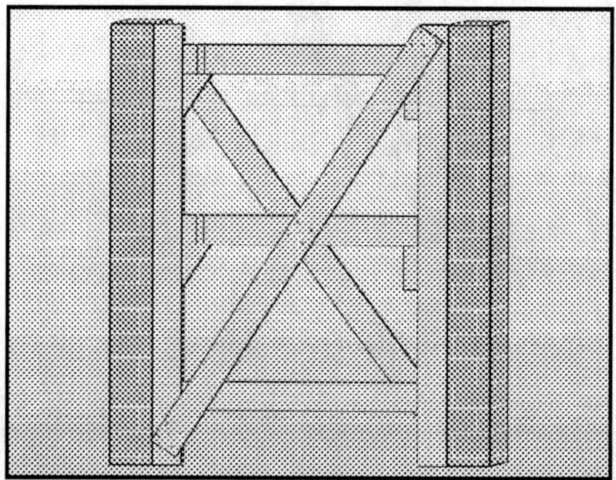

Fig. 4–85 Diagonal brace in the horizontal shore section.

Figure 4–86 shows a shore in place in a heavily damaged area. Place a strut at the site of the maximum amount of damage and deflection. The pictured shore utilizes three struts because of the extensive amount of damage. Shim or pad out from the wall plate to the damaged wall where the other struts intersect the wall. Struts need full bearing of material against the walls to be effective.

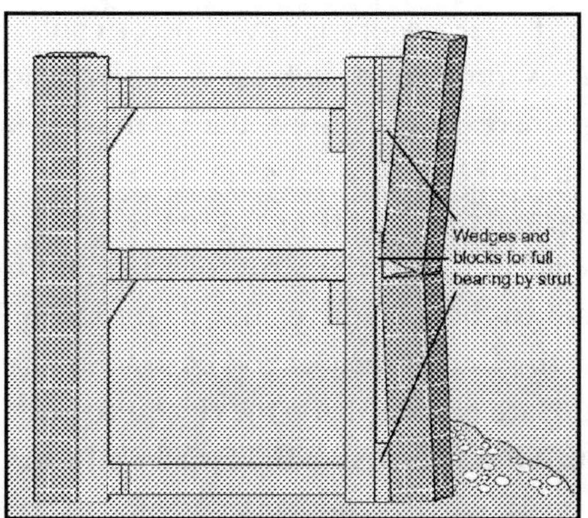

Fig. 4–86 Struts need full bearing of material against the walls to be effective.

Figure 4–87 shows a shore braced off with cleats and gusset plates for protection against vibrations from earthquake aftershocks. The wedge ends of the struts are capped at both top and bottom in order to prevent the wedges from popping out.

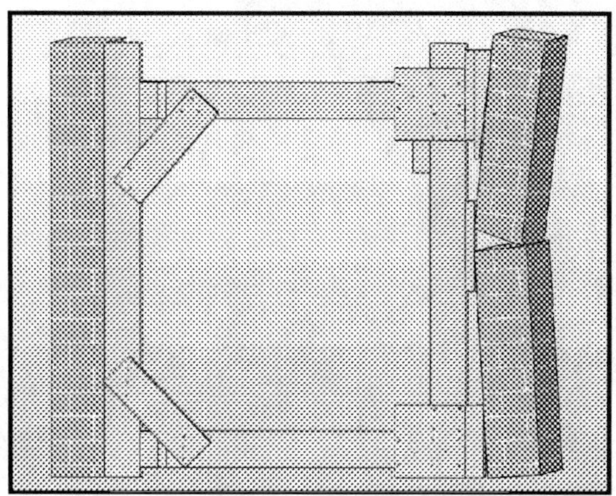

Fig. 4–87 Cap the wedge ends of the struts at the top and bottom with cleats and gusset plates.

After an earthquake, vibrations will occur continually. You must protect the shores from vibrating loose if at all possible. The wedges are the first element to be susceptible to movement. Figure 4–88 shows a method for keeping wedges in place. First, nail the wedges to the wall plate, and then nail the strut to the wedges with two nails. Next, place a 2x4 on top of the strut, overlapping the wedges and butting the cleat to the wall plate. The cleat must be a minimum of 16 in. long. Nail the face of the cleat on both sides into the wall plate, using 16d nails. Place one nail from the cleat into each wedge. Finally, nail the cleat into the strut, using the 5-nail pattern.

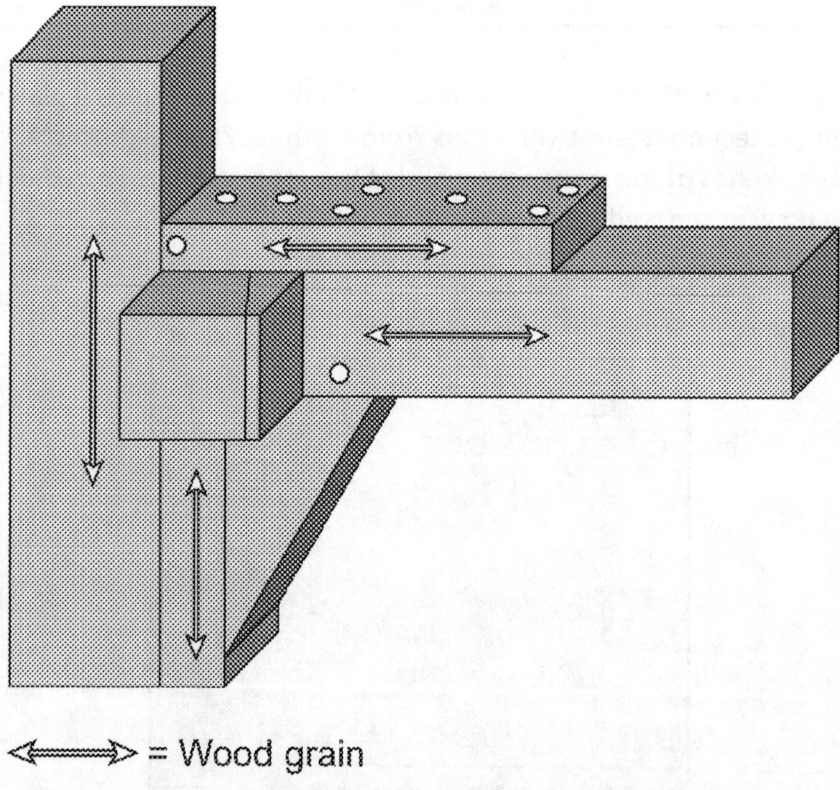

Fig. 4–88 A method for keeping wedges in place.

Another method for securing a shore in the aftermath of an earthquake is to use a 2x4, roughly 2 ft long, and nail it into both the wall plate and the strut. Use (3) 16d nails in both positions, keeping the 2x4 tight to the wedges. Make sure the wedges are also nailed into the wall plate.

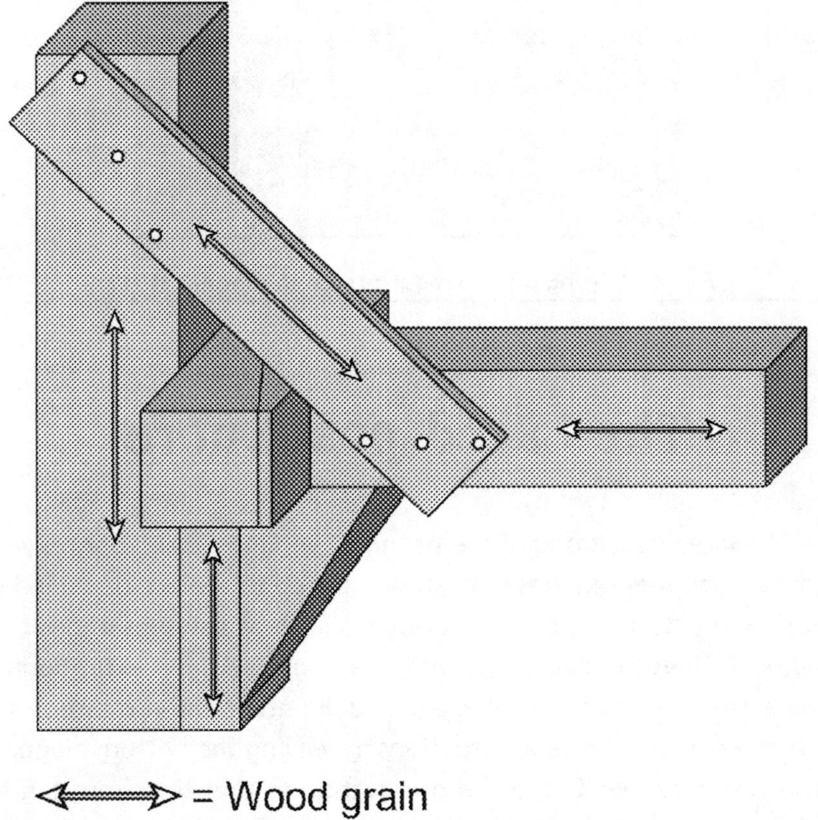

Fig. 4–89 Another method for securing a strut.

There is an advantage to using pneumatic struts for the shore's struts. It is that they can be quickly adjusted and provide superior strength in an earthquake situation. You need to do only minimal measuring with the shore, and the struts are simply expanded and placed in position. Pneumatic struts can be anchored to the wall plates with (4) 16d nails driven through the base plates. There is no need for gusset plates or cleats with these shores.

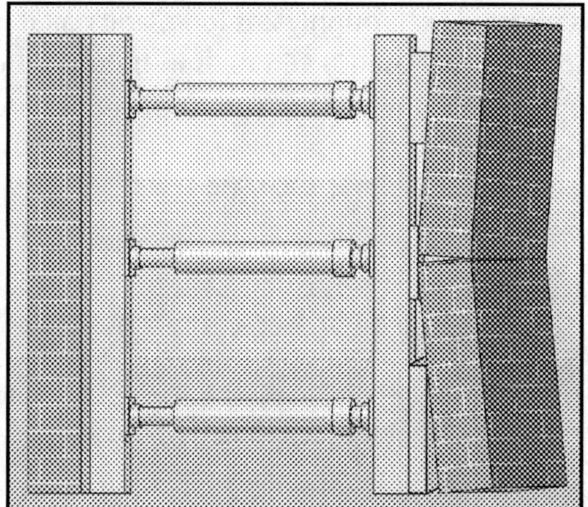

Fig. 4–90 Using pneumatic struts for the shore's struts.

The Laced-post Shore

The laced post is one of the strongest and most stable shores your rescue personnel can erect. It can be utilized to replace damaged or missing structural supporting elements such as columns and girder beams. This type of shore can easily stand by itself. It is exceptionally stable, making it very useful in earthquake situations. It may be used as a safe haven in unstable situations by leaving the bottom diagonal brace, or cross member, off in one section so that the shore can be accessed quickly if necessary.

Because the laced-post shore is quite stable and extremely strong, you can use one or a combination of several laced-post shores to create a protected area that you can rely on as a safe haven. This type of shore is also very efficient in the resupport of damaged or missing columns. The shore is easily erected around the damaged structural element without disturbing any part of that element. The damaged element then can be tightened up and properly positioned while the shore is in place. This makes for a very safe operation. The average width of this shore can be anywhere from 3 to 5 ft wide. If the laced post is erected with 4x4s then use 2x4s for your bracing. If your shore is erected with 6x6s then use 2x6s for all of your bracing, horizontal as well as diagonal.

One rule of thumb you should follow is to create a properly balanced shoring system. The height of the laced post should be roughly no more than three times its width. If you maintain this ratio, the shore is enough for the typical collapse situation, including earthquake situations where aftershocks may occur.

The height to which the shore is erected and the amount of debris it must support determines if the shore should be made with 4x4s or 6x6s. If the shore is to be made with 6x6s, then the bracing should be fabricated from 2x6-in. lumber. If you are concerned about the amount of weight the shore is to support and if you feel you may be taxing the limits of the post's rated capacity, it is a good idea to reinforce the header and sole plate by adding additional material to both. Or better yet, increase the size of the material.

Laced-post shore 3-to-1 ratio

As with any rescue-shoring situation, it is preferable to prefabricate as much of the laced-post shore as possible. To begin the prefabricating task, first clear an area in a safe location in close proximity to the shore's final destination. For example, clear an area about 6 ft wide and 4 ft longer then the length of the shore to be constructed. Once you know the total height of your shore, determine the height of your posts. To do this you have to account for the width of the sole plate, header, and wedges by deducting those values from the measured height. This gives you the exact height of your posts.

Construct the shore in two halves. Lay down two of the posts, keeping them parallel with each other. Measure and space them apart the predetermined distance. Cut the header 24 in. longer than the width of the shore. Center the header over the posts. You should have 12 in. of overhang on each end.

Laced-post shore step-by-step procedures

The following is a numbered list of the tasks that must be completed to build and place a laced-post shore in a collapsed building. The remainder of this section on lace-post shores provides detailed instructions for constructing and installing the various elements of the shore.

1. Survey the area and determine the load displacement and structurally unstable elements.

2. Clean the area to be shored.

3. Measure for proper lengths of shoring items.

4. Prefabricate the first two sections of the shore.

5. Place sole plates in position under the item to be stabilized.

6. Install the first prefabricated section under the damaged area, snug up both sets of wedges, and plum up shore.

7. Set the second prefabricated section in position and snug wedges under bottom of posts.

8. Check the shore for fit of materials and tightness to the opening then tighten all wedges.

9. Make sure the spacing between both prefabricated sections is accurate.

10. Place the center horizontal brace in position first, both sides of your prefabricated shore sections.

11. Install the top and bottom horizontal cross braces on both sides.

12. Measure and install the four diagonal braces, making certain that they are all running in the proper direction.

13. Double-check the shore's stability and tightness then anchor to floor and ceiling.

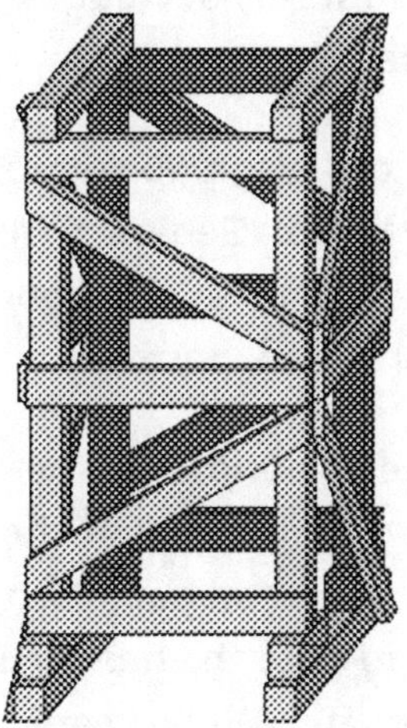

Fig. 4–91 Elements of the laced-post shore .

The laced-post shore elements

Headers. You need two headers. They can be 4x4s if you are supporting a lighter load, but generally if you are holding up a significant amount of weight, you should use 4x6s. That size lumber keeps the strength capacity consistent with the four posts.

Posts. The posts are the main support of the shore. Erect four of them, one in each corner. Posts generally are made of 4x4s, but at times they can be made of 6x6s or larger lumber.

Cross braces. There are eight cross braces in a laced-post shore. They are usually made of 2x4s with 4x4-in. posts or of 2x6s with 6x6-in. posts. There will be one in each center position. When the two main sections are tied in together, there will be one at the top under the headers and one at the bottom on top of the sole plates.

Diagonal braces. There are eight diagonal braces usually composed 2x4s for 4x4-in. posts and 2x6s for 6x6-in. posts. Two sets form an *X* pattern as you look through the shore, and two sets are parallel with each other.

Wedges. There are four sets of wedges; one set under each post, running parallel with the sole plates.

Sole Plates. There are also two sole plates, directly under the headers, posts, and wedges. Generally speaking, 4x4s are large enough to spread the load out to the floor for 4x4-in. posts, and a 6x6 sole plate is used for 6x6-in. posts.

After you determine the width of the shore, cut a header 2 ft longer than the width of the shore. If the shore is 4 ft wide, the header must be 6 ft. Using (4) 16d nails (two on the top face and one on each side) nail both posts to the header 12 in. from each end, the width of the proper overhang. While keeping the posts square to the header, go down to the center of the posts and measure 4 ft from the outside ends, and mark the place. The center horizontal brace will be positioned at the mark to keep the posts in position.

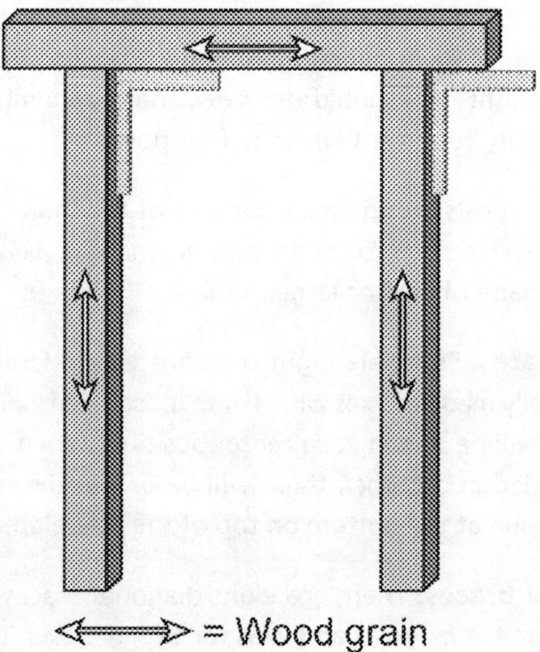

Fig. 4–92 Nail both posts to the header 12-in. from each end while keeping the posts square to the header.

After the header and post have been squared up, make a mark on the ground alongside both posts. Use this mark as a guide to the posts' position, indicating whether the shore has shifted when you are working on it. Always keep an eye on the marks when installing the braces. Cut two 48 in.-long 2x4s. Place one in the center of the post. Nail it flush with the outside edge of one 4x4, using (3) 16d nails in a diagonal row. Then take the other post and place it flush with the opposite end of the 2x4 in order to keep the width at exactly 48 in.—exactly what you want. Nail the joint down with (3) 16d nails.

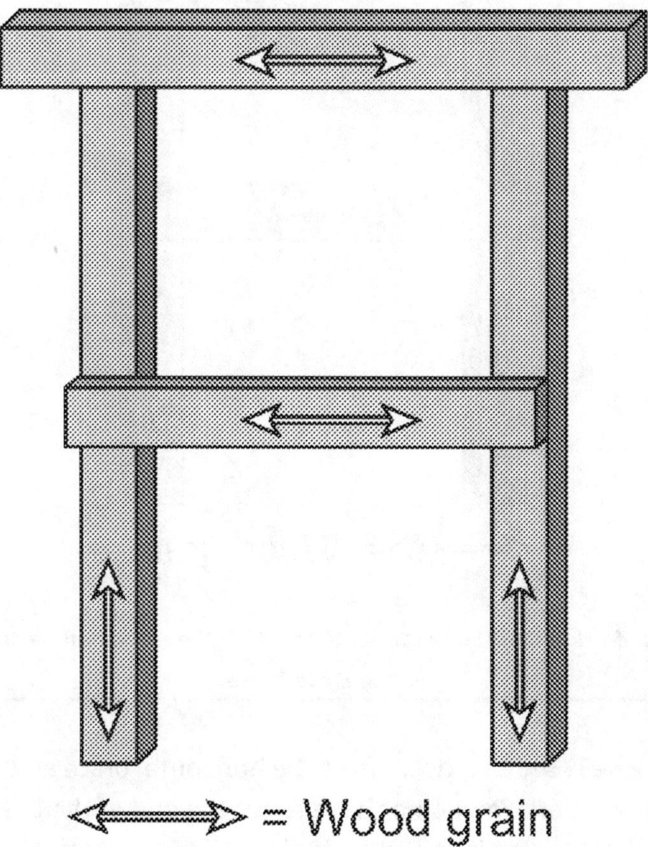

Fig. 4–93 Install the horizontal brace in the center of the post, nailing it flush with the outside edges.

Now measure from the face of the end of the header down to the outside face of the post. Measure this about 3 in. up from the top of the 2x4-in. mid-brace. You can cut the diagonal brace on an angle if you want, but that is not totally necessary. Nail the diagonal brace down on both ends with the same nail pattern as the cross brace.

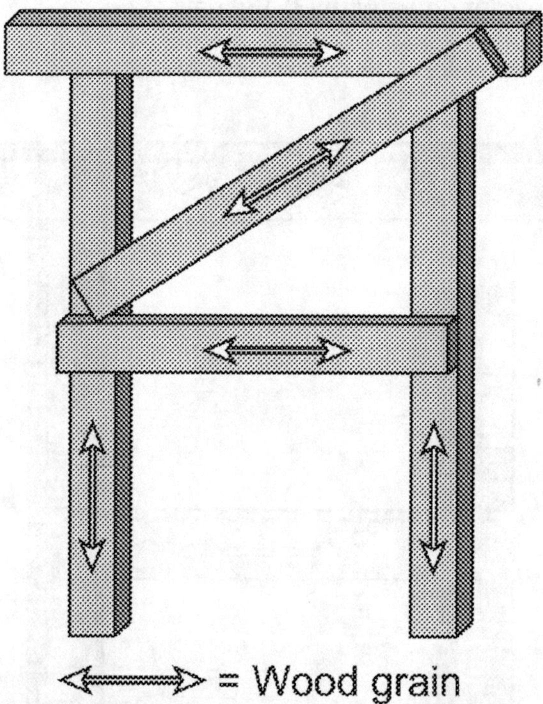

Fig. 4–94 Install the diagonal brace with the same nail pattern as the cross brace.

Figure 4–95 shows a close-up of the horizontal brace and diagonal brace intersection. You do not have to cut the end of the diagonal on an angle, but you can if you want. Nail both 2x4-in. braces with (3) 16d nails, keeping both braces flush with the outside edge of the post.

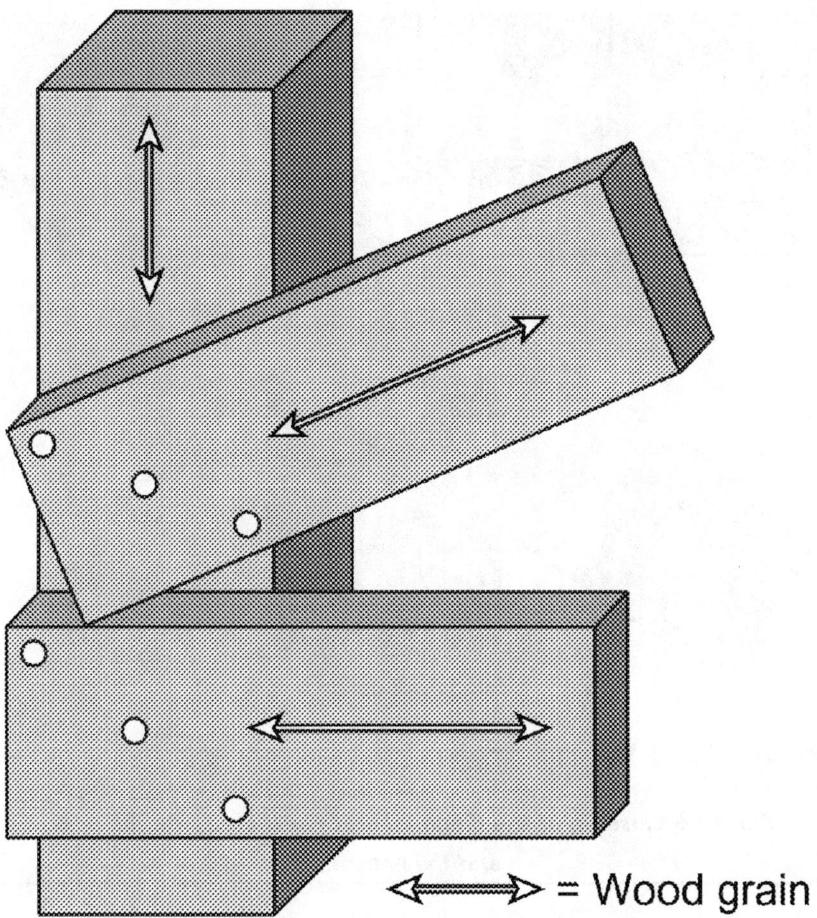

Fig. 4–95 Close-up of the horizontal brace and diagonal brace intersection.

Figure 4–96 shows the top of the diagonal brace anchored into the post and the header. This position is a *must*. The 2x4-in. brace must be nailed into both the post and header to be effective. Again, you may cut the angle into the brace, but it will not be necessary. Place (3) 16d nails into both the post and the header, making sure the brace does not extend past the top of the header edge.

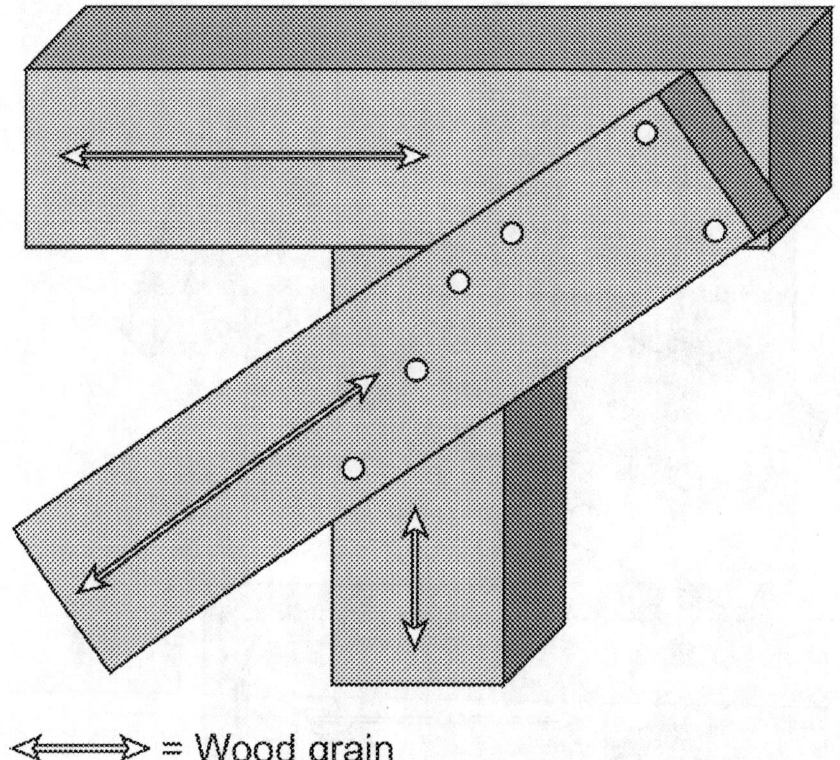

*Fig. 4–96 Close-up of the diagonal brace anchored into the post
and the header.*

Using the existing section you just completed as a template,
fabricate the exact same section in the same manner. You can lay the
lumber right on top of the preassembled section and place all the
pieces in the same, exact position as the bottom structure. With these
two sections assembled, you are now ready to erect the shore.

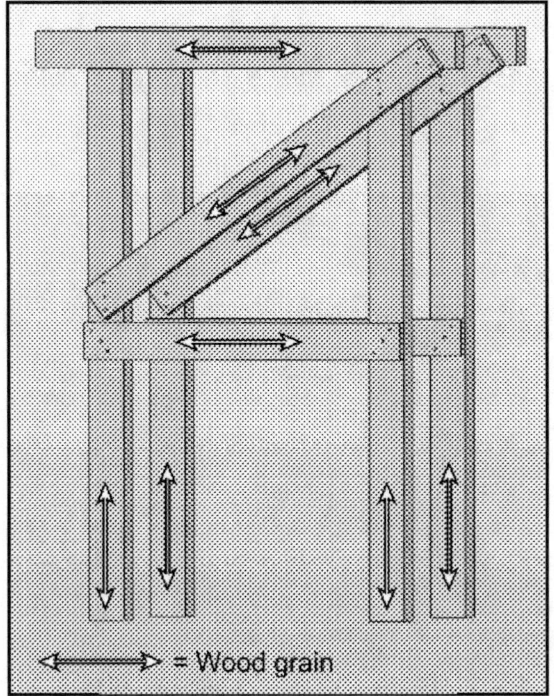

Fig. 4–97 Use the existing section you just completed as a template for the second section.

Place the first section on top of the sole plate, which must be at least 2 ft wider than the width of the posts. You need this space for the wedges. Be certain that the area where the shore is to be erected is clear. An area 8 ft x 8 ft is probably the minimum needed for safe operation. Place the upper section of the shore exactly where you want it, and then place the sole plate just under both posts directly in line with the posts. Lift up the section until it makes full contact with the ceiling. Insert a set of wedges under each post. Now, just snug up the wedges. You may have to adjust the shore a little before final assembly is complete.

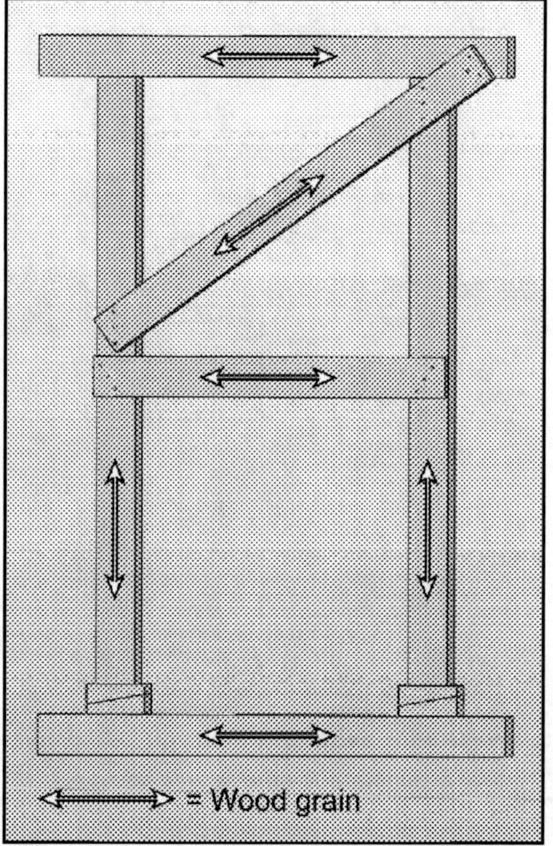

Fig. 4–98 The first section on top of the sole plate with wedges inserted.

Place the other prefabricated section into position. Center the wedges under the post and align the sole plate so that the shore fits evenly on it. Make sure the sole plate is parallel and in line with the shore. This positioning is important when the additional braces are installed. Plumb up the shore by eye. Make sure the two sections are in line with each other, plumb, and spaced properly apart. The outside face of each post on both sides must be 48 in. apart. Check this measurement in at least two places, generally at the bottom of the posts and at the top by the header. Make sure the two sections form a square shore. You don't want any fancy parallelogram shores.

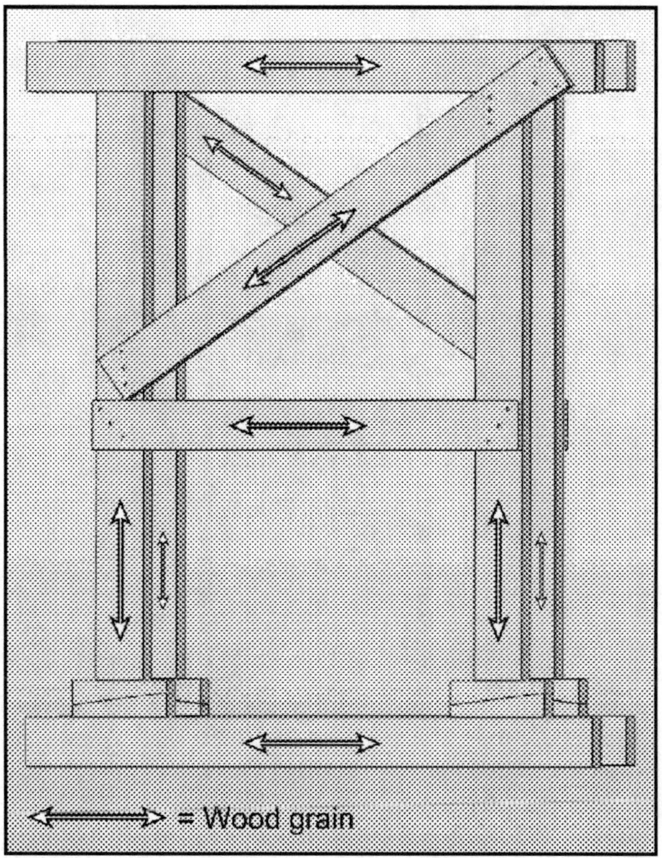

Fig. 4–99 Make sure the two sections are in line with each other, plumb, and spaced properly apart.

Next, tie the two shore sections together. Have your 2x4-in. cross braces precut for ease of installation. Place the middle brace first. It makes the other adjustments easier, and it keeps the shore upright and balanced. If the shore is 48 in. wide, you can make the braces the same size. They will fit flush with the edge of the posts. Anchor the brace in the standard fashion using (3) 16d nails.

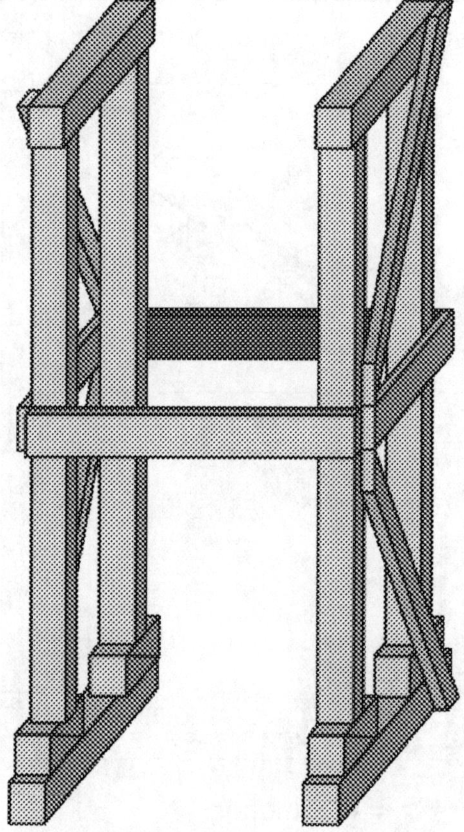

Fig. 4–100 Tie the two shore sections together with the middle brace first.

Figure 4–101 shows a close-up of the intersecting cross braces on the 4x4-in. posts. It is a good idea to overlap the ends of the brace so they are flush with the outside face of the other mid-braces. To do this, add 3 in. to the 48-in. measurement, making the braces 51 in. overall. You can place two nails into the other mid-brace at the end. Doing this ties the shore together securely.

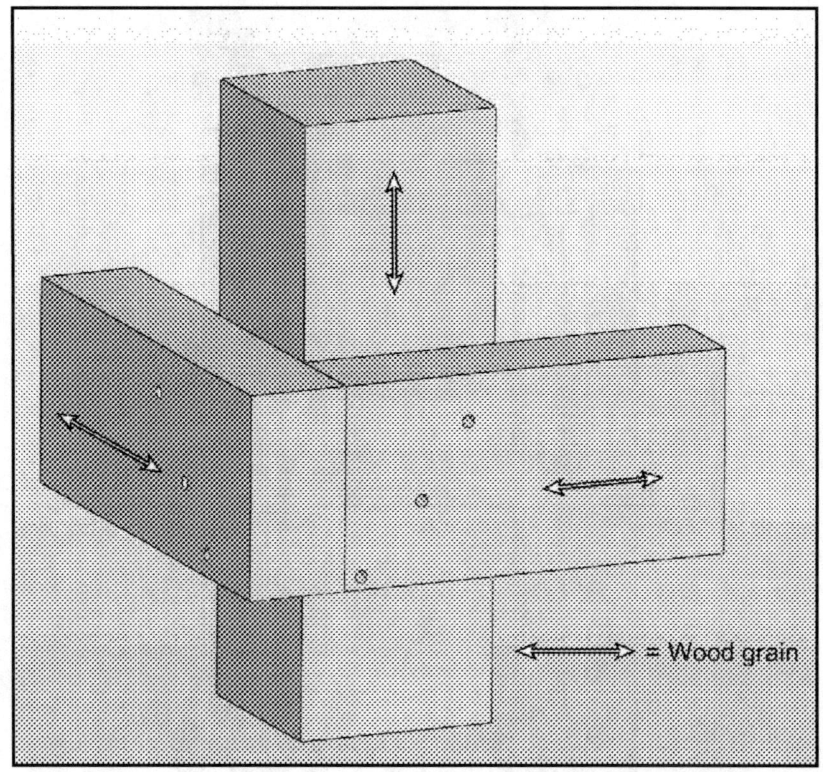

Fig. 4–101 Close-up of the intersecting cross braces on the 4x4 posts.

At this time, continue to install the 2x4-in. horizontal cross braces on the shore (Fig. 4–102). Do this on both sides! The nail pattern for a 2x4 is (3) 16d nails in a straight line or a triangle shape and five nails for a 2x6 brace. Install the top brace in second and the bottom brace in last.

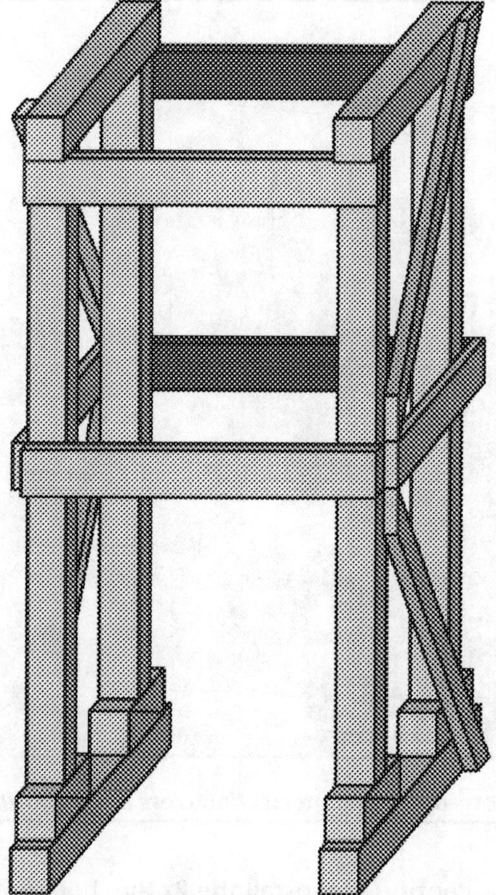

Fig. 4–102 Install the 2x4-in. horizontal cross braces on both sides of the shore.

Figure 4–103 shows a view of the shores tied together. When you have tied the shore sections together, you can then install the remaining diagonal braces. Make sure the top of the horizontal brace is flush and directly under the header overhang. The bottom horizontal brace is flush with or slightly above the base of the posts. The middle brace is the only one that overlaps the intersecting brace.

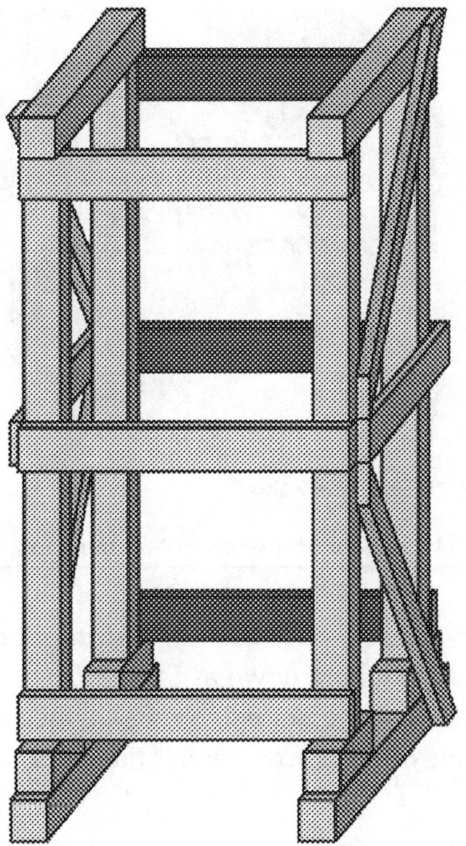

*Fig. 4–103 Install the remaining diagonal braces
once you have tied the shore sections together.*

It is important that the top 2x4-in. horizontal brace is up as high as possible. It should be flush with the ends of the posts and butt up under the two headers. This brace is installed on both sides. Anchor it with (3) 16d nails like the center cross brace. These braces are 48 in. long.

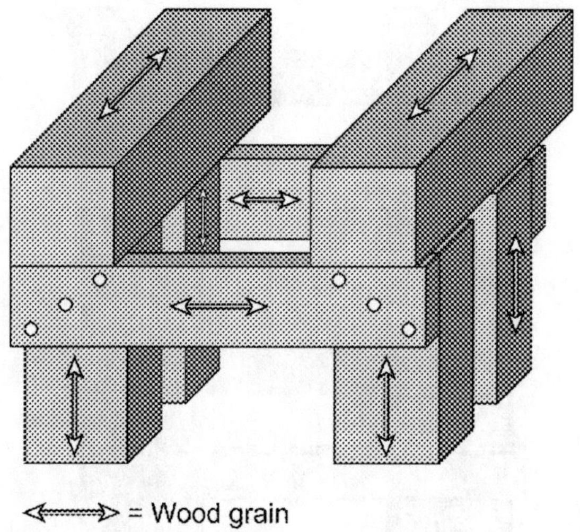

Fig. 4–104 Close-up of the top 2x4-in. horizontal braces.

Figure 4–105 shows a close-up of the bottom 2x4-in. horizontal brace. This brace should be down as far as possible on the base of the posts. You can have the brace sit on the wedges or have it slightly above. As with the other braces, keep it flush with the outside ends of the posts.

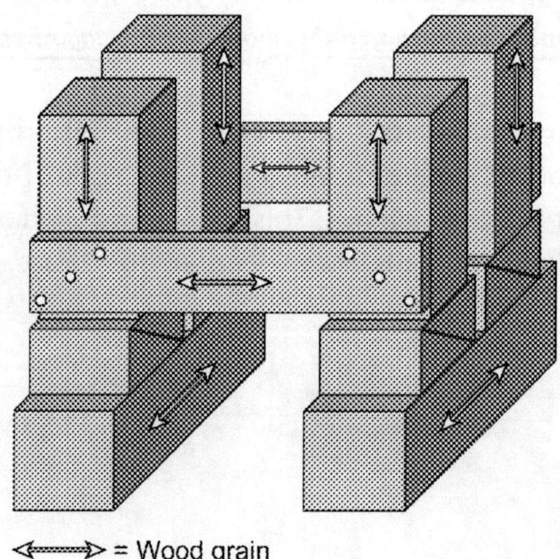

Fig. 4–105 Close-up of the bottom 2x4-in. horizontal braces.

On the two prefabricated sections, it is important for the bottom diagonal to extend to the ground and contact the face of the sole plate. This ties the plate into the shore and keeps it from sliding or shifting out of place. Anchor it with the 3-nail pattern into both posts and the sole plate. This diagonal is placed in the opposite direction as the one above it. When you build the shore sections on top of each other and follow the template, the braces match properly when the shore is erected. Each prefabricated section will look like a *K* when properly assembled. As you face the shore, after it is assembled, the braces form an *X* as you look through the shore.

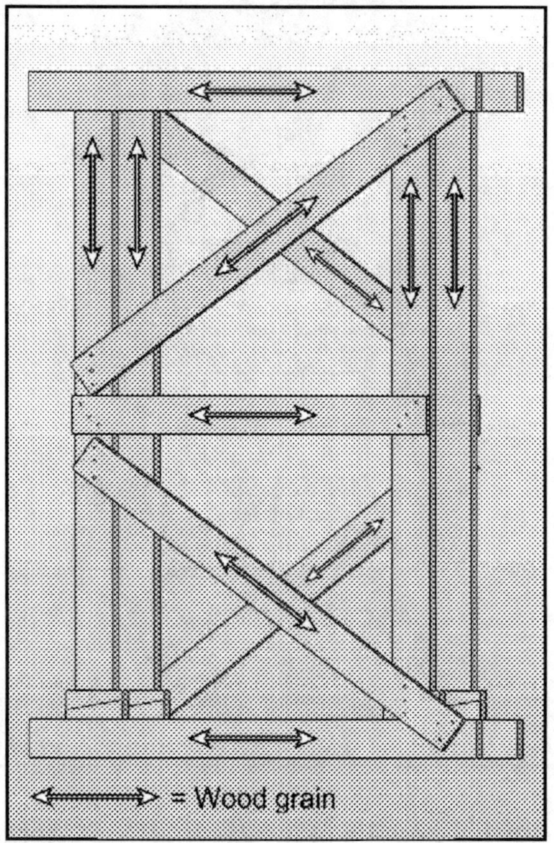

Fig. 4–106 Make sure the two sections are in line with each other, plumb, and spaced properly apart.

The next order of business is to install the side diagonal braces. Take the measurements the same as you did for the other diagonals. Install the top diagonal first, following the same angle as you did for the initial shore sections. Nail it in place. Install the bottom diagonal next. Both braces should just contact the horizontal braces. Take full advantage of all the useable space for bracing. The bottom brace follows the same angle as the top one. Note: These diagonal braces follow the same plane; they do not form an *X* when you look through the shore to the other side. This assembly helps to prevent lateral instability.

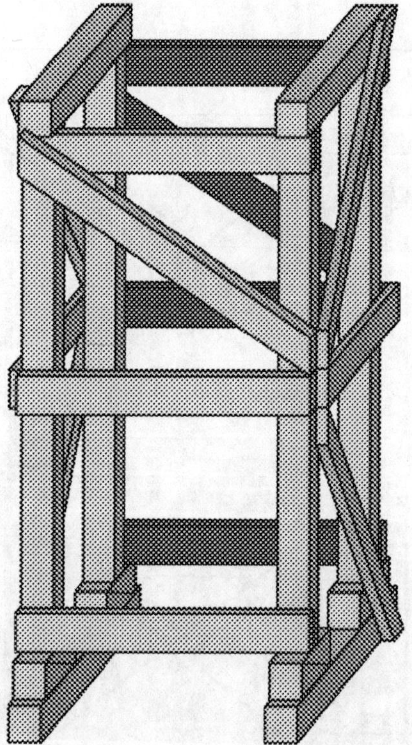

Fig. 4–107 Install the side diagonal braces, top brace first.

Figure 4–108 is a close-up of the 2x4-in. diagonal brace in the top bay. Nail it with (3) 16d nails. Try to keep the brace flush with the ends of the posts. It may overhang a little bit, but no more than an inch, so the overhang is not in the way.

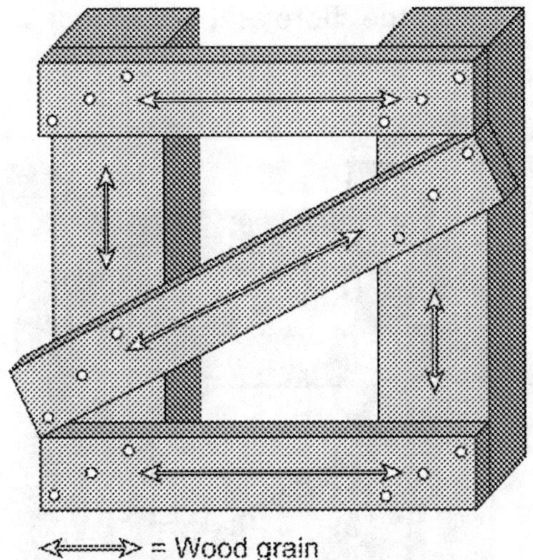

Fig. 4–108 Close-up of the 2x4-in. diagonal brace in the top bay.

Figure 4–109 shows how the diagonal braces would look if you cut them flush with the end face of the posts. This cut gives you some more surface bearing for nailing. Although this shore is somewhat fancier, basically it's an alternative and does not fill a critical need.

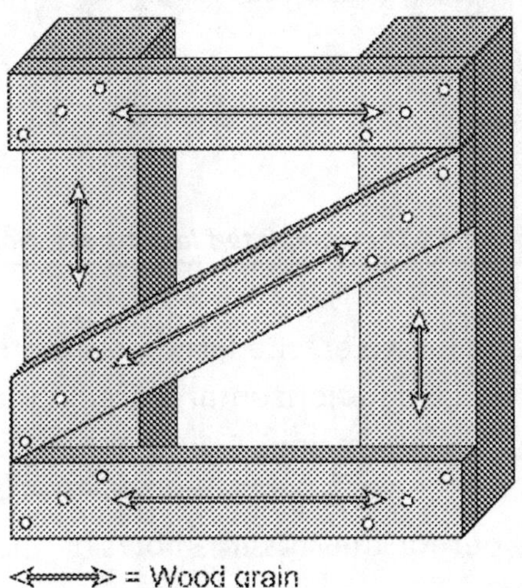

Fig. 4–109 Close-up of the 2x4-in. diagonal braces if you cut them flush with the end face of the posts.

Figure 4–110 shows the shore as it would look completed, ready for action.

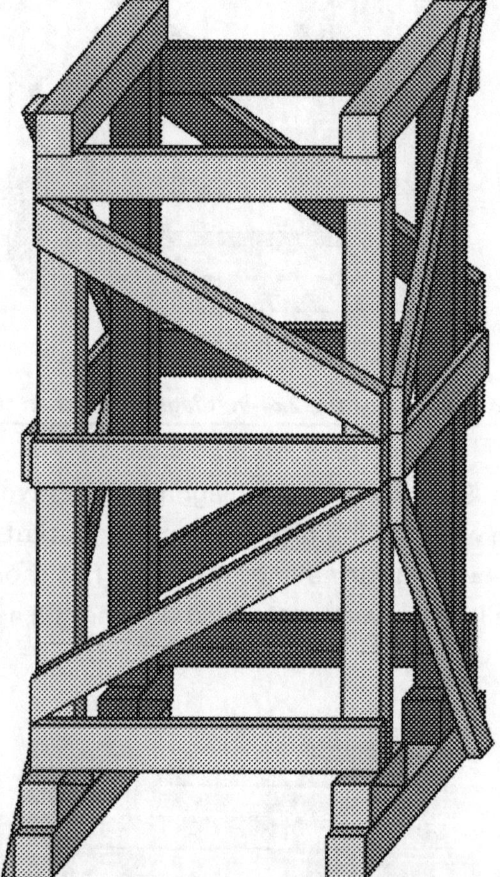

Fig. 4–110 The completed laced-post shore.

When your team has to erect a laced post more than 12 ft high, you need an additional mid-horizontal brace. In that instance, there are three bays to house bracing. Figure 4–111 shows the options available. On the left is the angle for all prefabricated sections. On the right is the layout for the cross-members. In the middle bay, you have the option of reversing the angle of the brace. The use of either option is fine. In any case, the center diagonal brace is to be in the same direction on both sides of the shore.

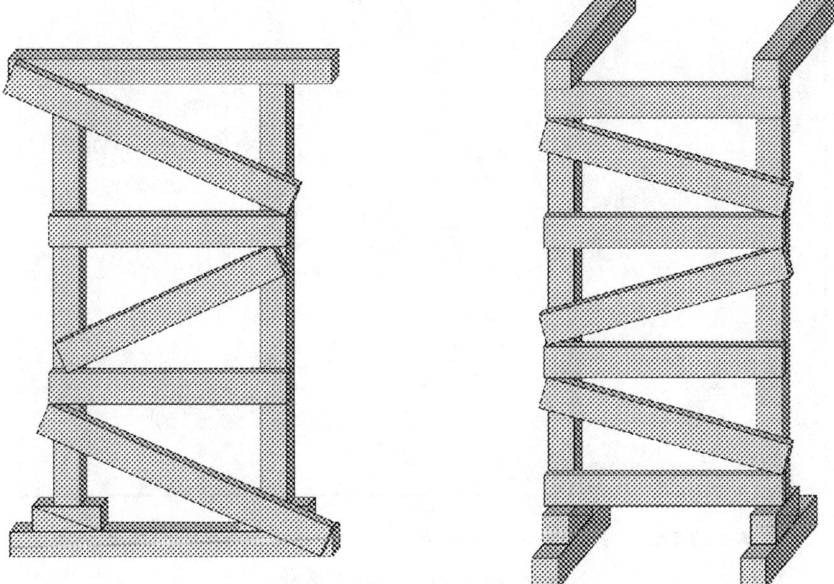

Fig. 4–111 There are three bays to house bracing when erecting shores more than 12 ft high.

To assemble this shore section, start as you would for the shorter laced-post shore. Overlap the header 12 in. on each side, nail the posts, and square them to the header. The difference from the shorter laced-post shore comes in this next step, the placement of the horizontal mid braces. Divide the length of the posts into thirds and place the centers of the horizontal braces on that point. Nail them just as you would a laced post of standard height.

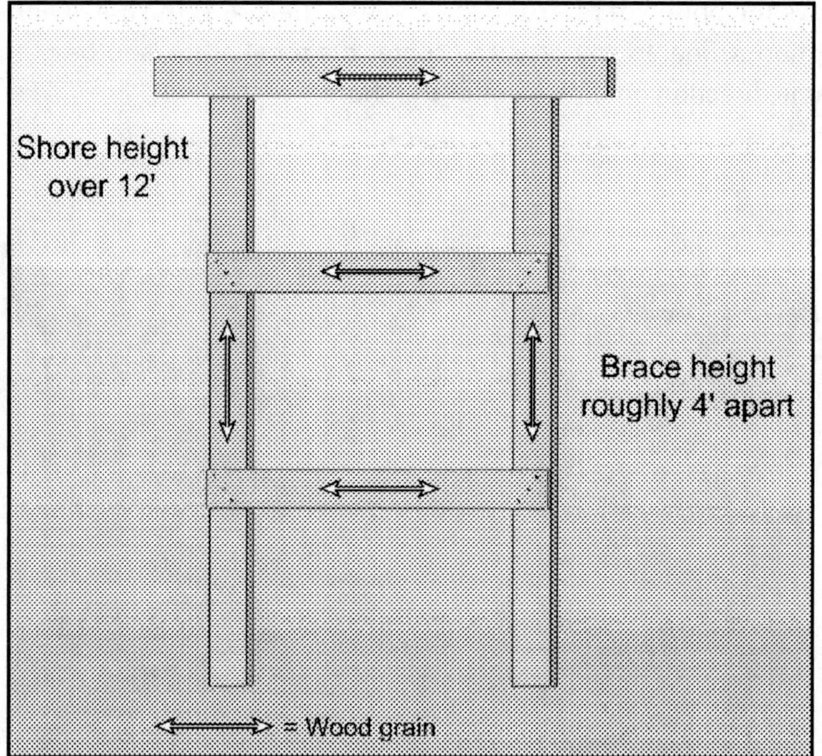

Fig. 4–112 The placement of the horizontal mid braces differs from the shorter laced-post shore.

In the case of a laced post more than 12 ft tall, install two diagonal braces instead of one. The top brace is the same as with a shorter laced post, and the center brace contacts both mid-braces as shown in Figure 4–113. Cutting the edges flush on the angle is optional.

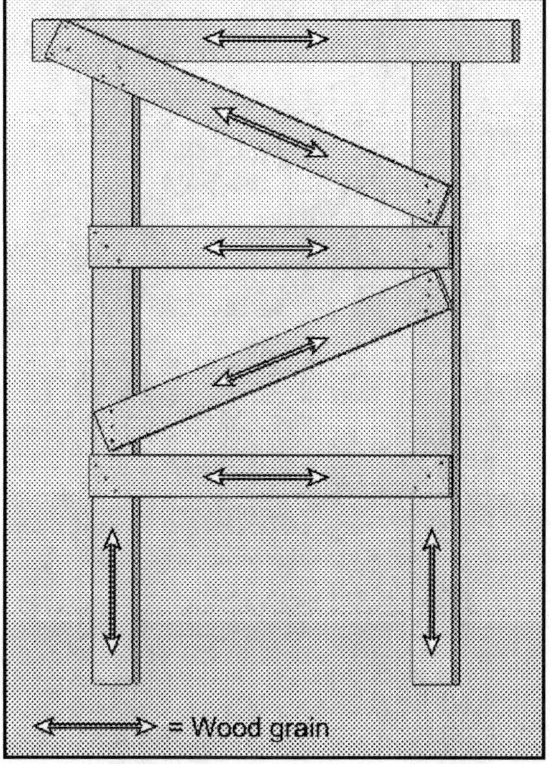

Fig. 4–113 Install two diagonal braces instead of one.

As you did for the first type of shore, you can use the first shoring section as a template for fabricating the second section right on top of the first. Keep the mid-braces in the same place and have the diagonals all going in the same direction.

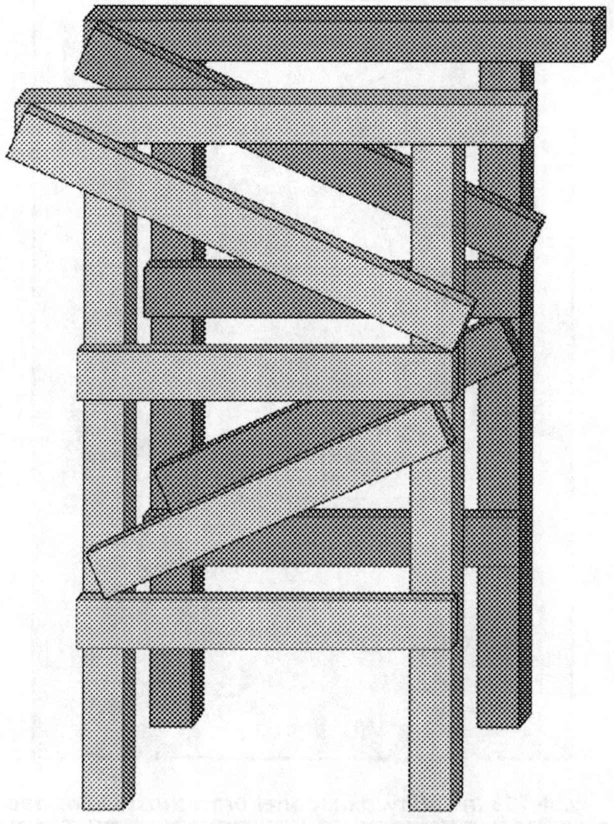

Fig. 4–114 Use the existing section you just completed as a template for the second section.

Use at least two firefighters and possibly more to erect this shore section. Place it on top of a set of wedges and a sole plate. Make sure both are situated directly underneath the posts and flush with the outside face of the posts and sole plate. All the elements must line up properly. There can be no haphazard overhanging of material.

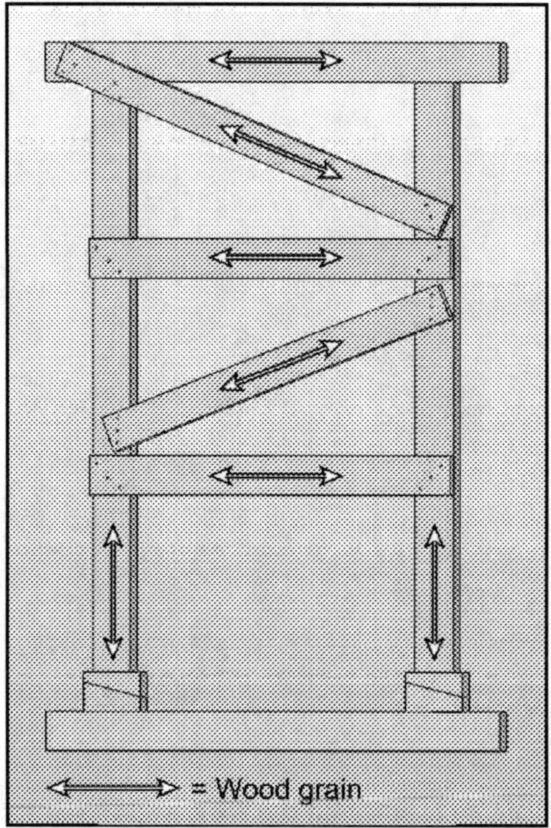

Fig. 4–115 The first section on top of the sole plate with wedges inserted.

While one or two firefighters hold up the first section, the other personnel put up the second section. Make sure it is in direct alignment with the first section and plumb, and the posts are the proper distance apart.

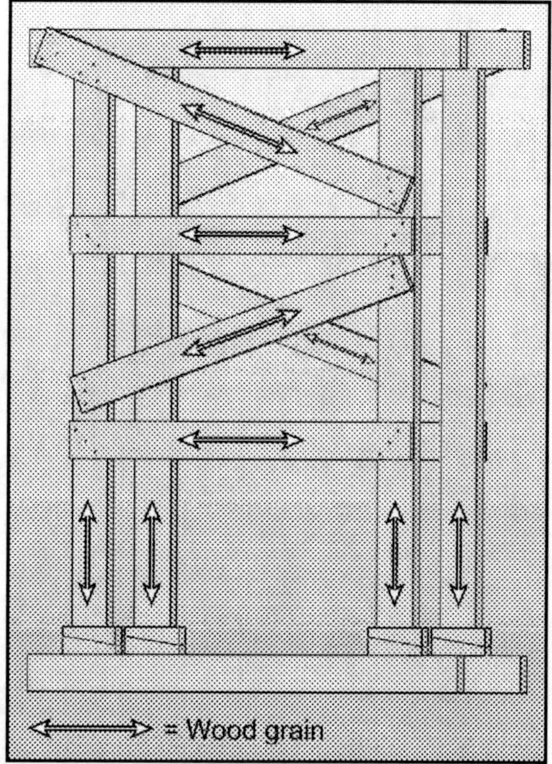

Fig. 4–116 Make sure the two sections are in line with each other, plumb, and spaced evenly apart.

As with the other shoring system, tie the two sections together. For a tall, laced-post shore, there are four horizontal braces instead of three. Begin bracing the shore in the middle and work your way to the top. Then install the braces down the shore to the bottom. Overlap the two middle braces on the outside face of the other braces and nail them in place.

Positioning of the four horizontal braces is important. The top brace contacts the bottom of the headers; the middle two braces are equally spaced through the posts' length and match up to the mid-braces of the two shore sections. The bottom brace sits on or slightly above the wedges. The top and bottom braces are flush with the outside face of the posts. The middle braces are flush with the outside face of the other mid-braces.

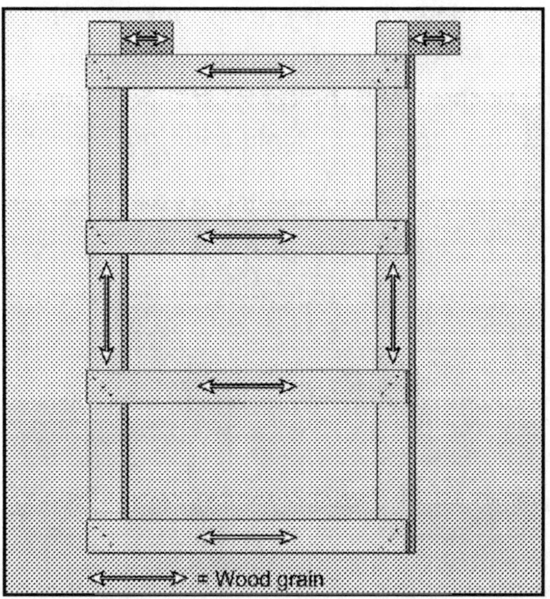

Fig. 4–117 Positioning of the four horizontal braces.

Figure 4–118 shows the laced post completed and able to support heavy weight.

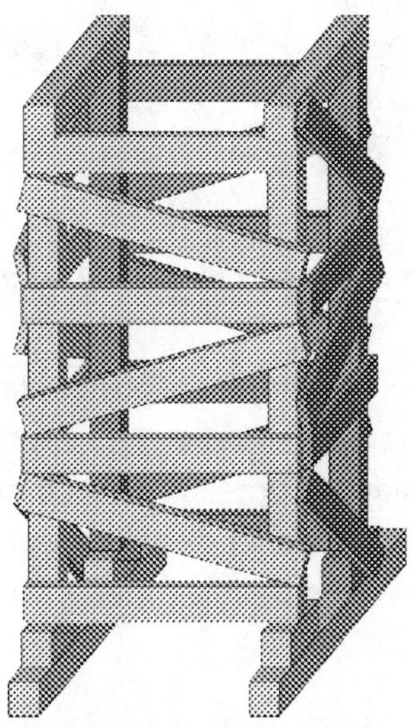

Fig. 4–118 Completed laced-post shore.

Figure 4–119 shows two of the options available to diagonally brace the shores. Make sure the top and bottom form an *X* when you look through the shore and the center section is running in the same direction. This gives the shore some additional torsional support in case of twisting.

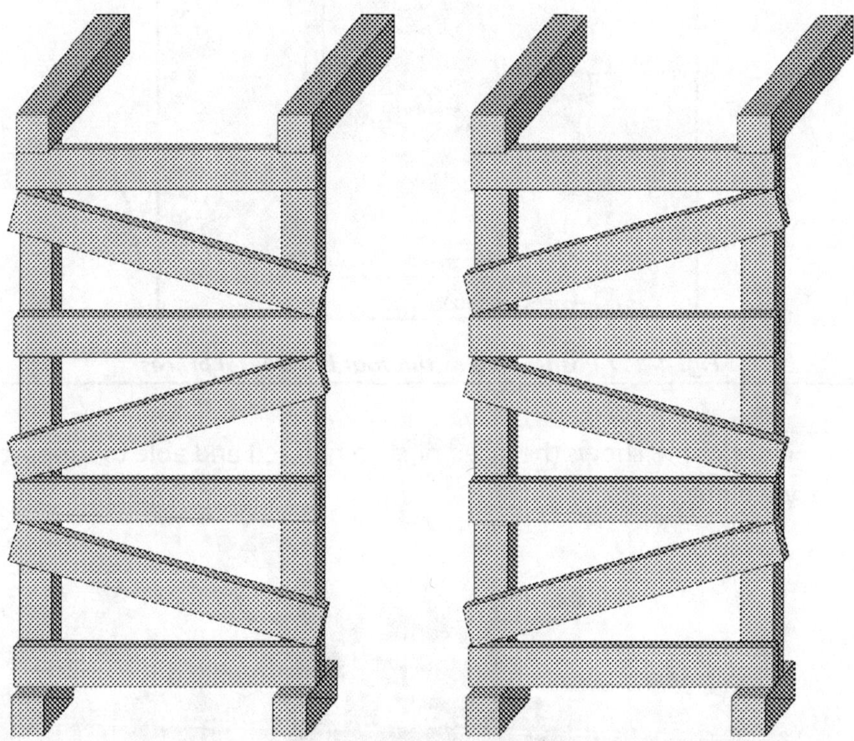

Fig. 4–119 Two options for diagonally bracing the shores.

Figure 4–120 shows the additional gusset plating needed to secure the shore in an earthquake situation. All four sets of wedges must be encased on both sides in order to keep them from vibrating out from under the posts. The header can also have a gusset plate in each corner opposite from where the diagonal brace is installed.

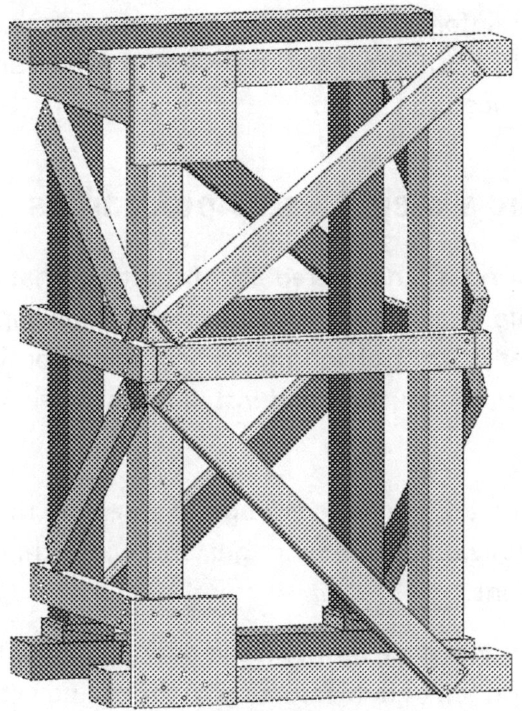

Fig. 4–120 Additional gusset plating needed in an earthquake situation.

The Vertical Shore

The vertical shore has also been called the *dead* shore mainly because it directly supports the dead load of a collapsed structure. The main purpose of the vertical shore is to stabilize damaged floors, ceilings, or roofs. It is one of the most common shores. It also can be used to replace missing or unstable bearing walls or columns if the need arises. In resupporting or replacing these damaged structural elements, some integrity is restored to the collapsed area. This affords rescue personnel some degree of safety while they perform search and rescue operations in the collapsed building. The two most common sizes of lumber used in vertical shoring are 4x4s and 6x6s.

Generally, 4x4s are more than sufficient for uses in the typical wood-framed or unreinforced masonry structure of average size (less than four stories). In much larger structures and concrete and steel buildings, you may need to use 6x6s.

Vertical shore step-by-step procedures

The following is a numbered list of the tasks that must be completed to build and place a vertical shore in a collapsed building. The remainder of this section on vertical shores provides detailed instructions for constructing and installing the various elements of the shore.

1. Clear an area suitable enough to assemble the shore safely. Generally an area 4 ft longer and wider than the shore is sufficient.

2. Lay the sole plate down on the floor and place the header directly on top of it. Measure up to the ceiling in three separate places—both ends and the middle.

3. Subtract the thickness of the wedges, generally 2x4 wedges for interior shoring. This measurement is the lengths of all your posts.

4. Install the end posts first and plumb up by eye. Place the wedges under the bottom of the posts and snug up.

5. Install the rest of the posts, snugging up with wedges. Anchor all posts to the header and tighten up all wedges. Nail the posts to wedges and the wedges to the sole plate.

6. Install the diagonal braces on both sides if necessary.

The vertical shore consists of six separate components.

Sole plate. This component is normally a length of 4x4 laid directly on the clean floor. It has to have a 1 ft overhang over the last post on each side. Make sure there is no debris under the plate. This plate must be directly under the area to be shored. Don't nail it down at this time.

Header. A header is made of a section of 4x4. In most cases, anchor the header to the ceiling joists.

Posts. These components are also usually 4x4s although they could be 6x6s. They will be installed at specific space intervals to be determined by the amount of debris weight above that they must support.

Wedges. Each post will have one set of wedges under it. Generally the wedges are 12 in.-long 2x4s. However, 4x4-in. wedges can be used, but they leave a large gap between post and sole plate—not always the best.

Diagonal braces. The last items to go into place are the 2x6-in. diagonal braces. They go on the outside face of both sides of the posts, header, and sole plate. The header, soleplate, and posts each have (5) 16d nails anchored into them through the 2x6.

Mid-point braces. When the shore is at its holding capacity and the height of the shore exceeds 9 ft, you may need to install 1x6s or pieces of ¾x6-in. plywood sections on both sides of the post in the center of the shore. The 2x6-in. diagonal braces go on the outside of the braces. Nail the mid-point braces in place with (5) 8d nails at each post.

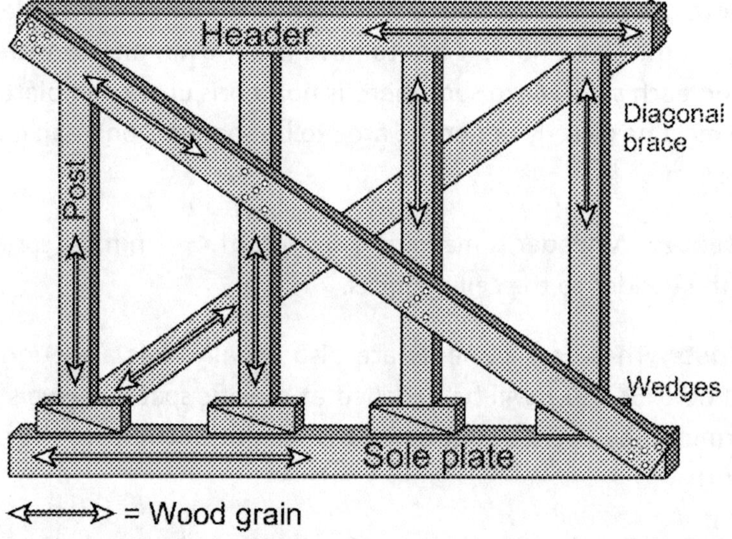

Fig. 4–121 Elements of the vertical shore.

The shoring officer and the structural engineer assigned to the team determine the location of the shore. Make sure the area is clear; you will need approximately 3–4 ft of clear space to work in. Lay down the sole plate. Do not nail it down to the floor at this time because you may have to shift it slightly to keep the entire shore plumb. Light carpeting need not be removed because the material will be compressed when you tighten the wedges; however, thick, heavy carpeting should be removed. If you are installing a shore on soft ground, for example in a basement, additional supports called *sleepers* or *mudsills* may be needed. They should be embedded in the ground and installed evenly so that the sole plate can be laid down as level as possible. Place the header on top of the sole plate, and measure from the top of the two pieces to the ceiling. This will give you the most accurate measurements. If you measure from the floor, you could have an inaccurate result if there are bellies in the floor. Measure in the center and at both ends of the shore. Take the smallest measurement, subtract the width of the wedges, and use the result for your post height.

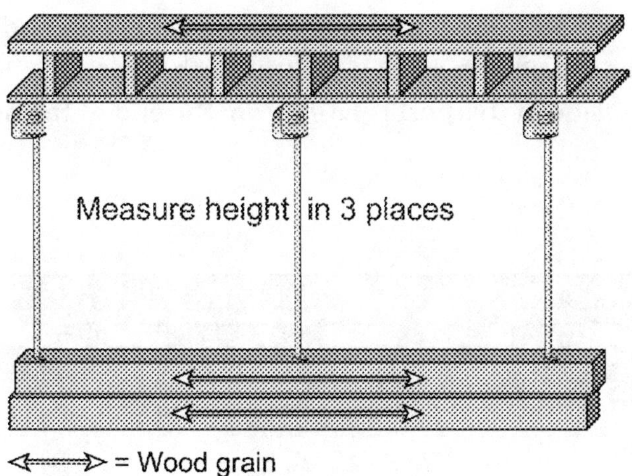

**Fig. 4–122 Place the header of top of the sole plate
and measure from the two pieces to the ceiling.**

Install one of the end posts first. Nail it to the header with (4) 16d nails, two on one face and one each on another face. Place a set of wedges under the post and snug up the post. Do not fit them too tightly because the shore may need some slight adjustment. Plumb the post up in both directions by eye. Make sure that the outside of this post is 12 in. from the end of the header and sole plate.

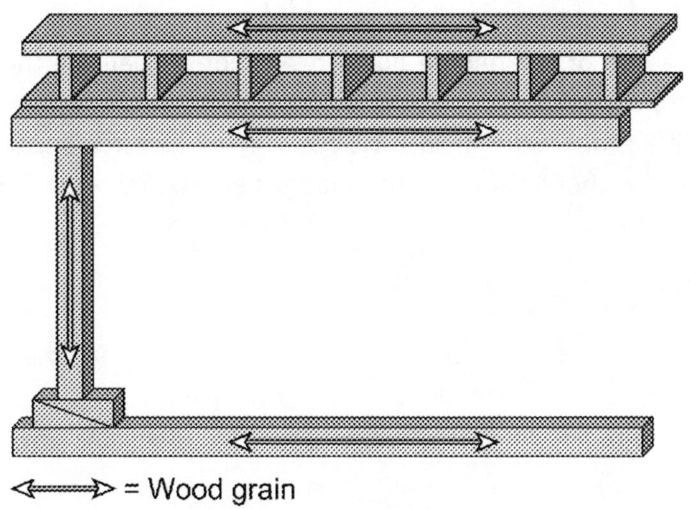

Fig. 4–123 Install one of the end posts first.

Install the other end post next, nailing it in place and snugging up the wedges. Plumb up this post in both directions. Make sure that the outside of this post is 12 in. from the end of the header and sole plate.

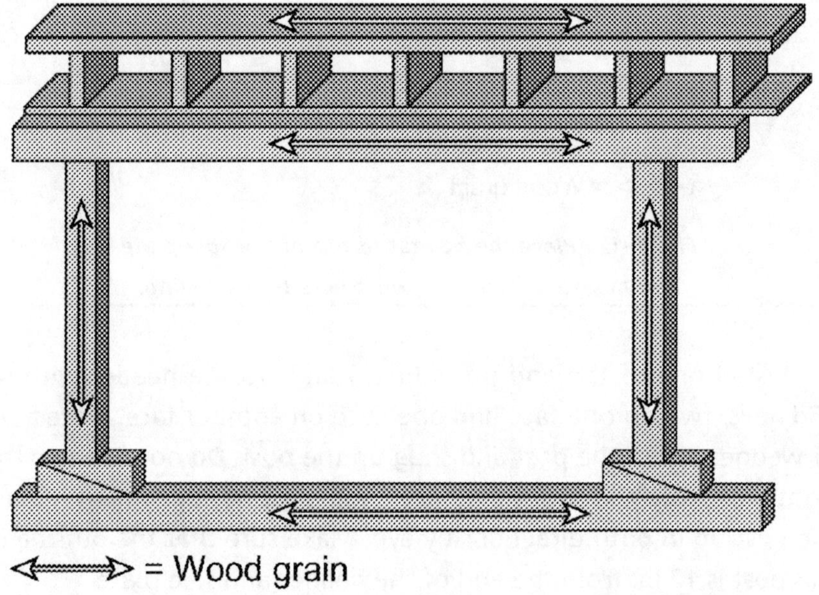

Fig. 4–124 Install the other end post next.

Place one of the middle posts in position to balance the shore. Nail the post in position. It should be directly under a floor beam if possible. This is the most efficient way of transferring the load through the shore. Snug up the wedges and plumb up the post in both directions.

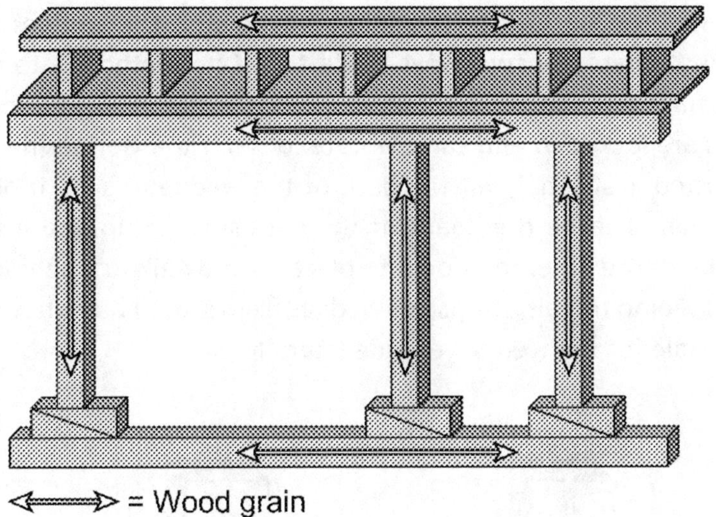

Fig. 4–125 Install one of the middle posts to balance the shore.

Place the remaining posts in position. Nail them in position under the upper floor's beams and snug up all the wedges. Be careful not to over tighten these wedges. Over tightening them can force the posts up. This loosens the other posts, causing the shore to be temporarily unstable. Recheck the tightness of all the wedges, making sure they are uniformly tight.

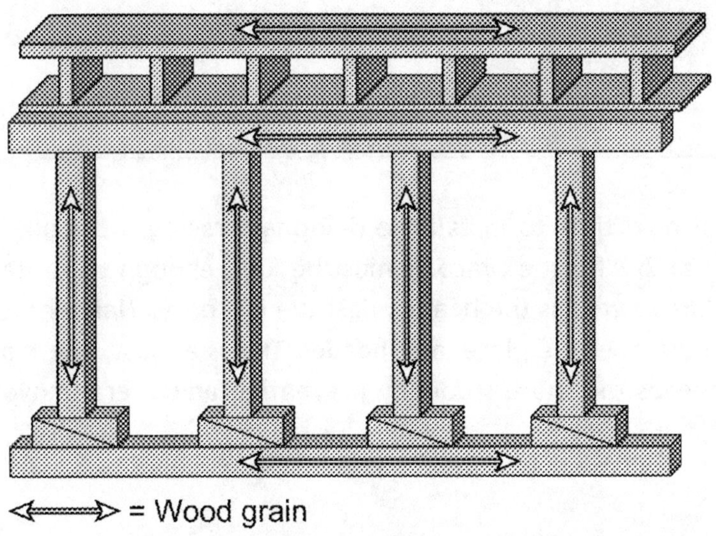

Fig. 4–126 Place the remaining posts in position.

Toenail the wedges to the sole plate to stop the wedges from loosening. There are two ways to do this. One method is to drive (2) 16d nails through the back of the wedge into the sole plate. The wedges are locked in and can't vibrate loose. If the wedges need to be adjusted, just simply hit the back of the wedge to slide it out of the toenail or leave the heads of the nails sticking up and just remove them. The other method is to place (2) 16d nails just behind the wedges. Doing this also stops the wedges from sliding back; however, it is possible for the wedges to slide laterally.

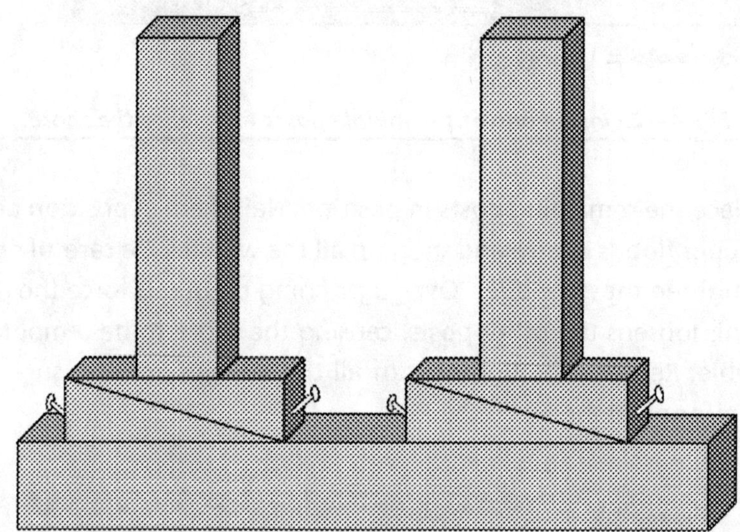

**Fig. 4–127 Toenail the wedges to the sole plate
to stop the wedges from loosening.**

It is now time to install the diagonal bracing. Normally, use a length of 2x6 for the brace. It must be long enough to contact the sole plate as well as the header, past the last posts. Nail a brace into every post, the sole plate, and header. This is a *must*. The diagonal brace keeps the shore stable by preventing any lateral movement.

For the proper anchoring of the 2x6, use the 5-nail pattern with 16d nails. If you do not have 2x6s, you can use 2x4s one stacked on top of the other. Anchor each one with (3) 16d nails into the posts, header, and sole plate.

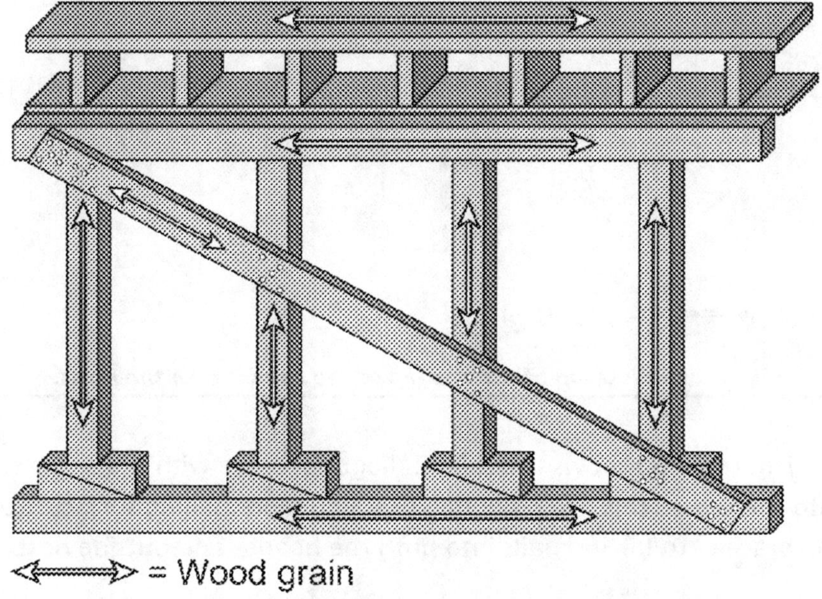

Fig. 4–128 Install the diagonal bracing.

Figure 4–129 is a close-up of the end lap of the header over the posts. This connection is important because it takes the main force of any lateral attack against the shore. As a rule of thumb, have the inside edge of the 2x6 intersect the top, inside edge of the post, right where it intersects the header. Place (5) 16d nails into the post and, just as important, nail (5) 16d nails into the header. The overhang is 12 in. from the outside face of the post to the end of the header and the sole plate. Cut the angle for the brace or just leave it square; either way is fine.

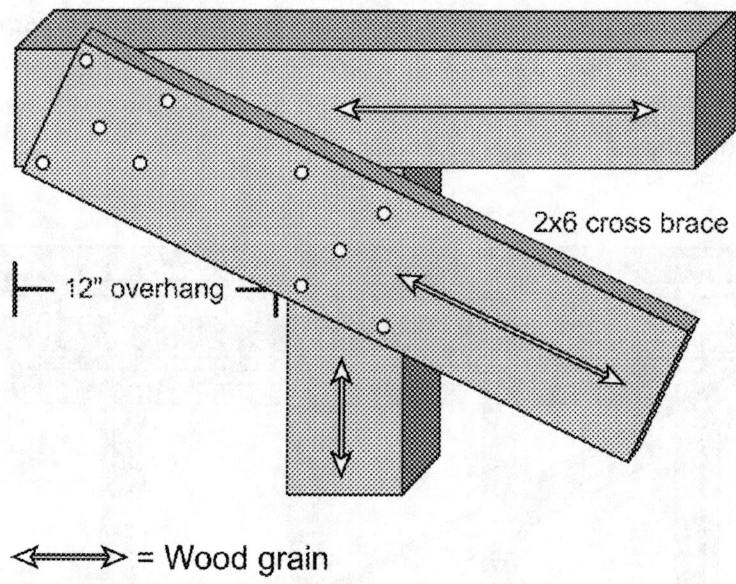

Fig. 4–129 Close-up of the end lap of the header over the posts.

Figure 4–130 shows the 2x6-in. diagonal brace with the angle cut into the brace. This size diagonal gives you more nailing surface. Nail the brace with (5) 16d nails into both the header and outside post.

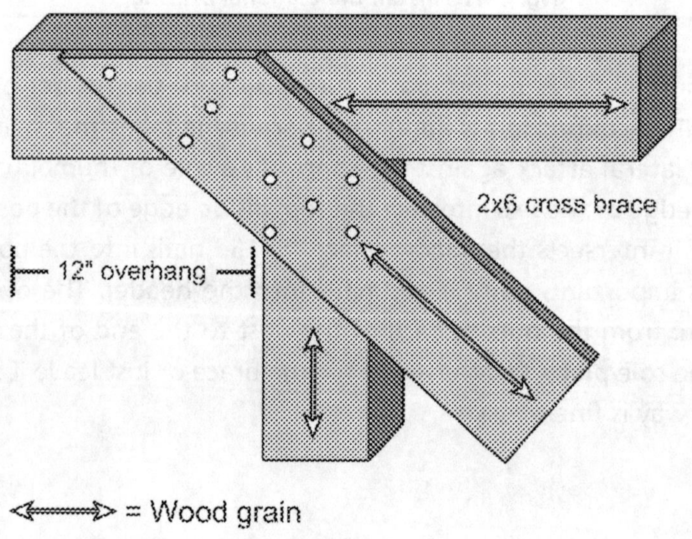

Fig. 4–130 The 2x6-in. diagonal brace with the angle cut into the brace.

Place the other diagonal into position at the opposite angle from the first brace. This is very important in keeping the shore from lateral pressure and racking in either direction. The diagonal braces must be able to withstand a pressure equal to 2% of the load being taken up by the shore itself. Your two braces in position form an *X*.

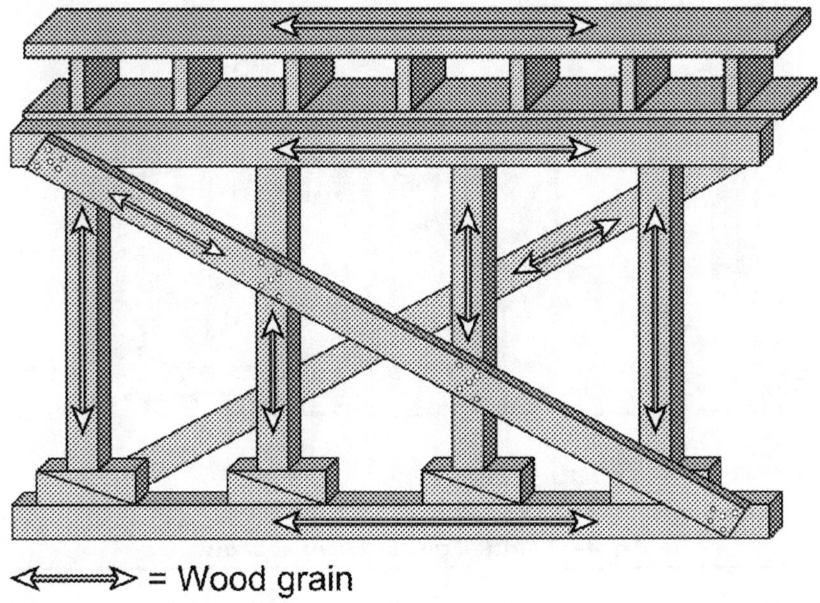

Fig. 4–131 Place the other diagonal into position at the opposite angle from the first brace.

If your team erects shoring in an earthquake situation, the shore must be gusseted after the diagonal braces have been installed. At the top where the posts meet the header, the outside posts don't need gusset plates. The diagonal braces hold them and stop them from moving. The interior posts need a gusset plate on one side only. That is enough to hold the connection in place. On the bottom by the sole plate, place plates on both sides of the posts. Do so because the wedges may dislodge. There is no direct contact between the posts and the sole plate.

At the bottom of the shore, place a gusset plate on each side, covering the wedges. Using 8d nails, nail the plates with the 5-nail pattern, five into the post and wedges and eight into the sole plate. The two end posts need a gusset only on the one side. The diagonal brace holds the other side in place.

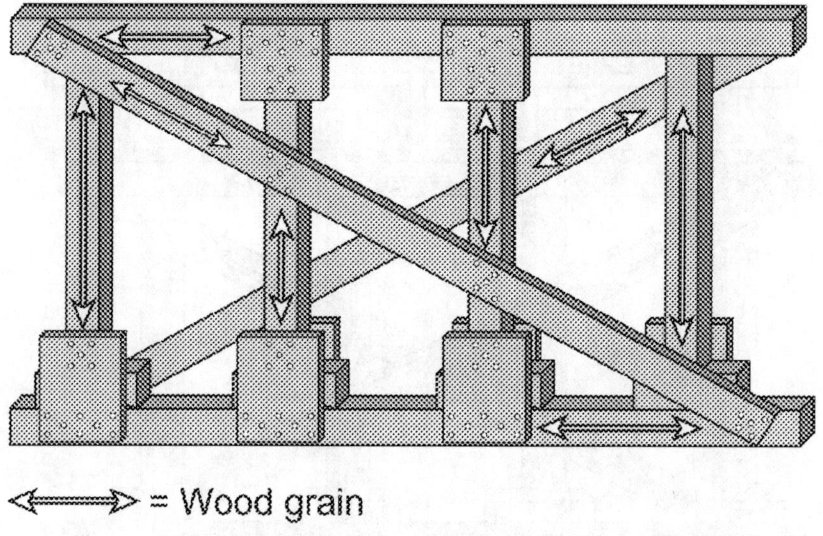

Fig. 4–132 Install gusset plates as shown.

Figure 4–133 is an example of a situation where there may be some vibration or movement affecting the shore. This could occur when the shore is erected in a high traffic area and may be hit by personnel and equipment. The gusset plates need only be applied on one side, at both the top and the bottom plates. At the end posts, there is no reason to gusset plate them. The 2x6-in. diagonal with (10) 16d nails in it most certainly will keep the post from separating. Make sure to toenail the wedges as previously described.

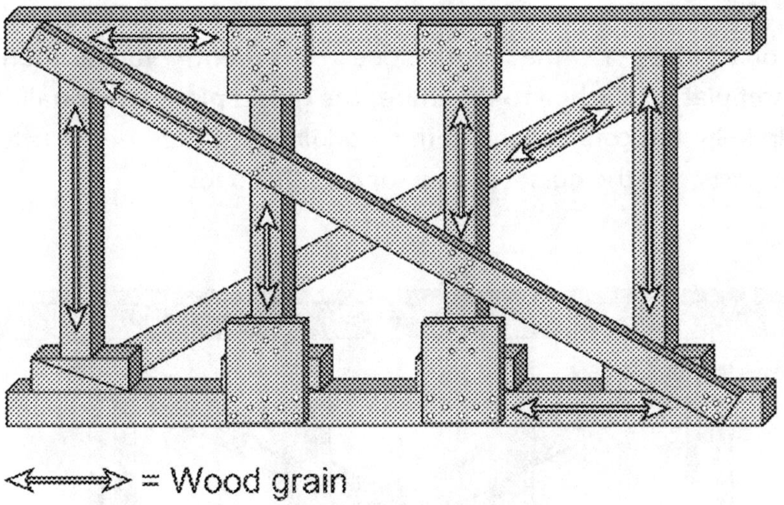

Fig. 4–133 In areas of high traffic,
the gusset plates need only be applied on one side.

Instead of using gusset plates, you could use sections of 2x4s. They can be 12 in. or 18 in.; however, 12 in. is the minimum length. If they were any smaller, they have a tendency to split when nailed with 16d nails. Therefore, drive (3) 16d nails into each post, header, or sole plate.

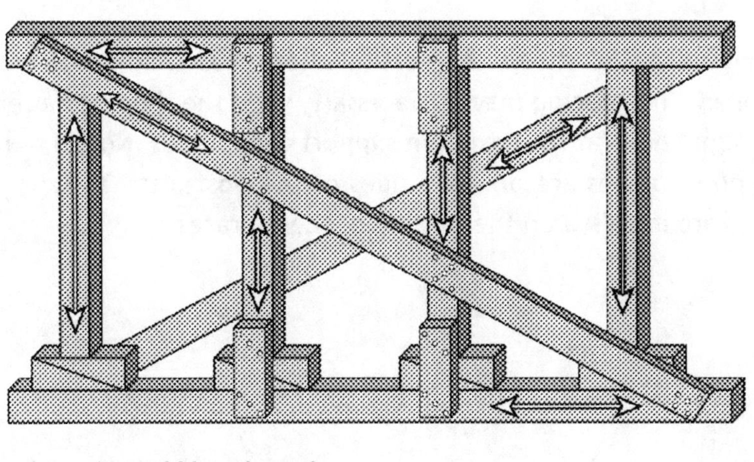

Fig. 4–134 Sections of 2x4s are an alternative to gusset plates.

If you erect some shoring in a non-earthquake situation, which accounts for 99% of the shoring done in this country, you do not need gusset plates attached to the shore. The gusset plates are installed to help keep the connections from dislodging. They are not structural. The posts, not the gusset plates, support the loads.

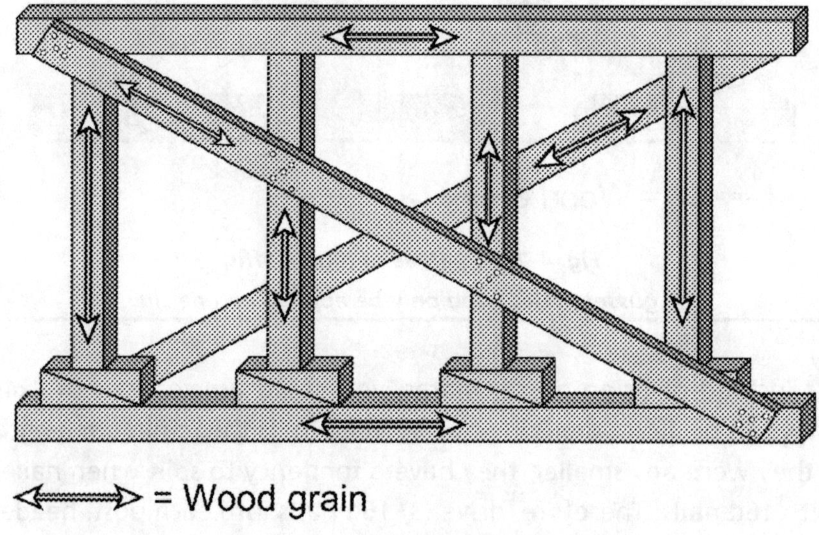

Fig. 4–135 Gusset plates are not structural and can be omitted in non-earthquake applications.

Figure 4–136 shows the vertical shore with mid-point bracing installed. This bracing may be necessary when the shore is more than 9 ft high and is at its maximum support capabilities. Notice that the mid-point braces are on both sides of the posts, and the diagonal braces are installed on the outside of these braces.

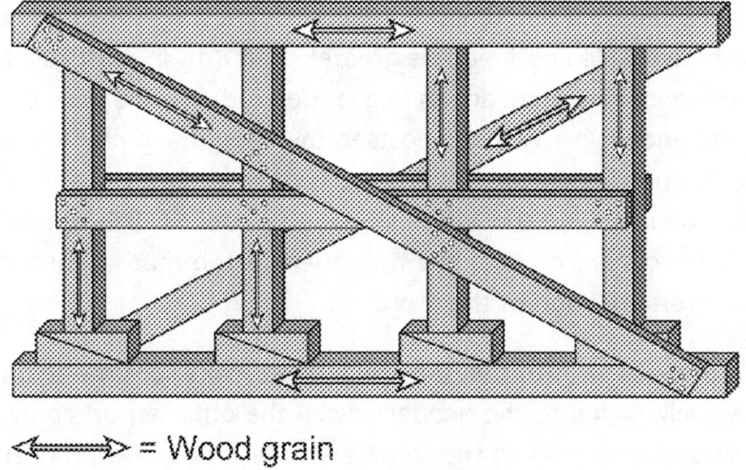

Fig. 4–136 Vertical shore with mid-point bracing.

Figure 4–137 shows a close-up of the mid-point brace to be installed if the post length is more than 9 ft. The easiest item to use for this is a piece of 1x6-in. pine. If that is not available, then a piece of 6x¾-in. plywood will do nicely. It has to be applied to both sides and anchored with 8d nails in the 5-nail pattern. This keeps the lateral deflection from occurring in the one direction.

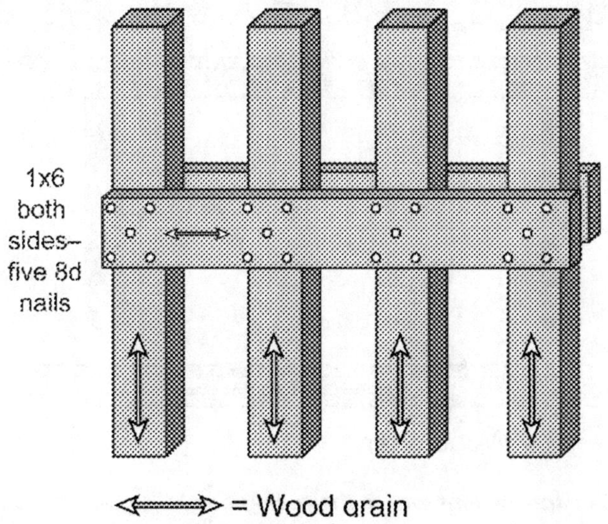

Fig. 4–137 Close-up of the mid-point brace for applications higher than 9 ft.

There will be many times when your team comes into an operation where the ceiling height is greater than 8 ft. In most cases, you will not have immediate access to a ladder. If that is the case, one option is to anchor the two end posts to the header and physically slide it into position. However, this technique only works if the area where you will be installing the shore is cleaned of debris. This option may work on occasion; most of the time there is too much debris in the way of prefabrication of the shore.

To proceed with this method, you must make sure the two posts are securely nailed to the header. Install the outside posts only; and with two people holding the shore and one calling the shots, slowly and evenly pick up the section. Keeping it steady and all movement to a minimum, swing the header up and walk the posts under the header. Keep the section as square and level as possible while you are doing this procedure. Together, raise the header into position against the ceiling then place wedges under the posts on to the sole plate. Tighten the wedges. Now you can nail the header above and install the rest of the posts.

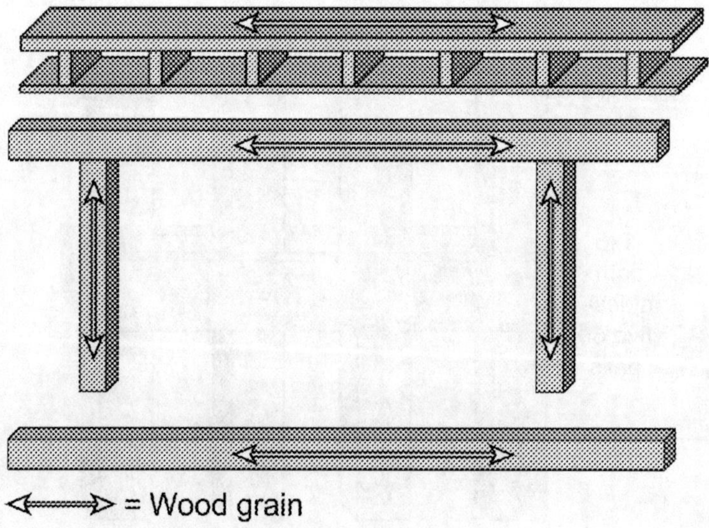

Fig. 4–138 In applications where the ceiling height is greater than 8 ft, one option is to anchor the two end posts to the header and raise it into position.

Figure 4–139 shows a close-up of one of the two outside posts. Keeping the spacing from the end of the header to the outside face of the post at 12 in., nail the post to the header with (6) 16d nails, placing two on the top face and two on each side face. Make sure these nails are driven flush, and the post is affixed tightly to the header. When both posts are anchored like this, you can raise the set-up and install from the ground.

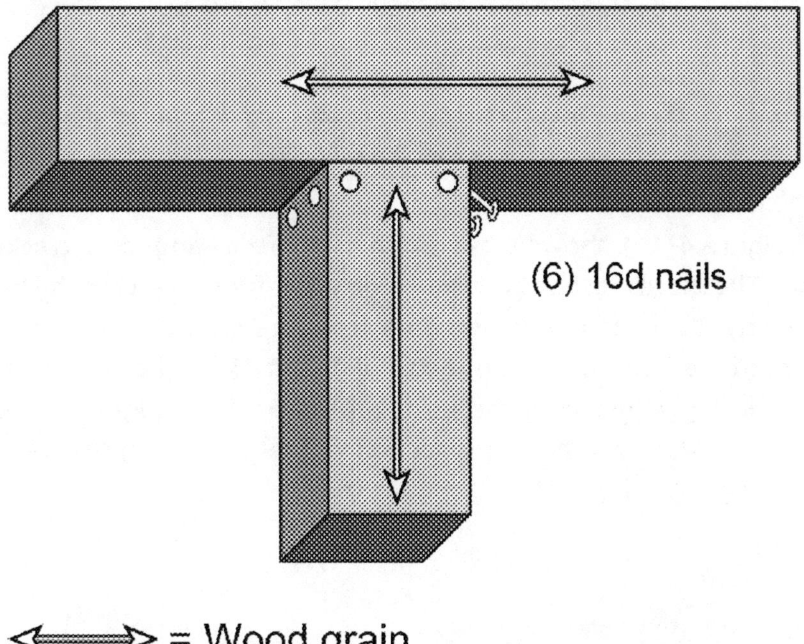

Fig. 4–139 Close-up of one of the outside posts.

With a vertical shore, the load is applied to the shore and collected. Then it is transferred through the posts and down to the next floor or to the ground. This is the principle behind the vertical shore. The header and sole plate keep the shore together as a system, and the header collects the load while the sole plate distributes it along the floor line. The diagonal braces keep the shore laterally stable.

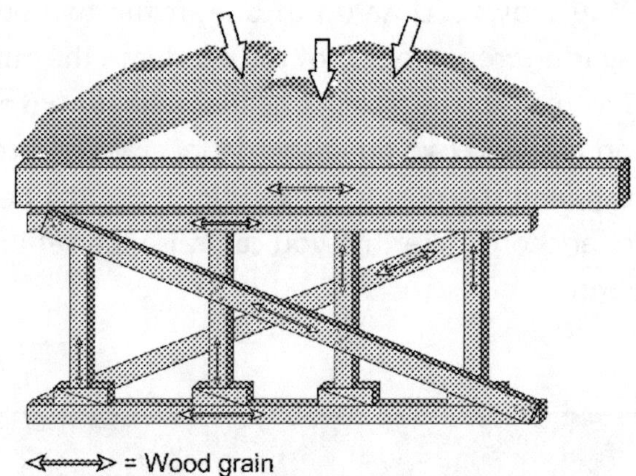

Fig. 4–140 The load is collected by the header and distributed along the floor line by the sole plate.

Figure 4–141 shows what to do to shore a sagged or cracked floor. The shore must be built square in order for it to accept the load properly. Your team must measure from the ceiling to the lowest point of the floor sag. This is the total height of the shore. Fabricate it accordingly. Install all the posts on top of the wedges as usual; however, when you have a space above the header, pressurize the shore slightly differently.

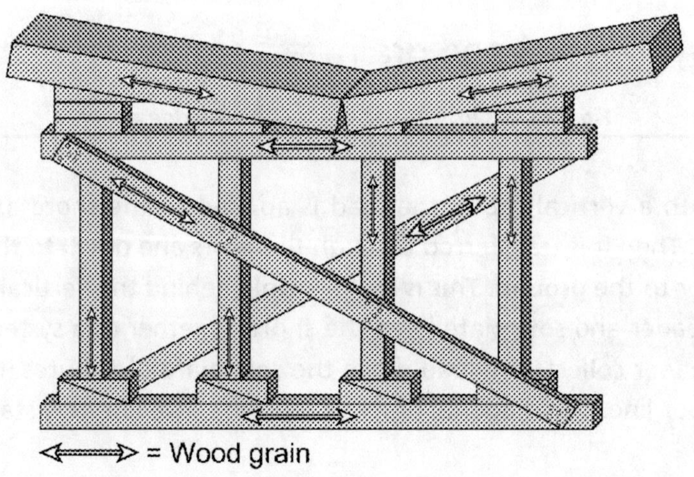

Fig. 4–141 Shoring a sagged or cracked floor.

In some situations, it may be easier to install one of the interior posts first in order to snug up the header to the center of the sag. After you have done that, start at one end and install the post on top of the wedges. Nail the post in position as usual. Before you fully pressurize the post, snug up the wedges until the header just moves slightly. Now, shim out above the header, directly over the post and to the floor beam above it. Tighten up the shims or spacers, nailing them to header and ceiling if possible. Now you can fully pressurize the wedges under the post.

In many instances, the vertical shore supports a frame floor or roof section. For the most part, the posts go under the joists. The spacing varies with the amount of weight overloading the floor. You may have to install a post under every joist or every other joist, depending on the weight of the material above. If the floor is sagged as shown in Figure 4–142, place shims, wedges, or blocks under the affected joists. Fabricate the shore to the lowest point and keep it square. Fill in the gaps under the joists. Do not pressurize the shims. Just place them in position and pressurize the wedges under the post to tighten up the shore. Anchor the blocks and shims to the header and the joist.

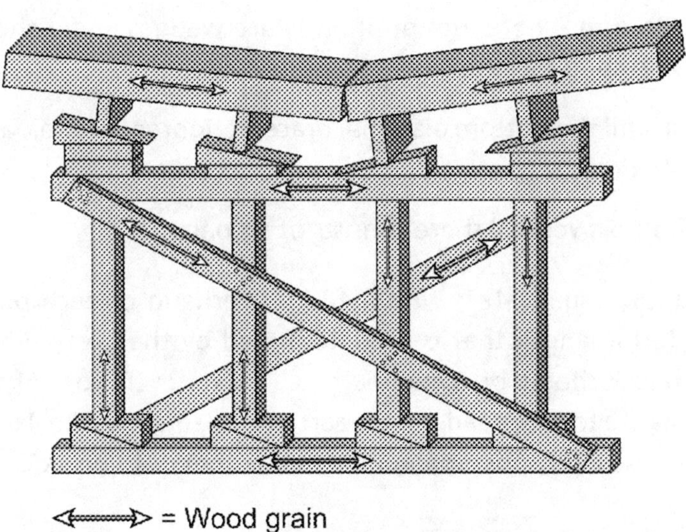

Fig. 4–142 If the floor is sagged as shown, install shims, wedges, or blocks under the affected joists.

Two-post Vertical Shore

A variation of the vertical shore is the two-post vertical shore. It comes into play when there is a large area of debris in a very unstable area. In this situation, you have to erect shoring rapidly, but clearing an area for a large shore may take too long, making the two-post vertical shore a good compromise. Although called a *two-post vertical*, it is constructed exactly like half a laced-post shore. Hopefully, in many cases you can prefabricate this shore and bring it into position, but this may not always be the case. Areas with large amounts of debris can cause access problems with fabricated shores.

Two-post vertical shore step-by-step procedure

1. Make sure the area where the shore is going to be placed is clear of debris.

2. Determine the size of and the width of your posts.

3. Prefabricate the header, posts, cross brace, and top diagonal brace if possible.

4. Place the shore into position, place wedges under the posts, and snug them up.

5. Install the bottom diagonal brace, retighten wedges, and nail it in place.

A two-post vertical shore consists of six parts.

Header. Usually 4x4s with a 12-in. overhang of each post, the header has a length that is predetermined by the post width. The header depth should be a minimum of 1 in. for every foot of span. If your posts are to be spread 3–4 ft apart, the header must be 4x4. If you decide to use a 5-ft post spread, the header needs to be 4x6.

Posts. Usually 4x4s also, generally spaced 3 to 4 ft apart, the maximum we can spread these posts would be 5 ft. The limit we can construct this shore would be to a height of 12 ft. Although this is the maximum, it will be rare for you to construct this outside and bring it into the collapse area. It's just too big to maneuver.

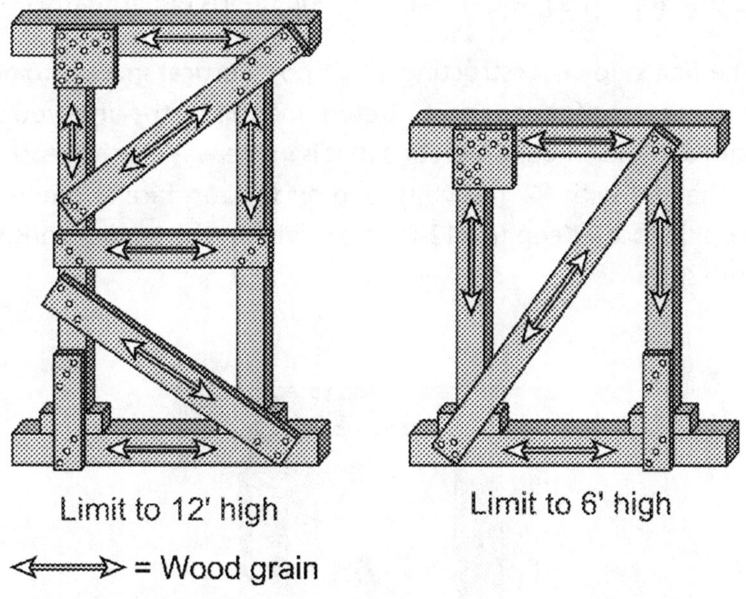

Fig. 4–143 Height options for the two-post vertical shore.

Cross brace. This will be a center brace between the two posts. Constructed from 2x4s, it is placed dead center of the posts and flush with both outside edges of the posts. It is anchored with (3) 16d nails at each post.

Diagonal braces. Two 2x4s are anchored to the shore on a diagonal from the header to the top of the center cross brace and from the bottom of the cross brace to the base of the sole plate. The diagonals are opposite each other, forming a large *K*. Anchor these with (3) 16d nails each point.

Wedges. For this shore, use one set of wedges under each post. You can use 2x4 or 4x4 wedges. Tighten them when the shore is placed in the final position.

Sole plate. Just like the header, the sole plate is a section of 4x4 with a 12-in. overhang of each post. It is placed in position before the preassembled section is raised up to the item or items to be shored. Make sure it is on a good surface with no debris underneath.

The first step in constructing a two-post vertical shore is to determine the spacing for your posts. Determine this by the area you want to shore and the amount of debris that is in the way. Anchor each post to the header with (4) 16d nails, two on the top face and one each on the side faces. Keep the 12-in. overhang on the header consistent on both sides.

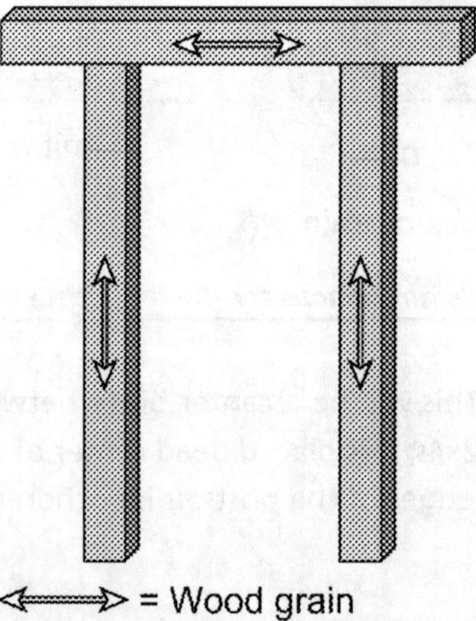

Fig. 4–144 Anchor each post to the header, keeping the 12-in. overhang consistent on both sides.

Square up both posts to the header. This is important! You do not want a crooked shore because it will not fit properly when installed. Use a 2-ft framing square to ensure that it is square.

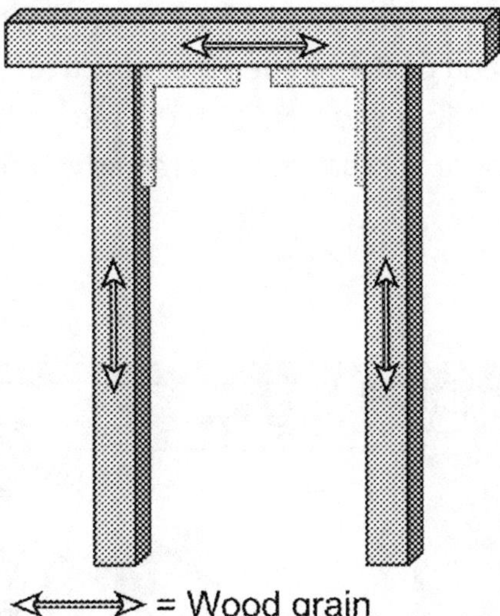

Fig. 4–145 Square up both posts to the header.

As an added measure to help keep the shore square, you can nail a 12x12-in. gusset plate to the top left-hand corner of the post and header. Find the center of the posts and anchor a 2x4-in. cross brace. Anchor it with (3) 16d nails to each post.

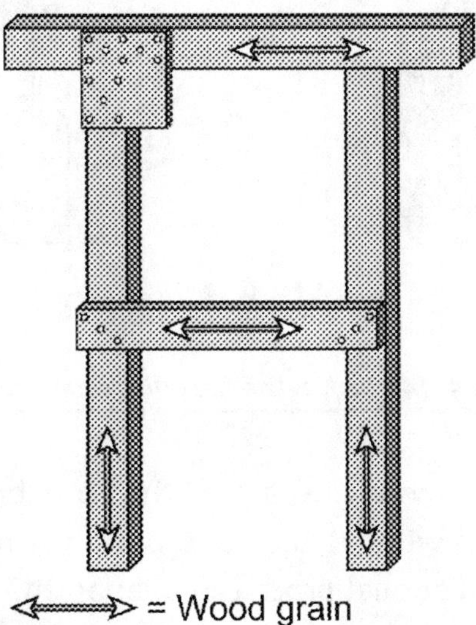

Fig. 4–146 A 12x12-in. gusset plate can be added to the top left-hand corner to help keep the shore square.

Next install the first diagonal brace. Span it from the top of the cross brace to the end of the header. Start from left to right as shown in Figure 4–147. Nail it in position with (3) 16d nails at each post and the header. The shore is ready to be erected if there is not a lot of debris in your way.

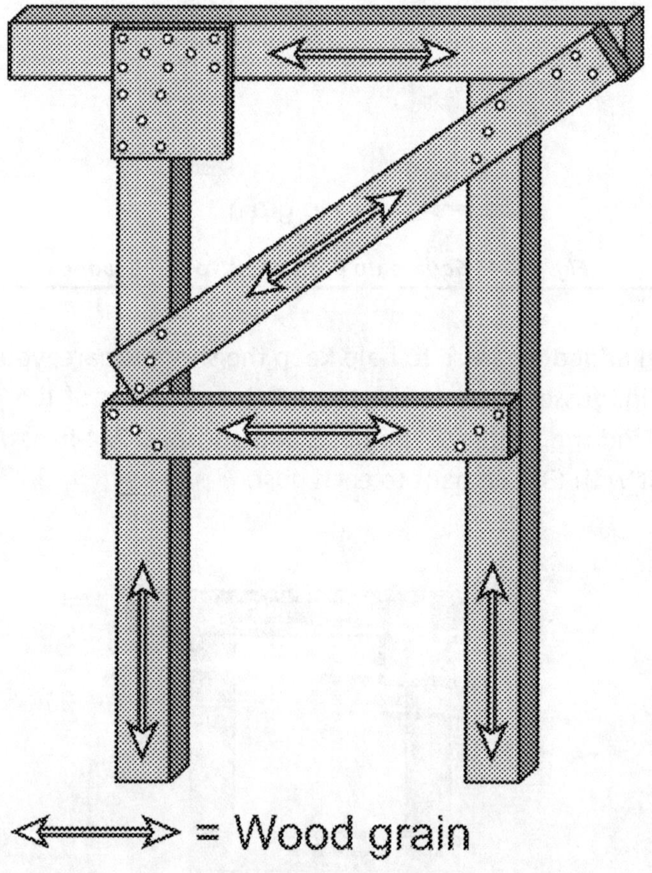

Fig. 4–147 Install the first diagonal brace.

Figure 4–148 shows a close-up of the cross brace and diagonal nailing. Nail all 2x4s with (3) 16d nails, using the nail pattern shown in the figure. The diagonal brace can overlap the end of the post a little or can be flush with it. Just make sure you have enough nailing surface for the three nails.

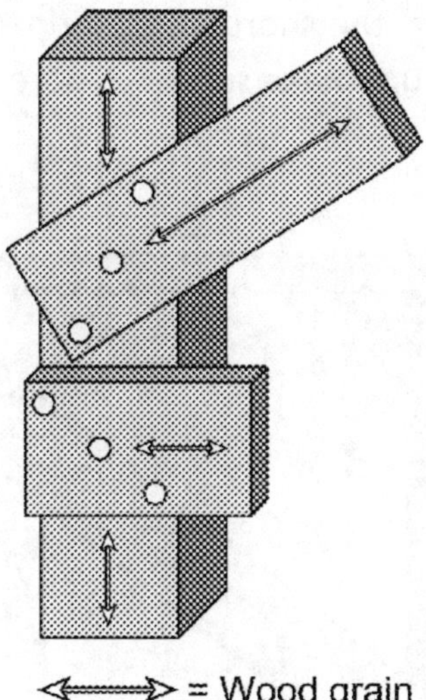

Fig. 4–148 Close-up of the brace and diagonal nailing.

Figure 4–149 shows a close-up of the top intersection of the diagonal brace with the other post and the header. Run the brace to the end of the header, place (3) 16d nails into the header and post as shown.

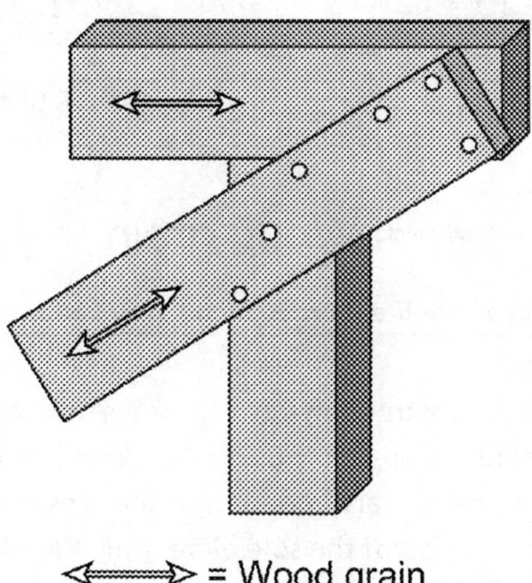

Fig. 4–149 Close-up of the top intersection of the diagonal brace with the other post and header.

Figure 4–150 shows the shore in position. Set the wedges under the posts and tighten up. Make sure the shore is plumb by eye.

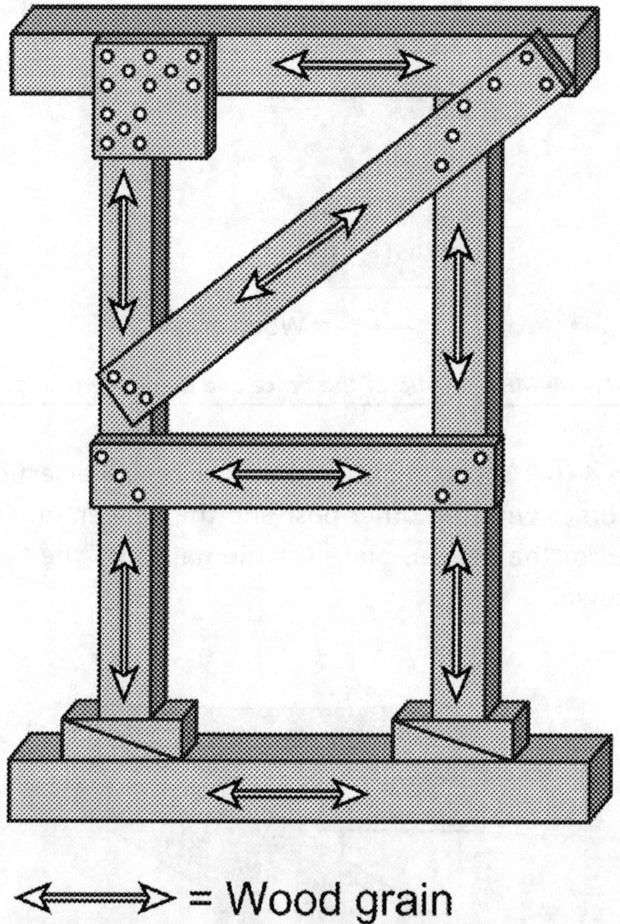

Fig. 4–150 Make sure the shore is plumb after you set the wedges.

The last step in constructing a two-post vertical shore is to set the bottom diagonal brace. Install it using the same procedure you used to install the top brace. Start underneath the cross brace then go all the way down to the end of the sole plate. Nail it in place with the (3) 16d nail pattern. Make sure the braces form the *K*.

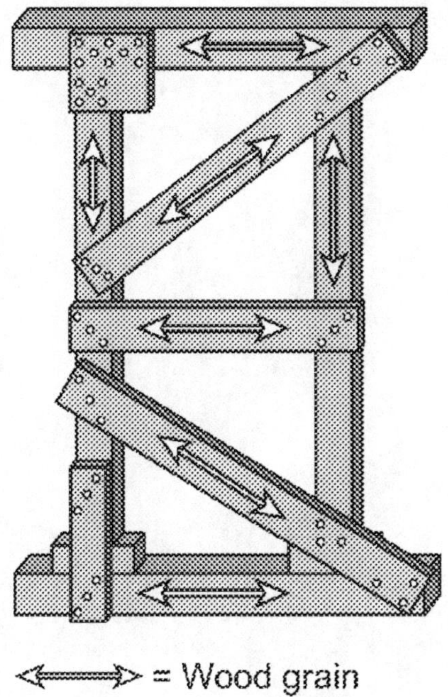

Fig. 4–151 Set the bottom diagonal brace.

If a shore is to be less than 6 ft high, you can fabricate it and just use one large diagonal brace. Make sure the brace contacts the header.

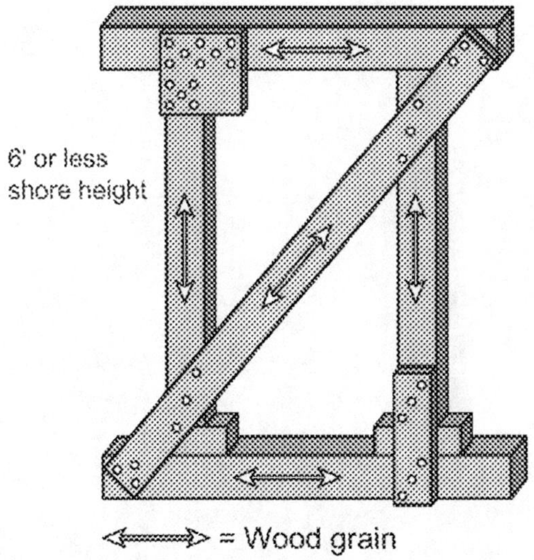

Fig. 4–152 Construction of a shore in applications less than 6 ft in height.

Exterior Rescue Shoring Procedures

Exterior Rescue Shoring

Some of the most difficult and complicated shores you will erect are exterior rescue shores. Exterior rescue shoring consists primarily of raker shores and exterior horizontal shores used to stabilize and resupport existing bearing or nonbearing exterior walls. These walls may be cracked, leaning, bulged, or in some other way damaged or not properly supporting their loads. In assembling exterior raker shores, you will work with lumber ranging from 4x4s up to lumber possibly as large as 12x12 in. in diameter. But generally speaking, most occasions call for 4x4s or 6x6s.

Assembling and installing exterior raker shores can be involved and exacting. These operations can be extensive and require large amounts of material. Make sure enough material is on hand to complete a given assignment. Keep in mind, even though the shores can be complicated, your team will find exterior rescue shores are not that difficult to construct after the team has had some practice building them. The flying shores are also complicated and can be erected without much difficulty after your team has practiced constructing them.

A series of fixed-raker shores properly anchored and braced together stops an unstable wall from moving outward. At least two should be installed in any given situation. Usually they are erected in a series for stability reasons. By connecting the individual shores together, you create a stable system of support that can safely handle extensive loads.

Exterior shoring size-up

Your team's exterior shoring size-up must cover the following factors:

- Type of construction

- Extent of damage

- Type and stability of the ground on which the shores will bear

- Secondary collapse potential

- Reason the building failed

- Height of the wall to be stabilized

Following is a more detailed discussion of these factors.

Bulged walls. Bulged, bellied, or leaning walls are signs of some type of structural instability occurring in the structure. Walls are designed to accept loads through their center axis when they are plumb. If for any reason the walls become eccentrically loaded, there can be drastic results, especially if those walls are bearing. As an eccentric load, the weight on top of the wall can quickly fail that wall. Any deformation in the wall indicates that the overall strength of that wall is compromised. The wall could possibly fail at any time, depending of course on how severe the deformation is. The majority of time erecting interior shores to accept the floor load from above is one of the safest ways to counteract possible problems.

Cracked walls. In concrete and masonry wall construction whenever forces are applied to the walls, there is the possibility of some type of cracking. This is especially true when forces are applied either laterally or horizontally to the plane of the wall. Although masonry is excellent under compression load, its lateral strength is not extremely efficient. As a result, lateral attacks against masonry walls can cause them to crack. In reinforced concrete, cracking is an inherent part of the curing process. Hairline or thin cracks in concrete do not mean anything. However, issues develop when the cracks are large and have depth and space, which means the sections of material have sustained heavy damage and may have separated from each other.

Another major factor to consider is the concrete adhesion to the reinforcement bar. Concrete keeps its structural integrity until the material itself has separated from the rebar. When there is no adhesion to the steel, the lateral strength of the concrete is severely compromised; and possible collapse situations are a real concern at this point. In masonry brick and block, cracking is much more of a concern than in concrete. The bond between the mortar and the masonry units is what keeps the wall's integrity intact. When a lateral force is applied to the wall and it fractures that bond, the structural integrity of the wall may be compromised. Thin cracks may not be much of an issue; however, large, long cracks with noticeable depth are of concern. In this situation, the integrity of the wall has positively been compromised.

Another situation that may develop is the appearance of an *X* pattern crack on the wall. This is not good. The *X* tells your rescue team that the wall has had stresses applied to it from two separate planes. Shifting or settling of the building is occurring in two separate directions. This is a serious concern, and a thorough size-up of the situation is called for.

Yet another key factor to look for is an indicator of where a wall crack began. For instance, suppose there is a wall crack with a large gap at its base. The crack runs from the base of the foundation laterally then goes up the length of the wall for an appreciable distance (say 10–20 ft) and terminates with a fine line at the top. This type of crack is a settlement crack that has occurred over time and indicates a foundation problem, affecting the stability of the wall as well as of the building. Major problems such as foundation issues can severely limit the possibilities of a rescue team helping to restabilize a structure.

Foundation issues. In a situation in which there is building instability due to the possible yield of the footings due to unstable soil conditions or water undermining the building, the shoring team has a significant problem. All of the rescue shoring to be installed must rest on a good bearing surface that can support the additional loads to be applied to it. In some cases, the good bearing surface can be another part of the structure; however, in many cases, it must be the ground or the basement level.

If the foundation of the building is somehow compromised, the chances of rescue shoring being effective are lessened. Support of the structure cannot be accomplished by the relatively simple installation of emergency building shoring. Major efforts must be considered in order to resupport a structure that has foundation problems. The resupport activities entail efforts not generally associated with the application of rescue shoring; it just requires too much time and commitment of resources. When your team responds to an incident involving a major foundation problem and after the structure has been evacuated, you may have to make decisions such as whether to turn the building over to a reputable contractor and let that party handle the operation and stabilization of the building.

Racked structure. On some occasions, mainly during natural disasters such as tornados, earthquakes and hurricanes, the entire structure may shift and become racked. In order to stop the building from shifting or racking any further, your rescue team can install raker shores at the corners of the building. A set of rakers installed at each corner, especially on a smaller structure, should provide enough support to arrest any further racking of the structure. Of course, the percentage of racking that has occurred to the building is one of the biggest considerations when deciding whether to attempt to restabilize the structure. The first consideration is whether there is anyone trapped in the building. If there is, every effort must be made to rescue the victim(s), and the structure must be shored up before rescue forces enter the building. By raker-bracing the corners, you lock in place the four corners of the building, stopping the structure from twisting any further.

Ground stability. Generally speaking, if you respond to a building collapse in an urban environment, you will probably be erecting your rakers on concrete or asphalt. On these types of surfaces, you should use solid-sole raker shores. This type of raker shore can easily be anchored to the hard surfaces, using any number of methods. In suburban areas where you are more likely to encounter bare ground adjacent to the damaged structures, the split-sole type of raker shore may be the easiest to use. However, if the ground is stable and firm, the solid-sole raker works fine. In each collapse situation, you have different anchoring options; choose the one that is the easiest and most efficient for your team to install.

Construction type. The building's construction type will help you determine the size of material to use and possibly the space between shores. Raker shores in a series normally should be placed no more that 8 ft apart. Buildings of lightweight construction, such as wood-framed structures, private homes, townhouses, and the like, usually don't generate a large amount of heavy collapse debris. Shores used when these buildings fail generally need to be constructed from smaller size lumber, such as 4x4s.

Larger structures are constructed of heavier materials, such as brick or concrete block, which are found in many commercial structures; and the weight of walls greatly increases. To handle the additional weight, use 4x6s and 6x6s for your rakers.

Deciding the size lumber to use can only be done on the scene. In concrete buildings and some much larger masonry structures, you may need lumber as large as 8x8s or 12x12s, which are not always readily available and can be a little more difficult to work with. In this case, using construction equipment to lift and place the material of this size makes the operation easier and faster to complete.

Adjacent structures. In many instances, the structures adjoining a damaged building can be utilized to help support the partially collapsed building after a careful survey of the adjacent structures to determine if they were affected in any way from the incident. The physical shape or the structural integrity of the building has to be thoroughly examined to ensure its potential use as a stabilizing force to help support the collapsed building. If it has been determined that the adjacent structure is capable of handling the additional loads that may be placed against it, then shoring operations can begin. On occasion, if the buildings are close together, then exterior horizontal shoring can be erected to support the remains of the damaged building. If the structures are farther apart, you may need to erect a flying shore system. In either case, the structure's specialist on the scene in conjunction with the incident commander will determine what is appropriate.

Building dimensions. A good rule of thumb is the bigger the building, the bigger the lumber. Building dimension determination is all part of your size-up. Most fire departments and technical rescue teams don't normally carry a lot of shoring material, generally due to space restrictions of their apparatus. When your team encounters an incident involving a large multistory structure and the building is constructed of heavyweight materials such as concrete and masonry or steel, you probably need to use larger dimensional lumber.

Typically, smaller structures such as townhouses, condos, wood-frame buildings, etc., need only 4x4s for normal shoring situations. However, in the large dimensional structures, 4x4s may not be adequate. In some situations, especially in structures that have been compromised and heavily damaged, large lumber will most often be necessary; 6x6s, 8x8s or even larger lumber have been used in the past.

On several occasions, steel has also been used for shoring. Pipe and box beam are two types used at some of the larger incidents.

Amount of damage. How extensively the structure is damaged dictates whether you do any shoring at all. As you size up the structure, determine whether the area is safe enough for rescue personnel to operate in it. Check the structural integrity for the following:

- Cracks or bulges

- Out-of-plumb walls

- Amount of the remaining structure relying on the wall for support

The presence of any of these conditions determines the amount of shoring needed. A pretty simple rule is that the more damage, the more need for shoring. Very extensive damage throughout a large building may dictate the use of multistory shoring systems in order to redirect the unstable loads to a good bearing surface, generally the ground.

Load Transfer

Most unreinforced masonry buildings are constructed in such a way that the interior weight is carried by the floor beams and transferred to the bearing walls, which in the majority of these buildings are exterior walls. An additional impact load transferred to these exterior bearing walls during a collapse can cause deflection and instability to occur. The purpose of the exterior raker shore is to help stabilize the bearing wall and help transfer additional loads to the ground. As the load is applied to the raker shore, the raker itself comes under compression, causing it to slide upward on the wall plate and away from the wall on the sole plate.

Figure 5–1 shows how the raker has the forces from the building applied to it. As the vertical force is applied to the raker and because it is on an angle, there are forces trying to push the raker up and away from the wall. The top cleat resists this force. When the vertical force is being applied to the raker, it also causes the raker to be pushed back from the wall. To counteract this horizontal reaction force, rescue personnel do two things: 1) anchor the sole plate so it cannot move; and 2) use a cleat on the top of the sole plate to resist the forces against the raker. By doing these things, workers enable the raker to transfer more of the load from the structure to the ground.

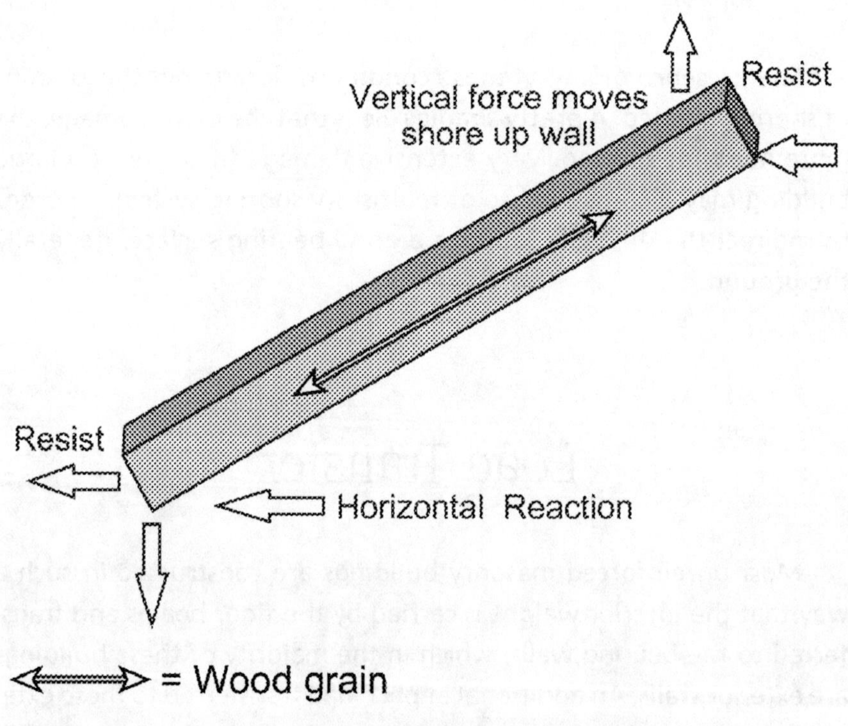

Fig. 5–1 Forces applied to the raker.

Figure 5–2 shows how the raker shore would be positioned against the leaning wall and pressurized. Counteract the vertical uplift forces and the horizontal lateral forces by using at minimum 24-in. cleats and by anchoring the raker against the wall and into the ground.

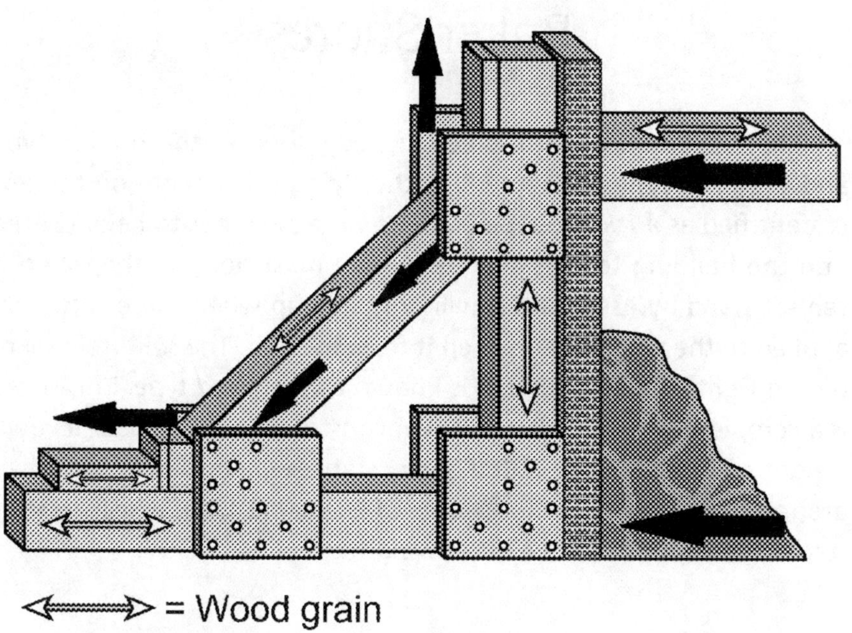

Fig. 5–2 A raker pressurized against a leaning wall.

Cleats placed against the raker stop that element from moving. These cleats must be a minimum of 2 ft in length and properly nailed. One way of doing this is by utilizing the 5-nail pattern. If done properly, the raker will not slide in either direction. In both wood-frame and URM construction, the floors are generally designed to support the building's main loads, transferring them to the exterior bearing walls through which they are then directed to stable ground. The main point at which rakers should intercept a building's load is at the center of the joists of the floor you want to stabilize. In general, if you come within 2 ft of this point (preferably lower than the joists center), you will be able to make full use of the shore's efficiency. With this in mind, you can round off your raker measurement to the nearest foot, making it easier to measure, lay out, and cut. If your raker insertion point winds up above the recommended spot, your shore will be less effective and may not be able to support the building load or prevent a secondary collapse from occurring.

Raker Shores

There are two general types of raker shores—the friction type and the fixed type. In Figure 5–3, the flying raker shore on the left is identified as a *friction* type of shore because it has to be anchored into the building to be effective. A wall must hold up the raker; it cannot stand by itself, and it will only stay up when there is friction applied to the wall plate to keep it from moving. The solid-sole raker on the right of the brick wall is known as the *fixed* type. This shore is a complete system by itself. When constructed, it can bear its own weight and stand by itself. It is a very stable type of shore; and when grouped with several other shores of the same type, it will be very stable and support quite a bit of weight.

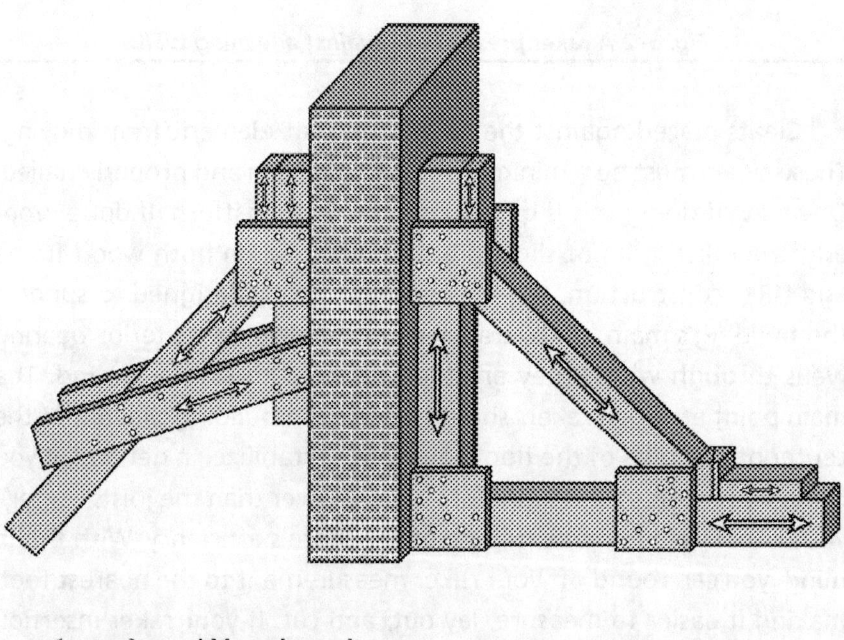

Fig. 5–3 The friction (l) and fixed (r) types of raker shores.

The friction-raker shore

Friction shores are used primarily in the construction industry. They can be quickly installed and use little material, making them attractive to contractors. A friction-raker shore relies on the compression force applied to the raker itself for its stability. It generally consists of a raker and some wedges or blocking at either the top or the base, sometimes in both places. The raker is generally installed against the object or wall to be supported and then wedged tightly into position. It can stay in position as long as it is under compression. In a rescue situation where fire department personnel are operating, this style of shore is not recommended for permanent shoring. If any movement, shifting, or secondary collapse occurs, the raker may loosen, slip, or fail entirely, placing the rescuers in jeopardy.

The vast majority of raker shores erected by contractors are friction-raker shores. When responding to a collapse at a construction site where friction shores have been erected, evaluate the shoring before committing rescue personnel to the area. In many instances, the shores may not have been constructed properly.

The fixed-raker shore

With a few exceptions, the majority of raker shores your team will construct and encounter in rescue situations will be of the fixed type. In a fixed-raker shore, all of the structural elements are connected, making the shore one integral unit when it is properly anchored to the wall and the ground. The shore itself is independently stable and can handle unexpected forces applied to it whether from secondary collapse or aftershock. Fixed raker shores are the number one choice for use in rescue situations.

Emergency service personnel normally deal with two types of fixed-raker shores: the solid-sole type and the split-sole type. Both have several variations. Some are adjustable, and most can be pre-assembled then moved into place. Some use more lumber than others.

Both the solid-sole and the split-sole raker shore can be used on either solid surfaces like asphalt or concrete or on bare ground. In general, the solid-sole raker is used in urban environments where concrete and asphalt cover the ground. The solid sole can also be used on bare ground with the addition of sleepers or ground pads. The split-sole raker shore is used primarily in suburban and rural areas where open ground is prevalent. Only after a survey of the structures in your response area and of the surface on which you will be installing will your team be able to decide which type of raker shore suits your department's specific needs. This is just one more example of why preplanning and surveying your response area are so important.

Constructing raker shores from rectangular lumber

Square lumber generally is your first choice of lumber from which to build your raker shores. However, if you only have rectangular lumber at your disposal, it can be use to construct raker shores when necessary. These shores have to be properly braced. Although the additional width of the material has some merit, it is not as much as you might think.

You can construct the shore with either the wide side of the rectangle or the narrower side facing the wall. It will matter; however, the bottom line is that the weakest point of the rectangular material will be at the narrowest side. Your team should run the wider part of the rectangular material parallel with the face of the wall for more surface contact. This is done for two reasons:

1. It makes the shore a little more stable with more surface contact available to the wall and ground.

2. The shore will be diagonally braced against the thinnest plane of the lumber. The diagonal braces are on both sides.

Two key points should be observed, however.

1. You should keep all the shore elements the same width. If the wall plate, the sole plate, and the raker are not the same widths, the gusset plates won't be able to hold properly, which is their primary function. The rectangular raker pictured in Figure 5–4 is constructed of a 6x6-in. raker and 4x6-in. wall and sole plates. This construction is a good example. You do not need 6x6-in. wall plates or sole plates because they are both part of the system. The wall and sole plates should be the same size (not one a 6x6 and the other a 4x6), and the 4x6 for both is sufficient. The 6x6-in. raker is taking all the weight and pressure. By using 4x6s for the two plates, you lighten up the weight of the shore, making it much easier to work with.

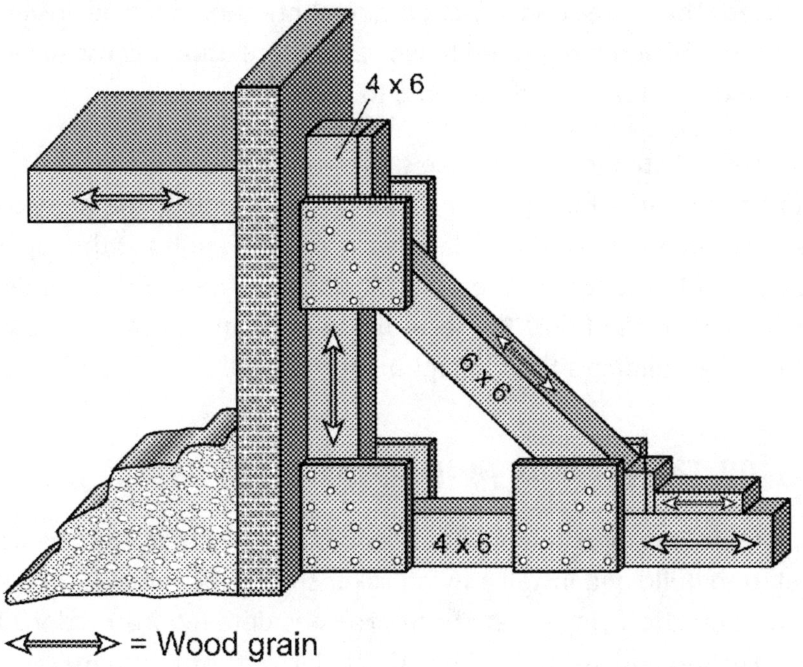

⟨⟹ = Wood grain

Fig. 5–4 Raker shore constructed out of rectangular lumber.

2. Make sure that the series of raker shores you install are properly braced; attach a diagonal brace to the center of each raker on both sides, and attach horizontal bracing across the top of the series of raker shores. With the wider part of the lumber on the horizontal plane, the top horizontal shores will have more contact with the shores, making them a little more efficient.

Flying-raker shore

The flying-raker shore is the least complicated of the raker shores. It is also the least stable. The main purpose of the shore is as a safety shore put in place before the main raker shoring systems are installed. You can quickly erect this shore and place it into position. Often, firefighters put up two of these as a temporary safety raker system. This will help stabilize the wall while the main rakers are being constructed. These are strictly friction type shores and must be anchored into the wall and the ground to work. They are usually constructed of 4x4s and set at a 45° degree angle to the wall.

The first step in erecting the shore is to determine the angle you want to use and the size of the shore. A 45° angle is the most commonly used. It is easy to calculate, and both the angles at the top and bottom of the raker are the same. The size of the shore depends on the height of the floor. The face of the raker must contact the wall where the building's floor beams meet the wall.

Flying-raker shore step-by-step procedure

The following is a numbered list of the tasks that must be completed to build and install a flying-raker shore. The remainder of this section on the flying-raker shore provides detailed instructions for constructing and installing the various elements of the shore.

1. Determine the size raker you want to use and the position on the wall where it is to be placed.

2. Preassemble the raker; clear an area 12 ft x 12 ft outside the collapse zone but close to the wall to be shored to assemble the raker.

3. Determine the wall plate length and the raker angle and then cut both pieces and assemble.

4. Nail a 2-ft 2x4 cleat to the top of the wall plate just above the raker.

5. Make sure the angle of the raker is accurate; double-check by measuring off the wall plate.

6. Nail the gusset plate to the joint.

7. Install the bottom cross brace at your predetermined location.

8. Flip the shore over and install the opposite gusset plate and bottom cross brace.

9. Install the shore and anchor it to the building.

10. Pressurize the base of the shore, using various techniques.

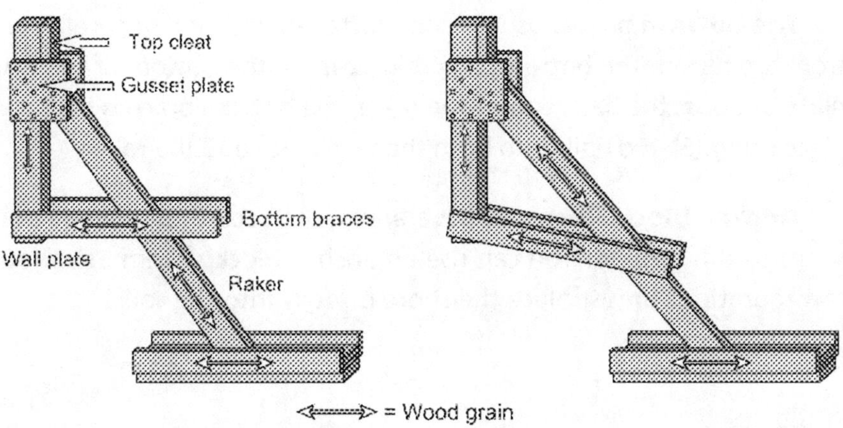

Fig. 5–5 Elements of the flying-raker shore.

Wall plate. The wall plate of the flying-raker shore is constructed of 4x4s usually not shorter that 4 ft and not longer than 6 ft. This is the typical size; but, of course, it can vary when necessary. The raker is also constructed of 4x4 material. Generally the angle is 45°, but, again, that can vary with the situation. Place the top cleat—a minimum 24 in.-long section of 2x4—just above the raker 2 ft down from the top of the wall and anchor it with (17) 16d nails.

Gusset plates. Place a ¾ in.-thick 12x12-in. plywood member on both sides of the wall plate and raker connection and cover the raker joint, the wall plate, and 2 in. of the top cleat. Use a 5-nail pattern with 8d nails for a total of 13 nails.

Raker. The main support element of the shore, the raker is also typically a 4x4. This section is cut to a specific length and has angles cut into both ends or can be square at the bottom when being installed in the ground.

Cleats. Cleats measuring 2x4x24 in. are installed on top of the raker and anchored to the wall plate. This is done in order to stop the raker from physically riding up the wall plate when pressurized. They are nailed down with (17) 16d nails each.

The bottom brace. Usually constructed of 2x6s and generally runs from the face of the bottom or within 12 in. of the bottom of the wall plate out past the back end of the raker, the bottom brace is nailed in place using (5) 16d nails into both the wall plate and the raker.

Anchor block. There are several ways to secure the flying raker shore to the ground. You can use an anchor block system of various configurations or just place the shore directly into the soil.

Step-by-step procedure

1. Place the wall plate on the ground and make a mark 2 ft down from the top of the plate. Your length of the wall plate can be anywhere from 4 ft up to 6 ft if you want. Keep the top of the raker return cut at the 2-ft line or just below it. This space is needed for the installation of the top cleat. Use (2) 16d nails and anchor the raker into the wall plate, making sure the face of the raker is sitting flush with the plate for a good fit. The raker must have full contact with the wall plate in order for it to be effective.

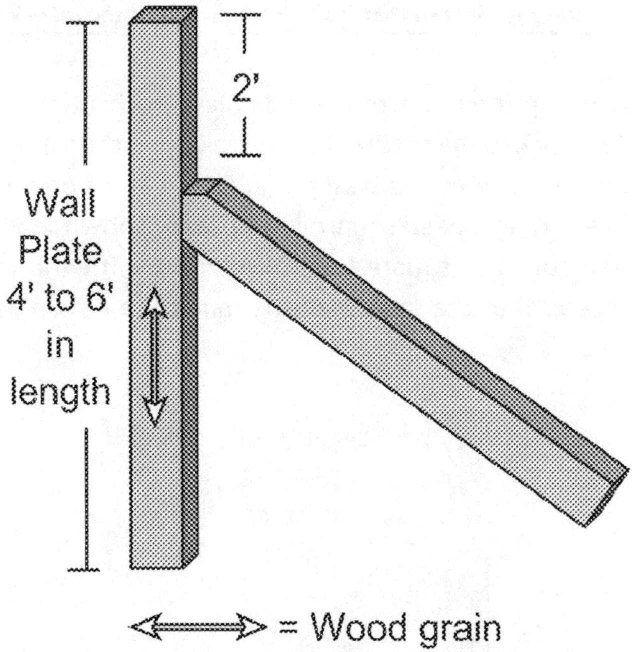

Fig. 5–6 The raker face must have full contact with the wall plate.

2. Install the top cleat. It is made of a 24-in. section of 2x4 and anchored with (17) 16d nails in a 5-nail pattern. Make sure the cleat sits flush on top of the return cut of the raker. This cleat stops the raker from riding up the wall plate when it is pressurized.

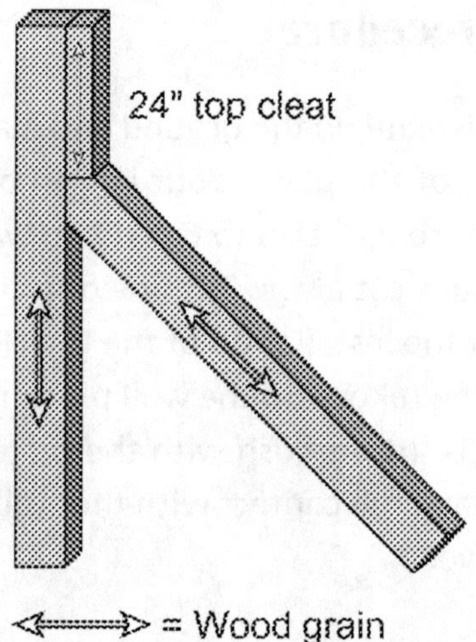

Fig. 5–7 Install the top cleat so that it sits flush on top of the raker's return cut.

3. Square up the raker to the wall plate at this time. Just to be safe, check to make sure that the raker is sitting at a 45° angle from the face of the wall plate. One way of doing this is to place a 12-in. speed square in the space shown in Figure 5–8. Make sure the square has full contact with the raker; this indicates that the raker is pretty much at a 45° angle.

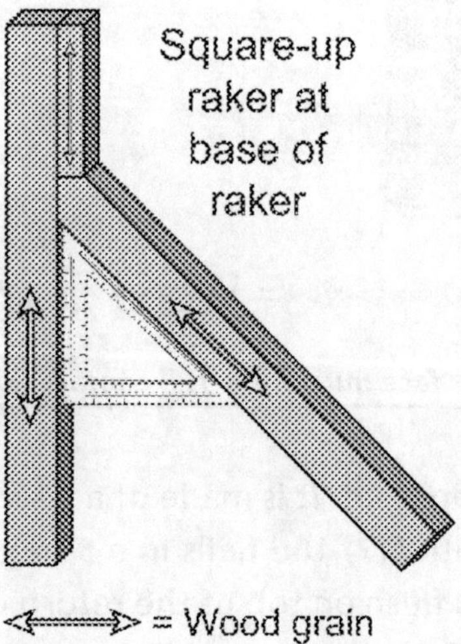

Fig. 5–8 Square up the raker to the wall plate.

Another and more accurate way of making sure the raker angle is 45° is by utilizing the *17 method*. From a predetermined point, usually the bottom intersection of the raker and the wall plate, measure down the wall plate 3 ft. To determine the 45° angle, multiply the number 17 by the 3 ft length you marked down the wall plate. Three times 17 equals 51. In this case, it will be the 3 ft that was laid out. Three times 17 equals 51. Along the bottom face of the raker and from the same point you made the previous measurement, measure down 51 in. Placing your tape measure at the 3-ft mark on the wall plate, measure up 3 ft to the intersecting 51-in. point on the raker. Now align the 3-ft and 51-in. marks. This will give you your desired angle.

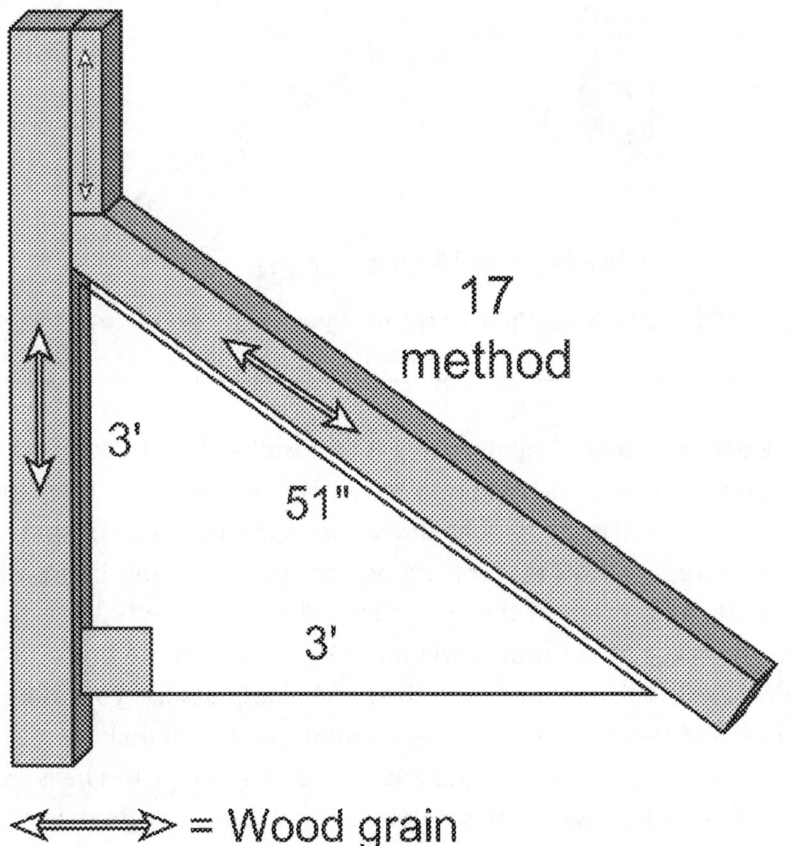

Fig. 5–9 Utilize the 17 method to make sure the raker is at a 45° angle.

After you have made sure the angle is right, next install a ¾ in. thick 12x12 plywood gusset plate at the joint where the raker meets the wall plate. Anchor the gusset with (8) 8d nails to the wall plate and (5) 8d nails into the raker. Use the 5-nail pattern for both.

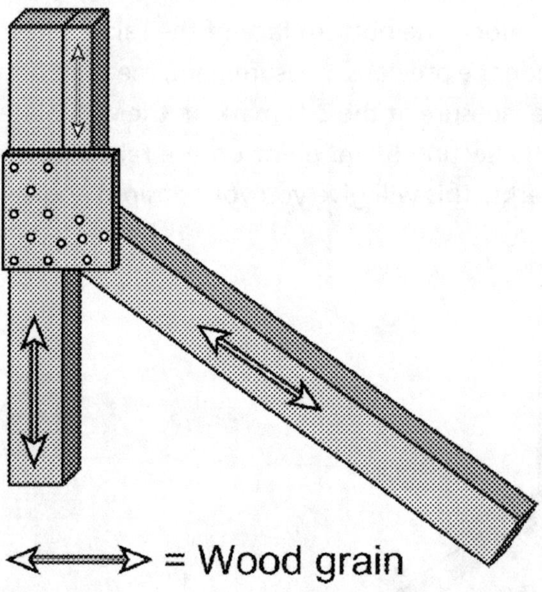

Fig. 5–10 Install a ¾ in.-thick 12x12-in. plywood gusset plate at the wall plate joint.

Bottom braces. The next step is to install a set of bottom braces. Because the flying-raker shore is temporary and needs to be maneuverable, the wall plate is much shorter than the solid-sole or split-sole raker shores. For this reason, the bottom braces should be installed down at the bottom of the wall plate. Make sure the brace extends past the raker. These braces will be 2x6s. Square up the brace to the wall plate. In this situation with the short diagonals, use your framing square. The measurement is close enough; you're not making pianos! Nail the brace to the wall plate and the raker with (5) 16d nails using the 5-nail pattern. When this is done and all the items are nailed, flip the shore over. If the raker insertion point is greater than 8 ft, you can put the bottom braces on an angle. This will help keep any deflection out of the 4x4 raker piece.

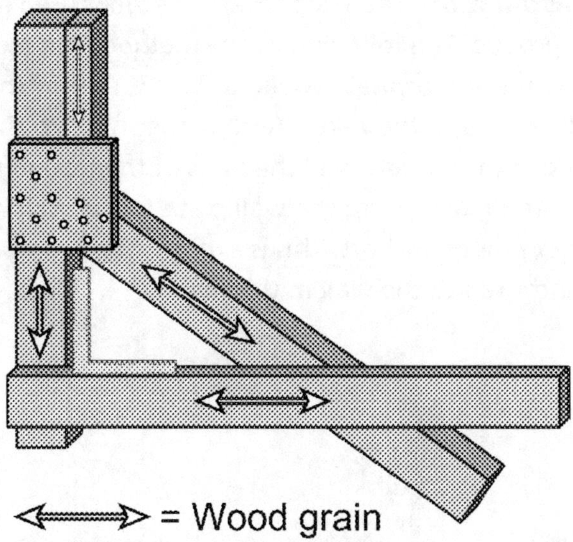

Fig. 5–11 Install the first bottom brace, making sure to keep it square.

After you flip the shore over, install the other 12x12-in. plywood gusset plate and the other 2x6 brace. Install them directly on top of the raker and in line with the other gusset and 2x6 brace.

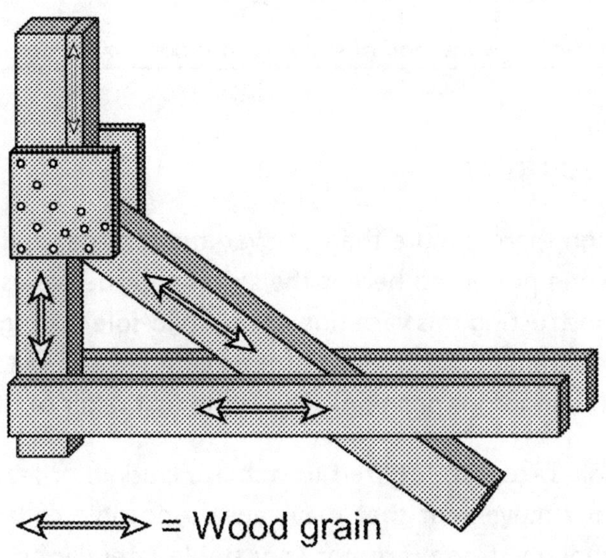

Fig. 5–12 Install the other 12x12-in. plywood gusset plate and the other 2x6 brace.

Shore installation. There are several methods used to secure the shore to the ground. The following is one method. This method works well on concrete and asphalt. Make a 2x6 trough in the sidewalk. Anchor a block behind the raker. Anchor the shore to the wall then nail the block to the trough. Nail the sides of the trough to the raker shore. Note: You must secure the wall plate to the wall with at least two ½ in. thick pins or anchors. This is a *must*! The shore will not hold unless it is anchored to the wall in this manner.

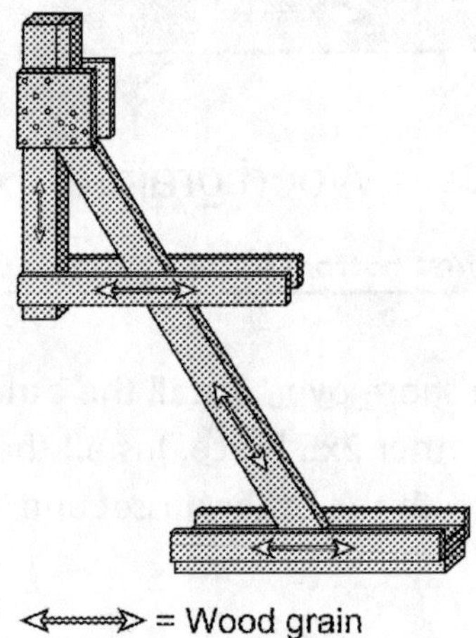

Fig. 5–13 One method of securing the shore to the ground.

Solid-sole raker

Although there is more than one variation of the solid-sole raker shore, the one presented here is the safest and the easiest to preassemble. Constructing this variation of the solid-sole raker goes a long way toward limiting the amount of time your rescue personnel have to spend in dangerous positions. This is the first raker of choice.

Solid-sole rakers are utilized to stabilize cracked or leaning walls, stopping any movement that may cause a possible collapse. These shores, although strong, are not very stable laterally; they must be erected in pairs. These pairs must be no more than 8 ft on center from each other. You will have to cross brace them with 2-in. material to keep them laterally stable. They can be installed along a wall face, using several rakers if necessary, or installed at the corners of buildings to stop the walls from racking or twisting in an earthquake situation.

Solid-sole raker step-by-step

The following is a numbered list of the tasks that must be completed to build and install a solid-sole raker shore. The remainder of this section on the solid-sole raker shore provides detailed instructions for constructing and installing the various elements of the shore.

1. Clear an area large enough to preassemble this shore; 20 ft x 20 ft would not be too big an area.

2. Determine the angle of your shore and the insertion point that you will be using.

3. Place the wall plate and the sole plate at right angles to each other, forming an *L,* and nail together.

4. Square up the plates and make your marks for the raker insertion point.

5. Anchor the corner gusset plate in position, keeping the 4x4s square.

6. Install the raker in the designated position and nail it down.

7. Install the top and bottom 2x4-in. cleats and nail them down properly.

8. Install the top and bottom gusset plates and nail them down with proper nail patterns.

9. Flip shore over, nail down the top, corner, and bottom gusset plates.

10. Install the shore into position and pressurize against anchor.

11. Pressurize raker to wall, anchor to wall, and install center diagonal braces.

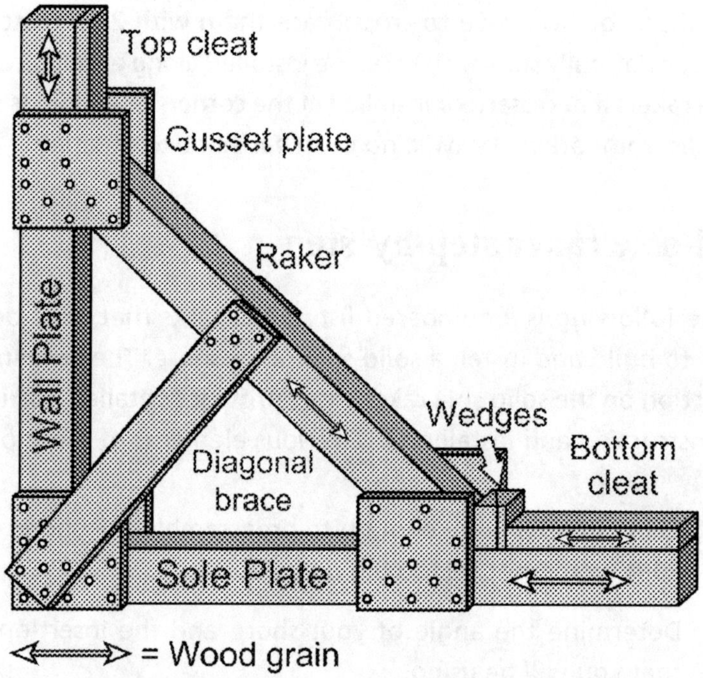

Fig. 5–14 Elements of the solid-sole raker.

Wall plate. Typically this wall plate is a 4x4 although 4x6s and even 6x6s have been utilized. In most instances, 4x4 by 12 ft long does the job. Make sure that the wall plate is at least 2 ft 6 in. longer than the designated shores insertion point.

Sole plate. The sole plate also is made of a section of 4x4, also generally 12 ft long. This piece gets nailed into the wall plate at the base and behind the wall plate.

Raker. The main support element of the shore, the raker is also typically a 4x4 or on occasion a 6x6. This section is cut to a specific length and has angles cut into both ends.

Cleats. Cleats measuring 2x4 by 24 in. are installed on top of the raker and the base of the raker. They are nailed down with (17) 16d nails each. With the use of 4x6 wall plates, 2x6 cleats are utilized.

Gusset plates. Gusset plates composed of 12x12-in. sections of ¾-in. plywood are used to lock all three connection points together. The gussets are anchored on both sides of the shore with (13) 8d nails each one in a 5-nail pattern.

Wedges. Usually either 2x4s or 4x4s, wedges are placed behind the raker and pressurized to keep the raker tight to the wall being shored.

Diagonal braces. The last items to go on the shore, diagonal braces normally are lengths of 2x6 lumber and are nailed to the outside of the raker. They are also to be nailed on both sides of the bottom corner gusset plates, using (5) 16d nails into the wall plate and sole plate.

When building a solid-sole raker, the first step is to lay down the wall plate, then butt the sole plate into the base of the wall plate. Make sure that the sole plate is behind the wall plate so that when the sole plate is anchored down, it holds the wall plate from being pushed out. To do this, lay the wall plate and the sole plate at right angles to each other. Make sure the ground is level, then butt the two together. If the ground is uneven, you can place a gusset plate under the joint if necessary to keep it flush. It is important that neither piece overlaps the other. Otherwise it is impossible to properly anchor the gusset plates. The 4x4s must be flush in order for the gusset plates to sit properly. Toenail the joint with (2) 16d nails; drive the nails flush. Drive the nails from the sole plate into the wall plate, following the grain of the wood so you do not split the lumber.

The next step is to measure and mark the sole plate and the wall plate at the point where the raker is to be. Always take your measurements from the inside joint. Because the raker is to lie inside the two plates, measurements taken from the outside will be off. It is quite a bit less confusing if you take all your measurements from this inside corner. In Figure 5–15, the insertion point is at 9 ft, and the marks are in place.

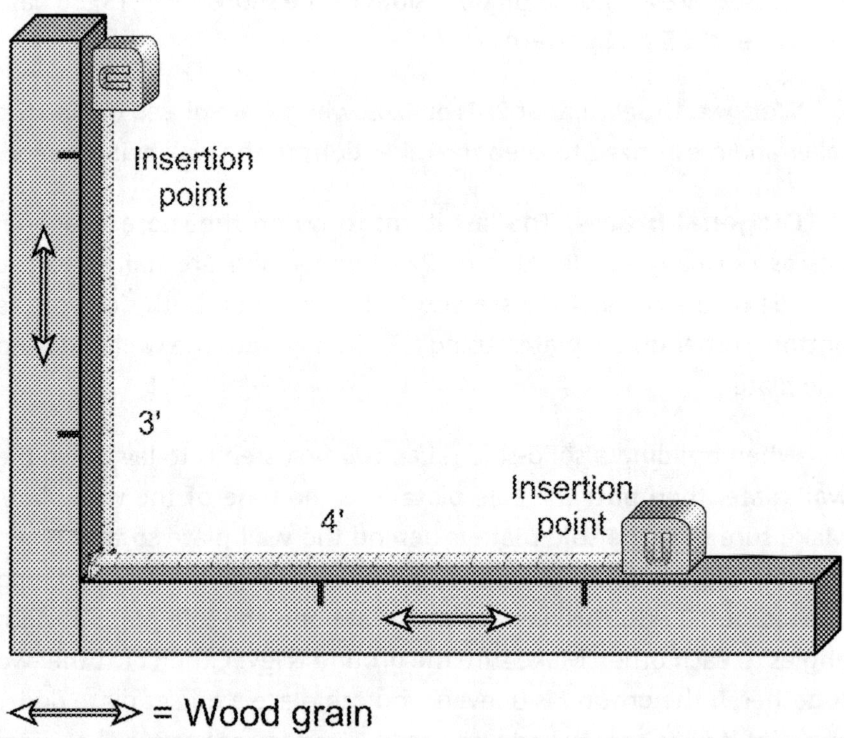

Fig. 5–15 Measure and mark the sole plate and the wall plate for raker placement.

Use the 3, 4, 5 method to square up the two plates. When you are taking the measurements for the raker, measure up the wall plate side 3 ft and measure down the sole plate 4 ft. Place a mark at each of these points. To square up the wall and sole plates to a 90° angle, measure off these two marks with the tape measure. When the space between the two marks is exactly at 60 in., the shore will be square (a right triangle).

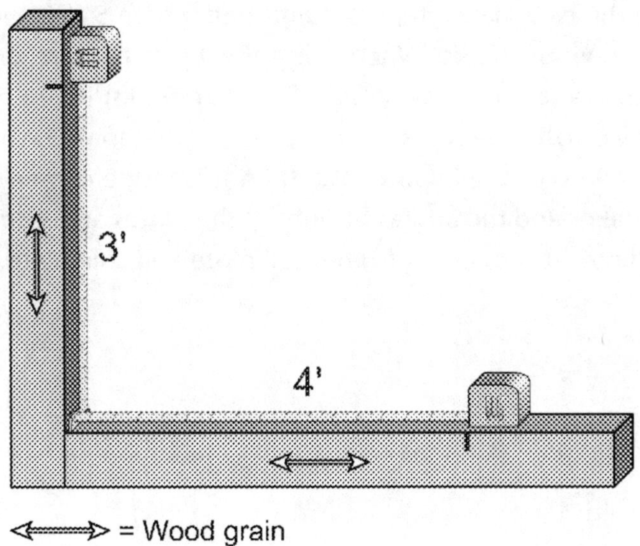

Fig. 5–16 Use the 3–4–5 method to square up the two plates.

It is very important to make sure when you start to erect the raker shore that the inside corner is square (90°). To square up the wall plate and the sole plate, toenail the two together. Measure out on one plate 4 ft and place a mark. Next, measure up 3 ft and place a mark.

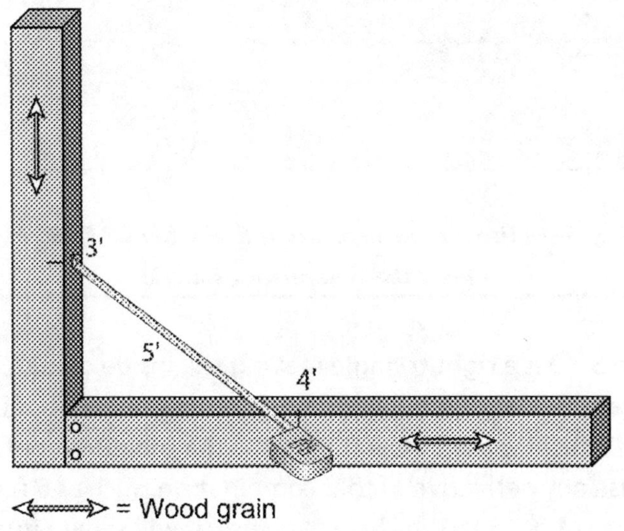

Fig. 5–17 Make sure the inside corner is square.

From the two marks, tape out and find where 5 ft (60 in.) meets the two previous marks. When using the tape measure, be careful to read the tape on the same side for both marks; if you don't, the corner will be off. When you have moved the material to where there is exactly 60 in. at the hypotenuse, then this inside corner is square. Now the raker and the angles fit right. If the plates aren't square, the raker will not fit properly into the wall plate and sole plate.

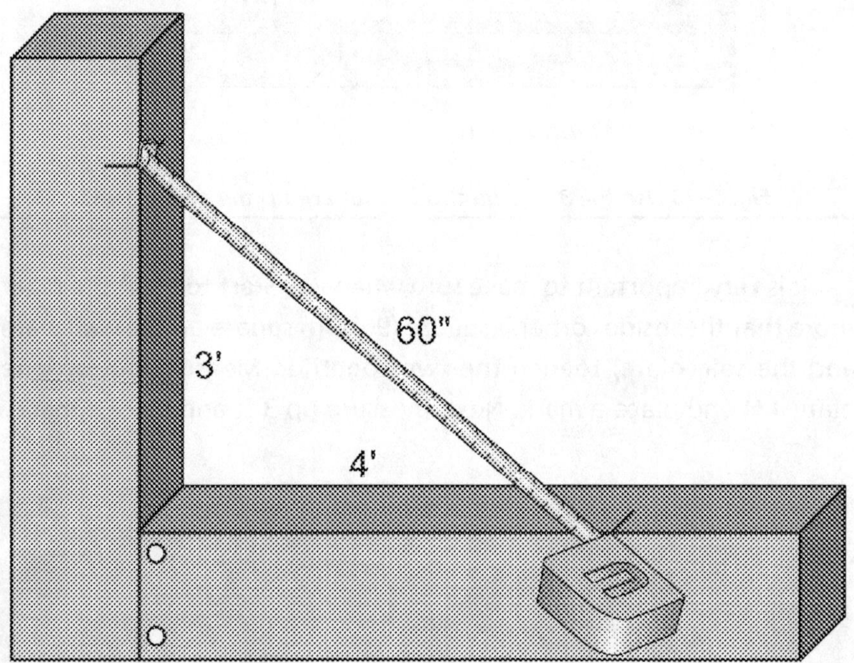

Fig. 5–18 Use a tape measure to find where 5 ft (60 in.) meets the two previous marks.

Figure 5–19 is a right triangle making the inside corner 90°. This is the *A squared + B squared = C squared* theory that most of us learned in school. Typically, use 3 ft, 4 ft, 5 ft for the measurements, but you can also use any derivative of that combination, such as 6 ft, 8 ft, 10 ft or even 9 ft, 12 ft, 15 ft. Whichever you use, it will square up the inside corner of the shore, making all the angles fit properly.

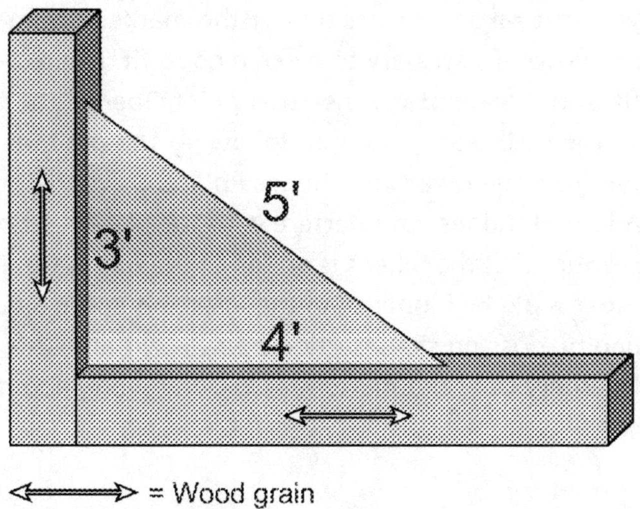

Fig. 5–19 A right triangle making the inside corner 90°.

After the plates are squared up, gusset plate the joint with a 12x12 ¾-in. plywood gusset plate. Use a 5-nail pattern, eight nails on the wall plate and five nails on the sole plate.

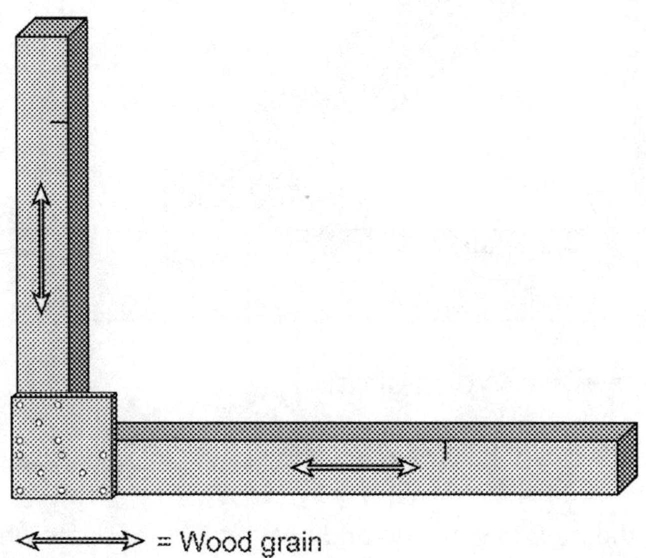

Fig. 5–20 Gusset plate the joint with a ¾ in.-thick 12x12-in. plywood gusset plate.

Lay your cut raker into position at the marked points; slide the raker up or down if necessary to make a good fit. Figure 5–21 shows a good fit of the raker at the insertion point. Toenail the top of the raker into the wall plate. One nail following the grain of the raker gets driven from the raker into the wall plate. Drive the head of the nail flush so that it does not interfere with the placement of the gusset plate. Notice that the raker sits on top of the sole plate. The raker's actual height is 3½ in. higher than the insertion point. Since you are not making pianos and the insertion point is within a 2-ft zone, the height difference does not matter at all. Don't worry about it!

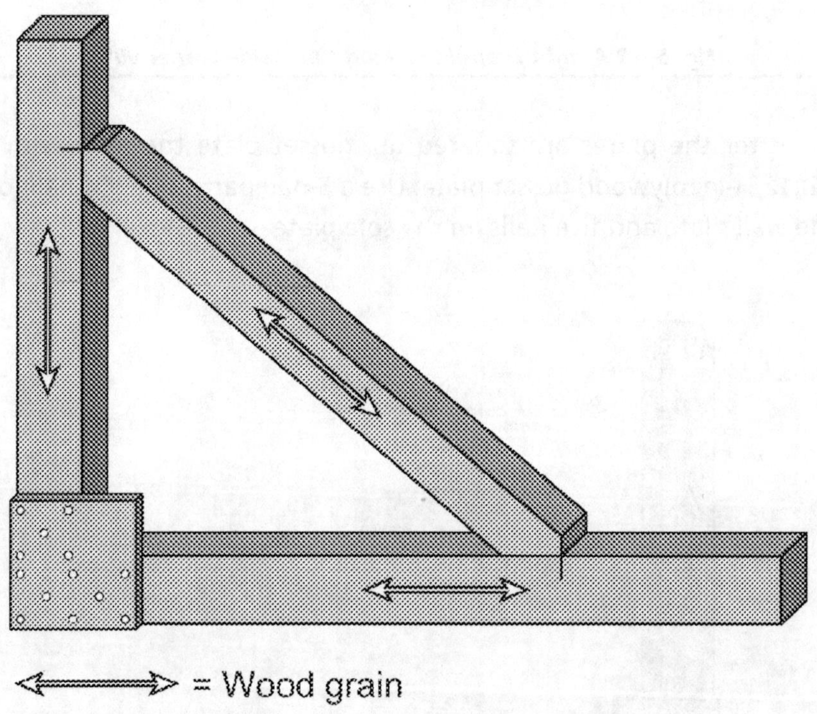

Fig. 5–21 A good fit of the rake at the insertion point.

Start the nails in your top and bottom 2x4-in. cleats while waiting for the raker to be cut. Doing so makes it much easier to anchor the cleat to the wall plate when you are nailing by hand. Nail the cleats in position when you're satisfied that the raker is fitting well. Nail the top cleat in position first, using (17) 16d nails in the 5-nail pattern. When nailing the bottom cleat, make sure you leave a space equal to the thickness of the wedges. If you are utilizing 2x4-in. wedges, leave a 1½-in. gap between the base of the raker and the face of the 2x4-in. cleat.

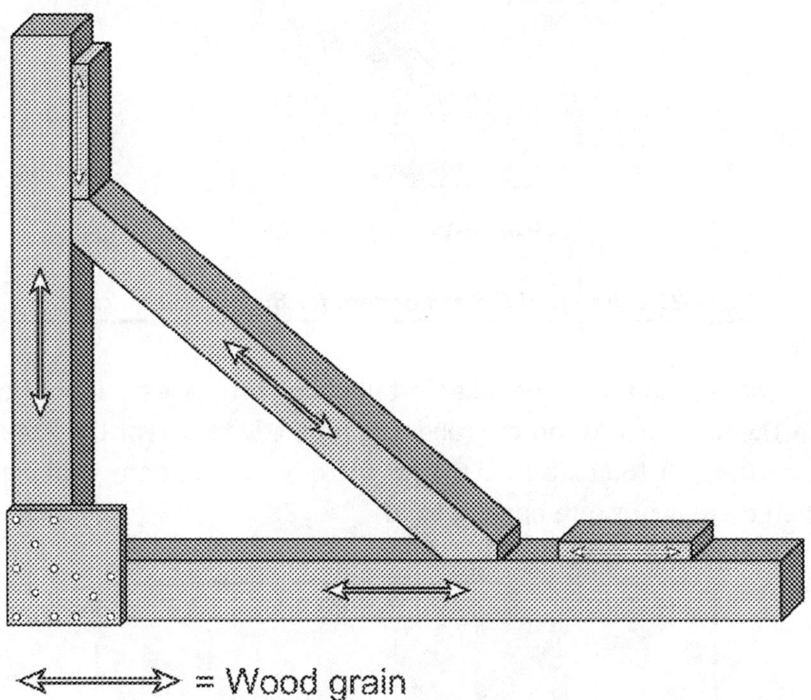

Fig. 5–22 When nailing the bottom cleat, leave a space equal to the thickness of the wedges.

Figure 5–23 shows a close-up of the top 2x4-in. cleat. Use (17) 16d nails in the 5-nail pattern anchor to the wall plate at the raker insertion point and keep it flush with the face of the plate.

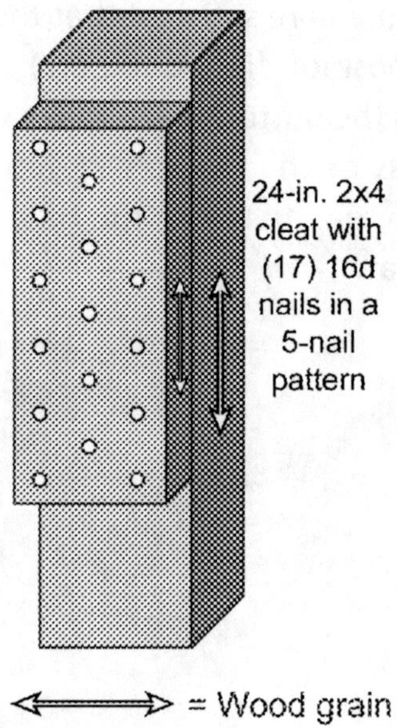

Fig. 5–23 Close-up of the nail pattern for the top 2x4-in. cleat.

Figure 5–24 shows the raker installed under the cleat. Notice how the 1½-in. return cut on the raker sits perfectly to accept the 2x4-in. cleat. The (17) 16d nails hold the raker from sliding up the wall plate when there is pressure applied to it.

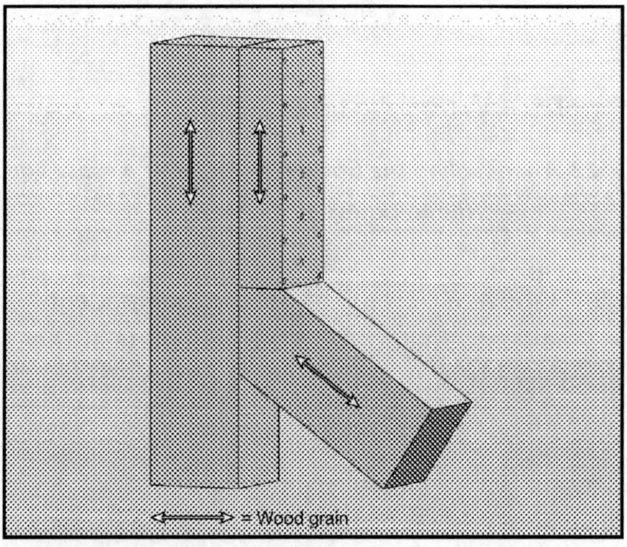

Fig. 5–24 Close-up of the raker installed under the cleat.

After the cleats are nailed, finish this side of the raker by anchoring the top and bottom gusset plates. Use the same nail pattern as the other gusset plate, 8 and 5 nails.

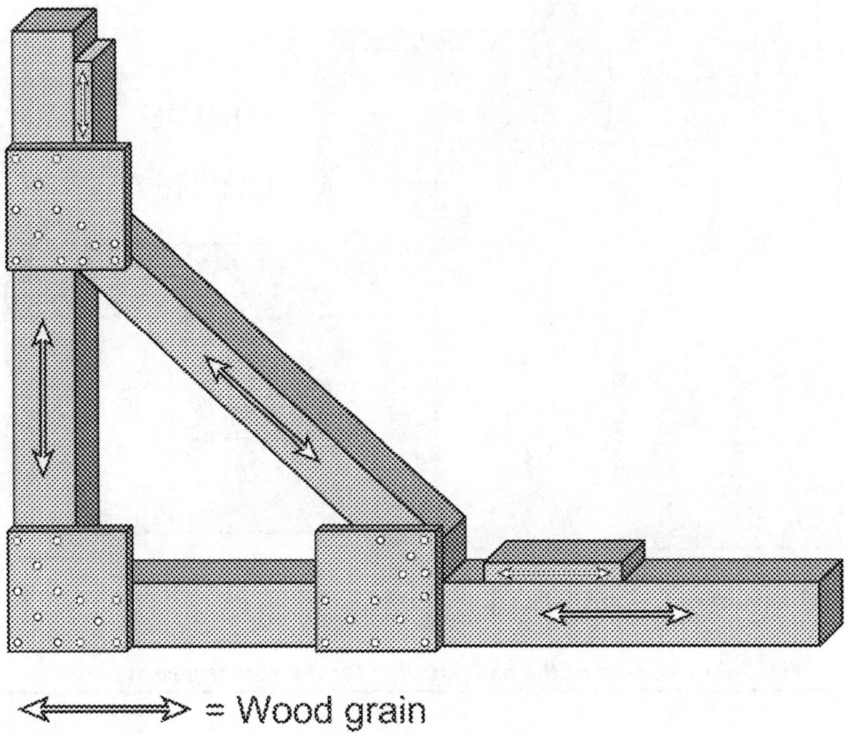

Fig. 5–25 After installing the cleats, finish this side of the raker with top and bottom gusset plates.

Figure 5–26 shows a close-up of corner gusset plates. Notice the 8-nail and 5-nail patterns. The gussets must be even with or slightly inside the face of the 4x4s, or the plywood may keep the shore from full contact with the wall or the ground or both. Place the 8-nail section on the wall plate and the 5-nail section on the sole plate.

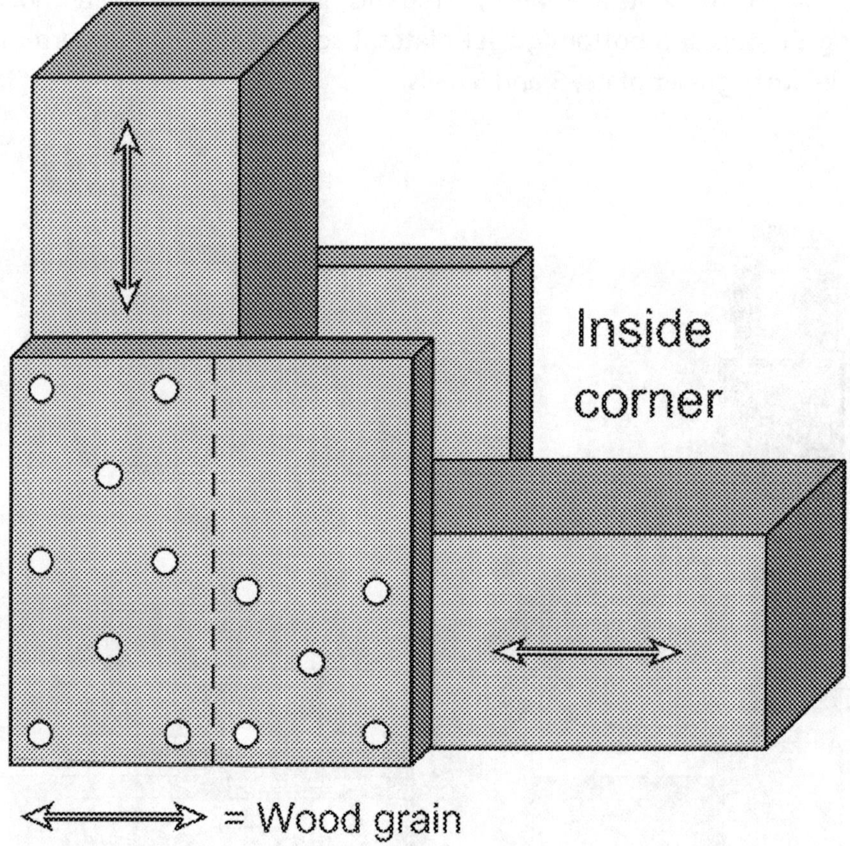

Fig. 5–26 Close-up of the nail patterns for the corner gusset plates.

Figure 5–27 shows the area where the top gusset plate is to be placed. As you can see, the gusset plate covers the entire joint and several inches above it. Lap this joint with the gussets in order to lock the joint in position so it cannot move or separate. A typical gusset plate with the proper nail pattern measures 12x12 and is secured by (8) 8d nails on one side and five on the other. As a rule of thumb with the 45° angle, you can place the corner of the top gusset plate in the center of the 4x4-in. raker. This will place the gusset plate about 2 in. above the end of the raker—right where you want it.

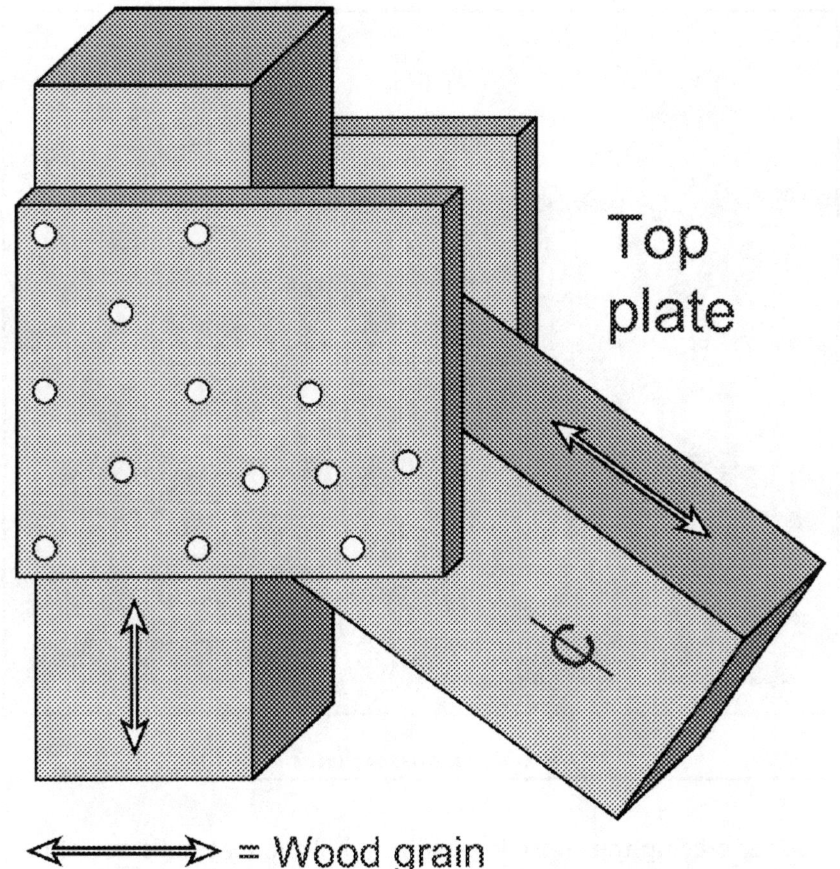

Fig. 5–27 Close-up of the nail patterns for the top gusset plates.

Figure 5–28 shows the bottom gussets in position. Leave the plate forward about an inch off the back of the raker. Nail the gusset into the raker, using the 5-nail pattern with five nails. When assembling the shore, put two nails into the bottom of the gusset by the sole plate. They will be pulled out when you pressurize the wedges against the raker.

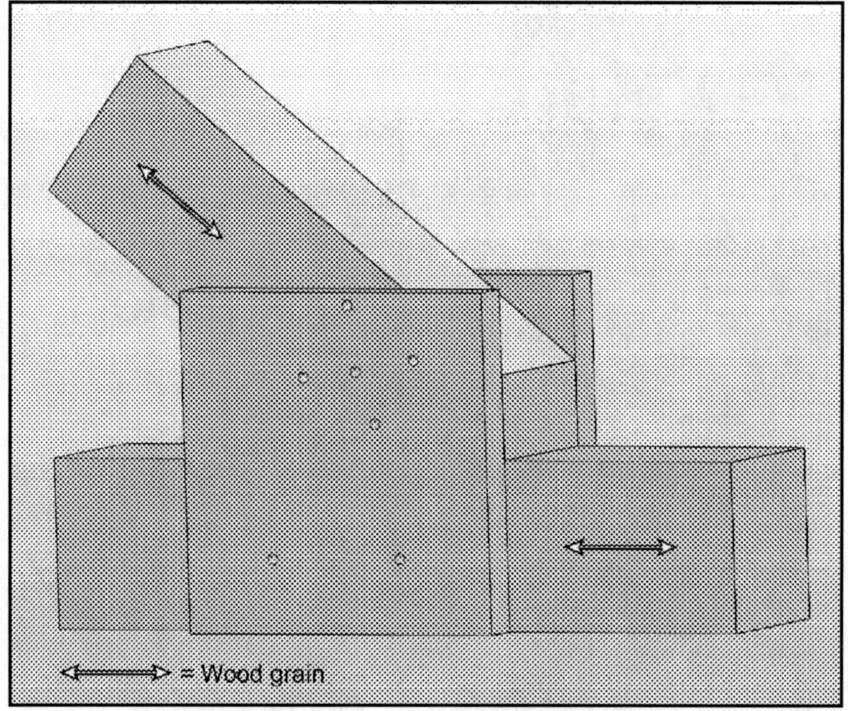

Fig. 5–28 The bottom gusset plates in position.

After placing the shore in position, pull the two nails out of each side at the sole plate and tighten up the wedges to the raker. After the shore has been set, nail it, using the regular 5-nail pattern shown in Figure 5–29 with the eight nails. Remember to make sure the gusset plates don't pass or overlap the bottom face of the 4x4. The sole plate must have full contact with the ground.

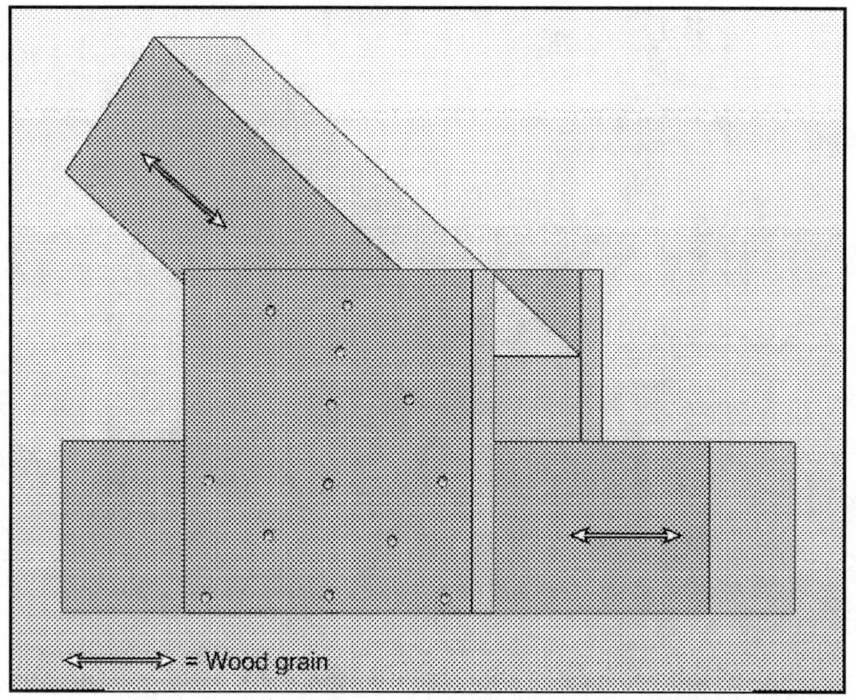

Fig. 5–29 Use the regular nail pattern on the gusset plates once the shore has been set.

At this point, flip the shore over and anchor the gusset plates on this side. The exterior raker shore joints must be gusset-plated on both sides. By doing this, you lock the joints together, helping to make the shore more efficient. Nail the gussets directly over the existing ones, covering the raker joints; use the same number of nails and the same 5-nail pattern.

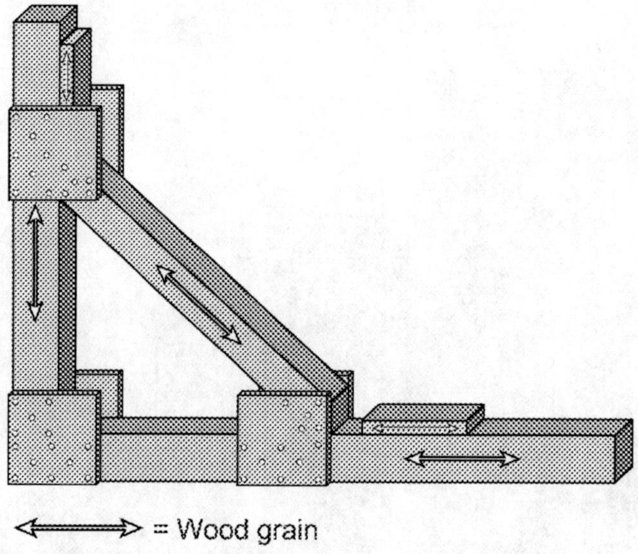

Fig. 5–30 Flip the shore over and install gusset plates on the remaining side.

The 45° solid-sole raker shore is now complete. After you carry the shore into position and place it on the ground, pressurize it against the anchor block system you installed. Then install the wedges on the raker and pressurize them to finish off the shore.

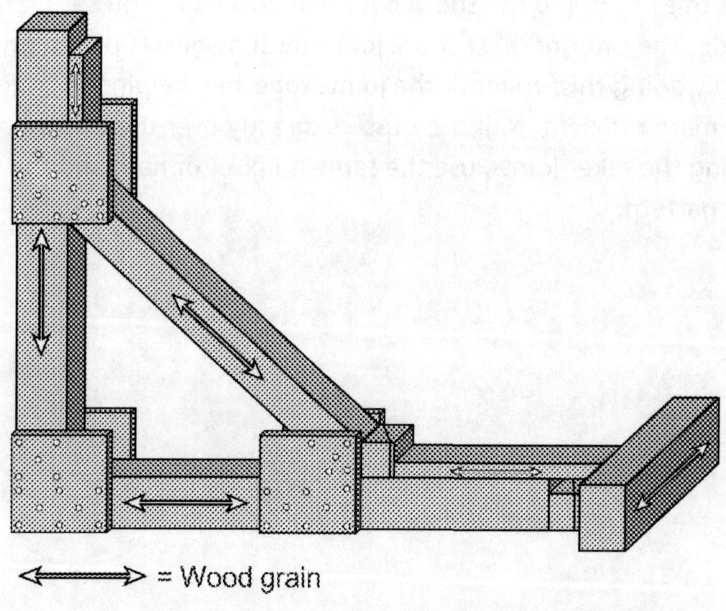

Fig. 5–31 The completed 45° solid-sole raker shore.

Figure 5–32 is a close-up of the wedges against the bottom cleat and pressurized up against the back of the raker base. These wedges must be tight. They pressurize the raker against the building, enabling the raker shore to transfer the load to the ground.

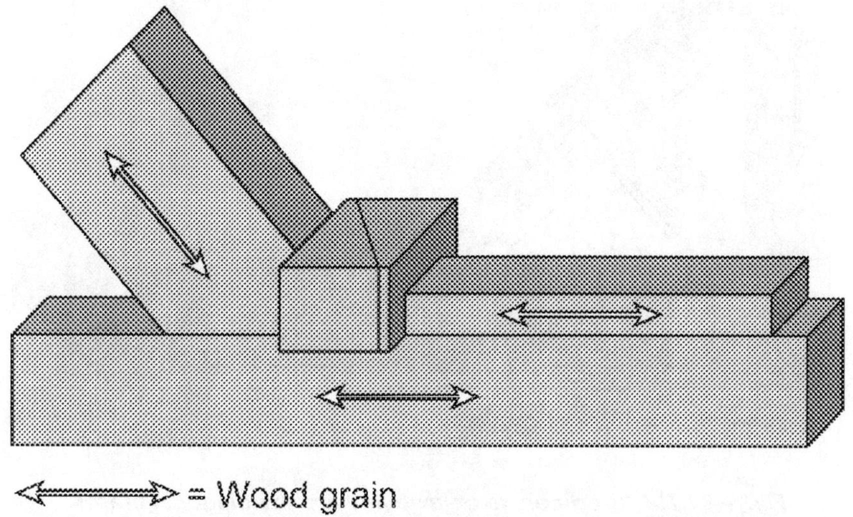

Fig. 5–32 Close-up of the wedges against the bottom cleat, pressurized up against the back of the raker base.

The last pieces to go on are the diagonal braces. Place two 2x6s on top of the corner gusset plates and right against the raker itself. Set the braces from the bottom corner into the center of the raker. These braces take out any possible deflection of the raker when it is under load, thereby increasing the efficiency of the raker substantially. Use the 5-nail pattern and utilize 16d nails as in Figure 5–33.

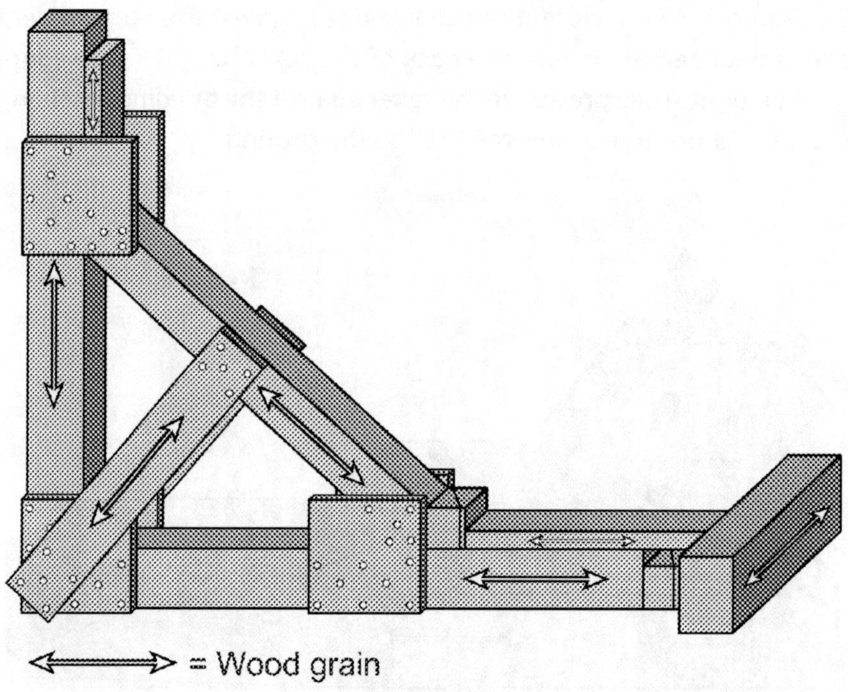

Fig. 5–33 The last pieces to be installed are the diagonal braces.

When cutting the raker, use the number 17 as a multiplier. The result is the exact length needed for the total hypotenuse of the right triangle. Because the shore starts from the ground and the insertion point goes straight up, the exact measurement is from the ground. This is the A point. However, because you wind up placing the raker on top of the sole plate, you actually elevate the raker 3½ in. higher than the exact insertion point (point B). This is fine because, as said before, you are not making pianos! The insertion point has an acceptable range. If your team wants to make it exact, it can do so.

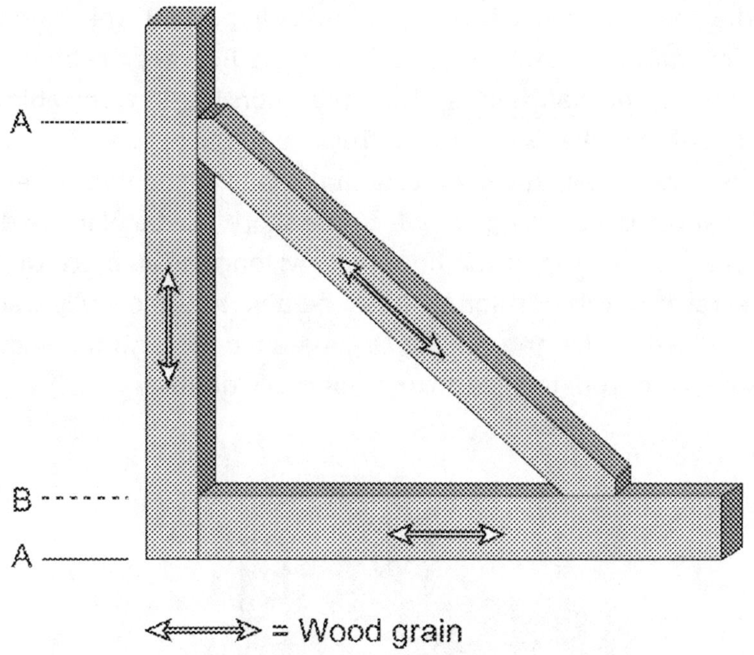

Fig. 5–34 Exact raker lengths for 45° rakers.

The following list shows some of the most common wall insertion points calculated by multiplying the number 17 by the wall height in feet. The result is the length of the raker in inches. To measure the raker length, begin at the outside edge of the angle cut (the longest point) and measure to the opposite end of the raker. The outside edge of the angle will be placed and cut to fit at this point. The measurement will be exact.

If you want to be exact with the measurements for some reason, that's ok, you can do it easily. All you have to do is deduct 5 in. from the length of the raker, and the wall height face of the raker will be exact. And you have your piano!

Raker Lengths for 45° Rakers

- 6 ft — 102 in.
- 7 ft — 119 in.
- 8 ft — 136 in.
- 9 ft — 153 in.
- 10 ft — 170 in.

Exact Raker Lengths for 45° Rakers

- 6 ft — 97 in.
- 7 ft — 114 in.
- 8 ft — 131 in.
- 9 ft — 148 in.
- 10 ft — 165 in.

The insertion point of your raker is very important. The raker must transfer and support the loads applied to the floors of the building in question. As the wall starts to lean, the floors become unstable. The purpose of the raker is to redirect the load to the ground. In order to do this, it must have full contact with the wall at the floor level. You do, however, have a range in which the shore can be placed—down within 2 ft of the top of the floor joist. As long as you place the face of the raker in this position, you will be able to successfully transfer the overload to the ground. For this reason, deducting the width of the sole plate from the exact cut is not really necessary.

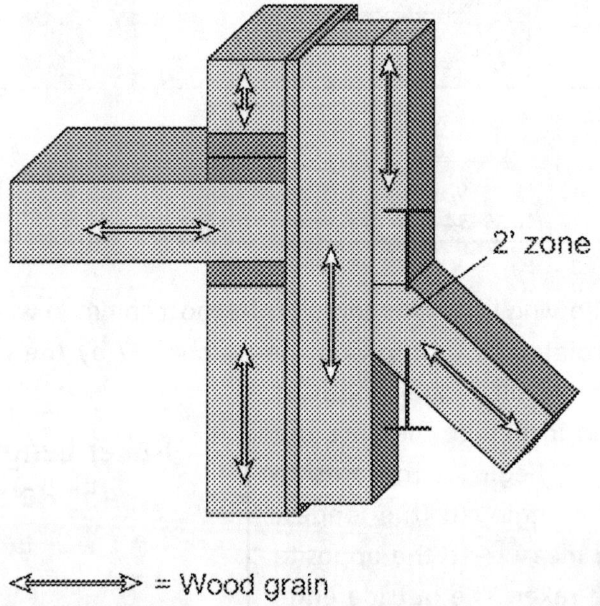

Fig. 5–35 The acceptable range for shore placement—
within 2 ft of the top of the floor joist.

Figure 5–36 shows a view of the raker shore at the proper wall height. The raker face is just even with the face of the floor beams.

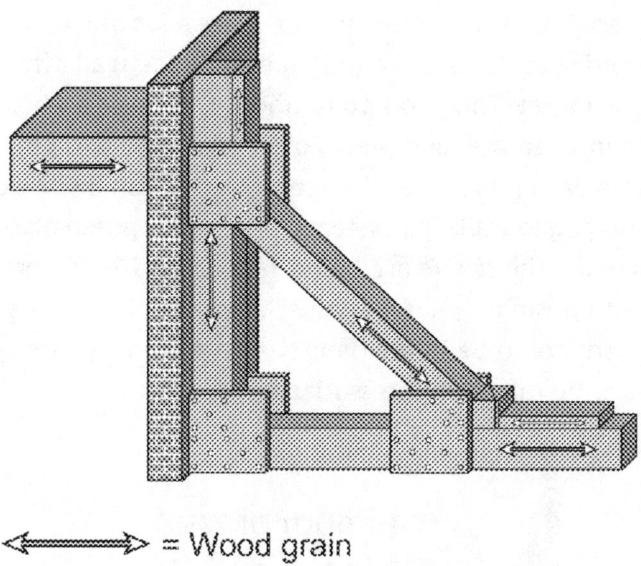

*Fig. 5–36 The raker at the proper wall height,
even with the face of the floor beams.*

Raker shore anchoring methods

Based on a number of factors—the material available, the type of shores you construct, the ground conditions, and your team's experience—you can choose one of many methods of anchoring your raker shores. There are several uncomplicated techniques described here. In most cases if the anchoring you are using can resist a force of 3000 lb, you will be fine.

This is probably the most common method used out in the field. This method can be used on bare ground as well as on asphalt or concrete. The raker is backed up by a piece of lumber at right angles to the raker. The lumber can be the same size or larger than the raker. A minimum of two 1-in. steel pins are driven into the ground or street behind the anchor block, roughly 12 in. apart. This is necessary to have the proper amount of resistance needed to keep the raker from backing off the wall.

Between the raker and the anchor, place a set of wedges that will be tightened up to pressurize the raker into the wall after the steel pins have been set. The good point about this raker anchor system is that it can incorporate long pieces of lumber with two pins driven in place where each raker goes. This construction can all be done away from the damaged wall. This system can be safely put in place on most occasions while the rakers are being fabricated. The rakers are then brought into position and set against the finished anchor system, allowing personnel to be in the danger area for a very short period of time before the rakers are pressurized.

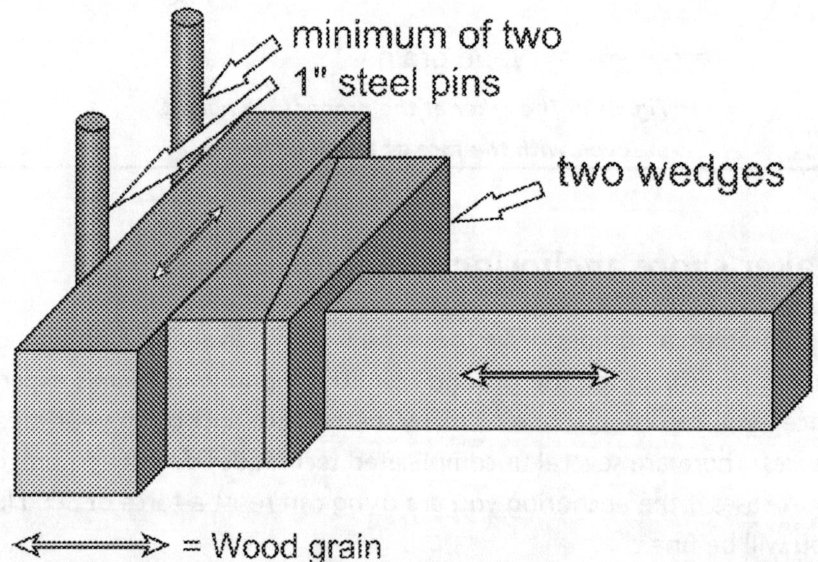

Fig. 5–37 Place a set of wedges between the raker and anchor, to be tightened after the pins are set.

In an earthquake situation where aftershocks are common, this variation is better suited for your anchoring. Driving the pins through the lumber (after predrilling the holes) prevents the anchor from vibrating loose with small aftershocks. Drive these pins roughly 12 in. apart for best distribution of the load.

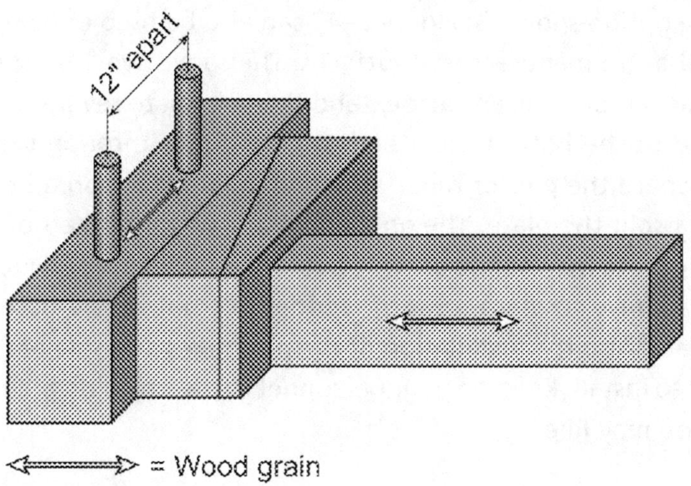

Fig. 5–38 In earthquake situations, drive the pins through the predrilled holes.

Figure 5–39 shows an anchor variation that can be used in hard manufactured surfaces such as concrete, or asphalt; typically streets and sidewalks. Place the anchor block and wedges behind the raker the same way you would place your pins. However, instead of using two steel pins, use other fasteners that are designed to hold lumber to hard surfaces. Place a 2-in. piece of dimensional lumber, usually about 2–3 ft long (the minimum should be 2 ft). The width of the wood should be 10–12 in., and its thickness should be 2 in. There are many types of anchor bolts on the market today, and you can use whatever one you want as long as you use enough of them. If you are going to use shots and pins, you need at least ten. Two rows of five will be fine.

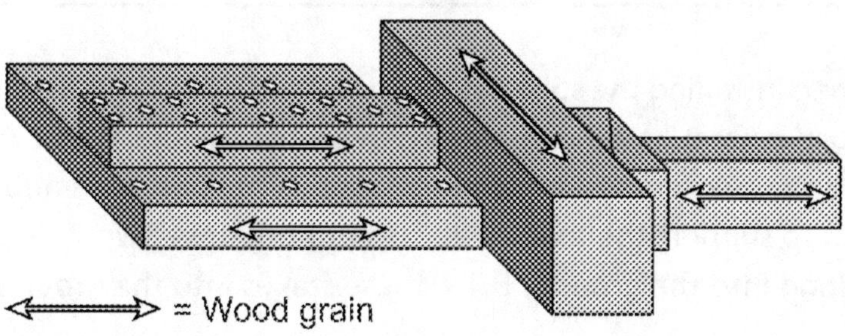

Fig. 5–39 An anchor variation for hard manufactured surfaces.

The option shown in Figure 5–40 can also be used either on bare ground or on manufactured surfaces. The same two 1-in. steel pins or rebar can be utilized. Just extend the sole plate several feet past the end of the bottom cleat and drive the pins through predrilled holes. Space the pins or rebar about 12 in. apart to spread the load and not split the plate. The only drawback with this method is that the sole plate may not be as tight against the wall as you would like. You may have to place a set of wedges in front of the base of the sole plate. Also the installation of the pins takes a couple of minutes longer to install, keeping your personnel in the unstable area longer than you may like.

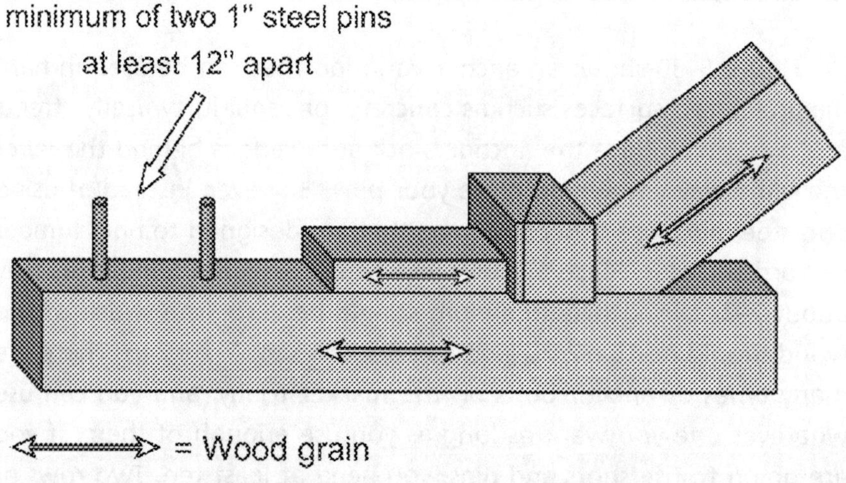

Fig. 5–40 Another anchor variation for bare ground or manufactured surfaces.

When installing the solid-sole raker system directly on the ground, there are several anchoring options available to your team. You can still use the two-pin system if the ground is hard enough and stable enough to support the load. With this option, drive two 2x4s at least 36 in. long into the ground. Drive these stakes into the ground as far as they will go; do not pass the top of the anchor block. If the stakes are difficult to drive all the way down, then they should be driven down enough to hold.

On occasion the soil may be a little soft or loose. In this case, drive two sets of stakes into the ground, keeping them about 3 ft back from the first set. Depending on the soil conditions, these stakes can be 2x4 or 2x6. After you have driven the second set into the ground, place two 2x4 braces from the front set to the back set. Place them as shown in Figure 5–41. The braces should contact the two front stakes right at the top of the anchor block. The back of the 2x4 braces should contact the second set of stakes at the base by the ground (See Fig. 5–42). Nail both ends with at least (2) 16d nails.

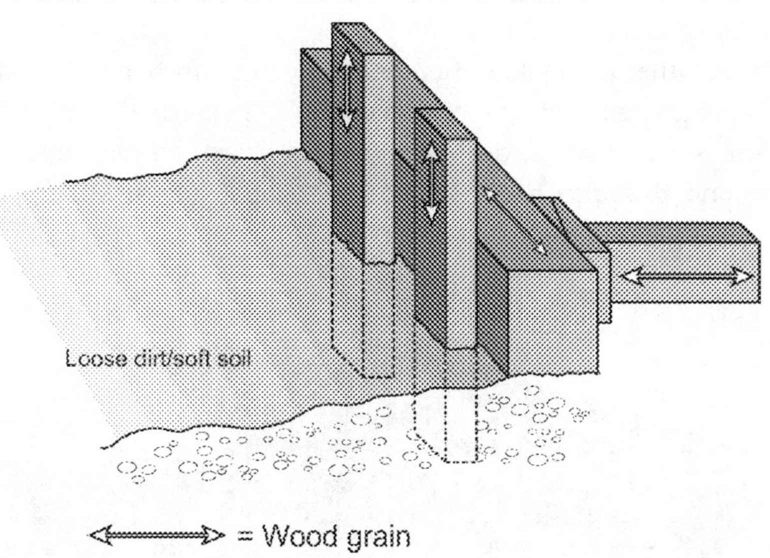

Fig. 5–41 The braces should contact the two front stakes at the top of the anchor block.

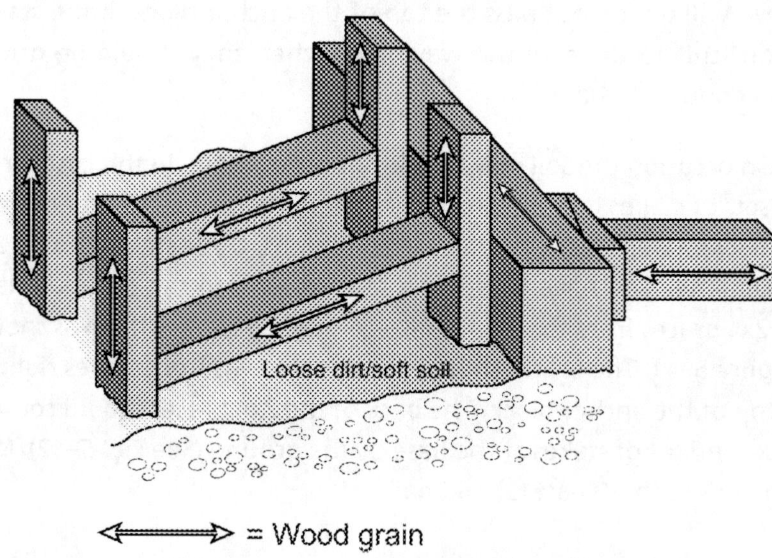

Fig. 5–42 The back of the 2x4-in. braces should contact the second set of stakes at the base by the ground.

Another available option enables you to anchor larger rakers to concrete or asphalt. This option uses a 36-in. piece of steel with three 1-in. holes on each side. When you place 1-in. anchor bolts into the ground, this raker has an anchor block capacity that will support any load applied to the larger rakers.

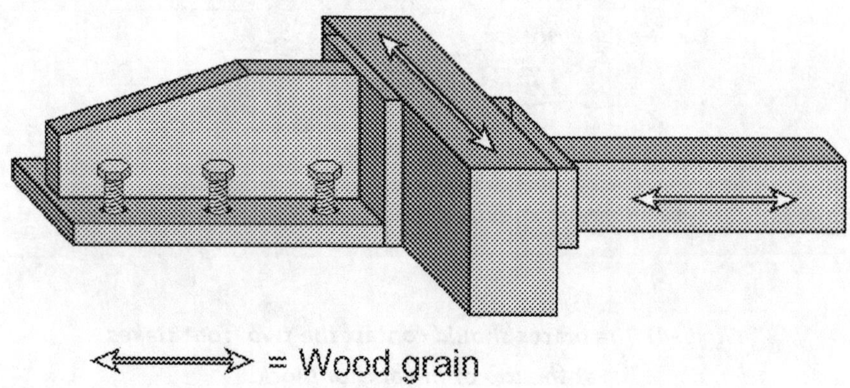

Fig. 5–43 An option for anchoring larger rakers to concrete or asphalt.

Many times when you place a raker against the wall and use an anchor system, you do not wind up back that far from the base of the raker in reference to the base of the sole plate. When this is the case, extend the bottom cleat to the anchor block itself. By doing this, you have solid contact with the cleat, which is better; and you do not have to place all those nails into it. In this case the block is holding the pressure, not the nails. All you would need is just (5) 16d nails set into the whole cleat to keep it anchored down to the sole plate.

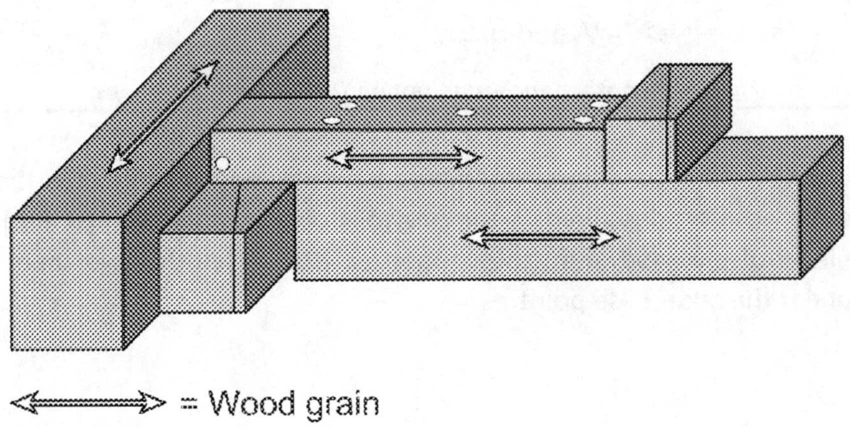

Fig. 5–44 Extend the bottom cleat to the anchor block itself as an alternative to heavy nail placement.

Figure 5–45 shows another method to help support the raker in soft soil conditions. If you feel the ground is a bit soft, to be safe, pad out the base of the raker. Along the bottom face of the raker where it contacts the ground, place three 3-ft long 2x6s under the shore. This will help redistribute the weight over a larger area, giving the shore more resistance force.

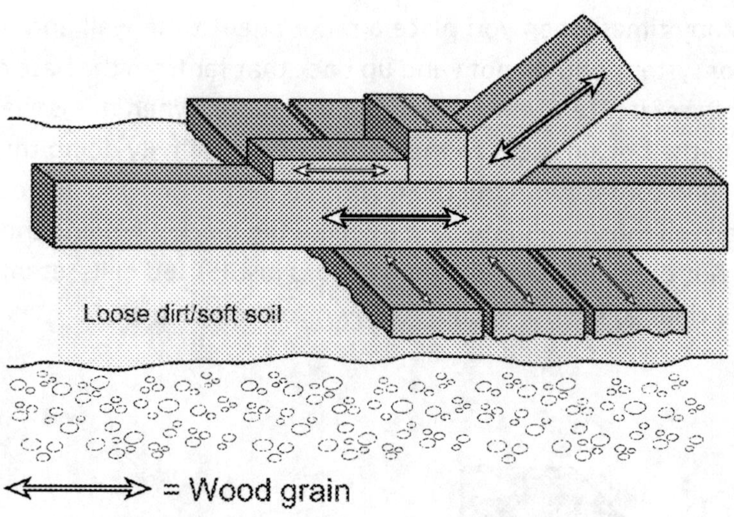

Fig. 5–45 Another anchoring option for soft soil conditions.

Along the same lines, this next option also will work fine. In the same position described in the previous option, nail two 18x18-in. pieces of plywood ¾ in. thick together. Now place the two sheets under the raker base point.

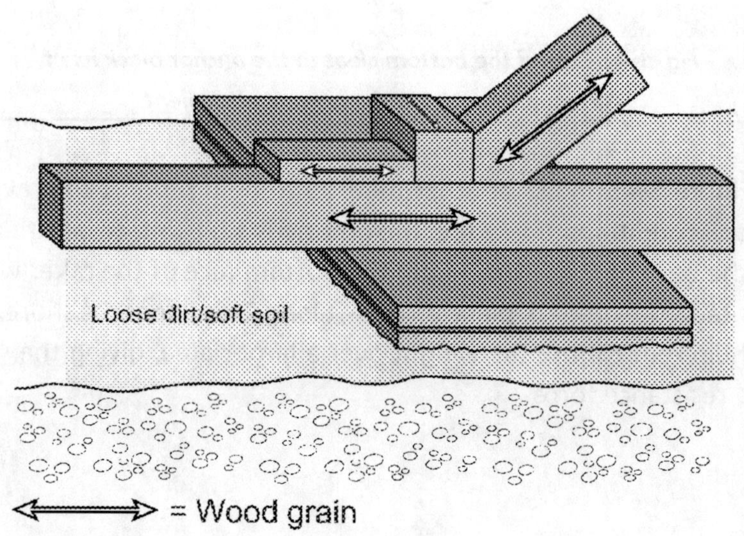

Fig. 5–46 A variation of the previous option for soft soil conditions.

The raker also must be anchored to the wall in order to resist any upward force being applied to the shore. Typically two ½-in. pins are all that are necessary to secure the raker to the wall. They can be placed anywhere within the wall plate. Generally, drill two holes into the wall plate and wall and insert the pins. The bolts don't have to be very tight because as the shore is pressurized, the pins lock up. You can use anchor bolts, *J* bolts, or rebar for the pins.

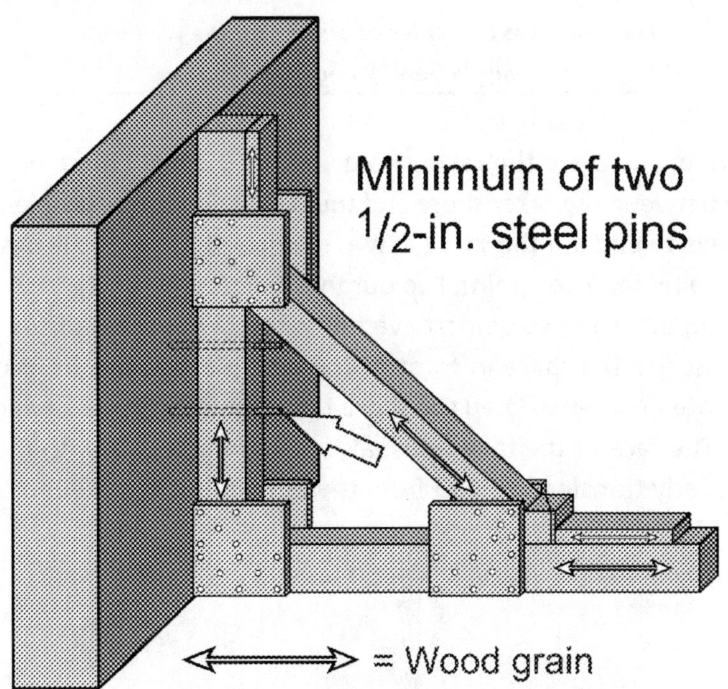

Fig. 5–47 Typically two ½-in. pins are all that are necessary to secure the wall plate.

Figure 5–48 shows two more examples of ways to anchor the wall plate. On the left is a section of angle iron on the top and an angle piece made from two 2x4s nailed together that can be anchored or nailed to the wall and the wall plate. On the right is a section of plywood that is nailed to the wall plate before installation. Bear in mind, the plywood needs to be secured by a minimum of (12) 8d nails into the wall plate to be effective.

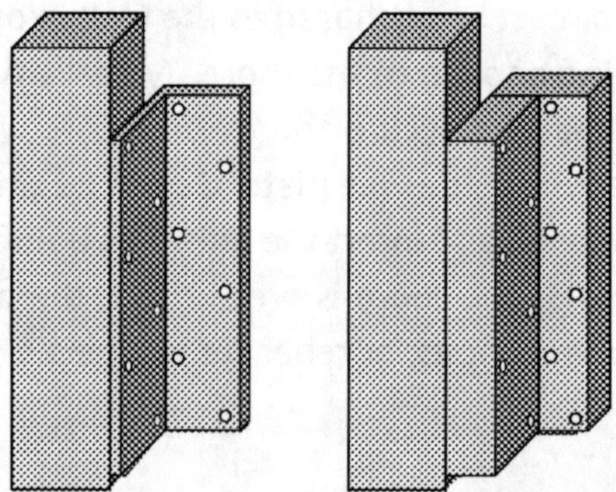

Fig. 5–48 Two examples of anchoring the wall plate; angle iron (l), angle 2x4 (r).

When you place the raker shore against the wall, there may be a space between the raker shore and the wall. This is not a problem; you will have to pad out the shore in two important spots: at the base and at the raker insertion point. Pad out these spots with wedges or cleats, ensuring that the two points have full contact with the wall. You have to pressurize the space in front of the wall plate at the front of the sole plate in order to keep the shore tight against the wall, which is a *must*. The face of the raker must also have full contact with the wall to properly transfer the load from the building to the ground.

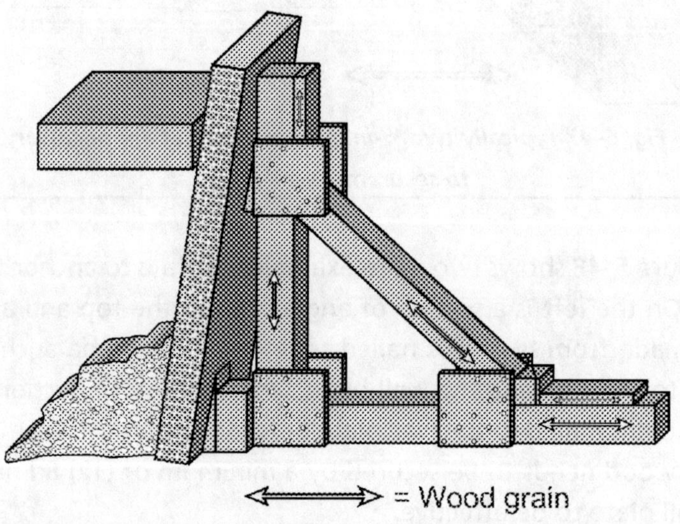

Fig. 5–49 The face of the raker must have full contact with the wall.

There is another way your team can adjust the raker so it fits flush with the leaning wall. If your team decides to use this method, leave off the bottom cleat temporarily. Place the shore in position and pull out the two nails on each side of the bottom gusset at the sole plate. Now the raker bottom is free to move. Pull back on the raker; it will slide and bend the wall plate slightly, which is fine. Push the sole plate forward as you pull on the raker. The raker can come back as much as 10 in. if necessary.

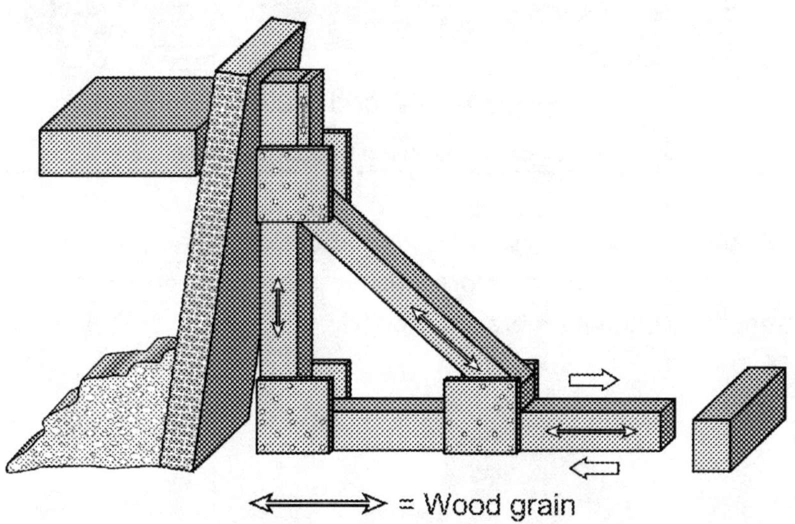

Fig. 5–50 This method allows the raker to travel up to 10 in. if necessary.

In Figure 5-51, the wall plate is flush with the wall. Now pad out the sole plate to the anchor and pressurize your raker as you normally would. Many firefighters prefer this option to the one shown in Figure 5–50.

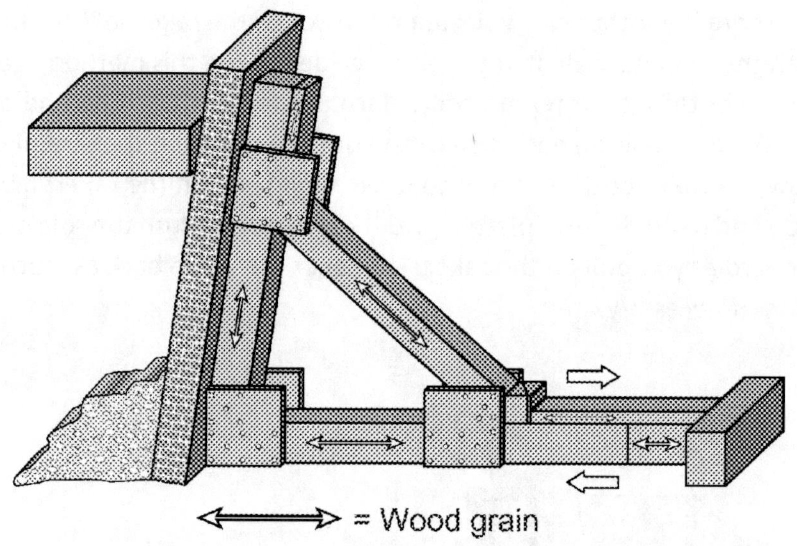

Fig. 5–51 An option involving padding the sole plate.

Figure 5–52 shows a set of pneumatic rakers against a wall. These can be assembled much more quickly than wooden ones and are extremely strong when braced properly.

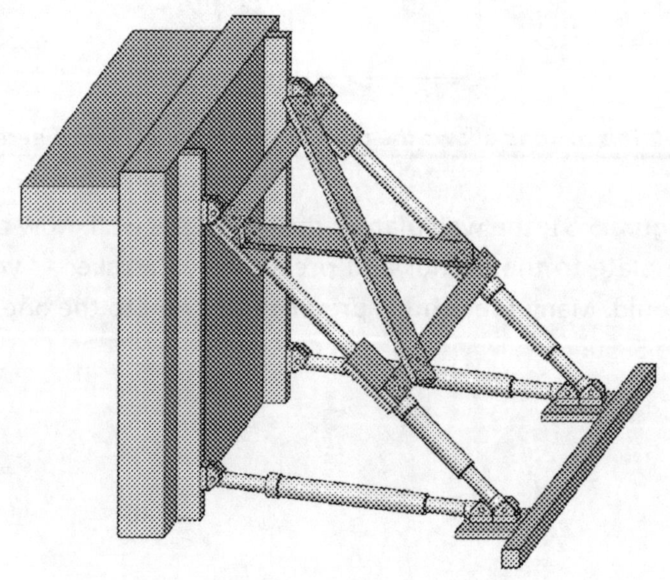

Fig. 5–52 Using pneumatic rakers in place of lumber.

Split-sole raker shore

Split-sole rakers generally should be placed into the ground at a 60° angle for safety reasons. This will cause more of the force of the raker to bear directly into the ground, while having less lateral force pushing against the soil, possibly causing the shore to move. Always think about the available options your team has. Size-up is the most important part of any shoring operation. The more effective and thorough it is, the more it reflects on the overall efficiency of the shoring system. To eliminate any possible movement problems that may occur in soft ground, always keep the raker angles imbedded into soil at a 60° angle.

Split-sole raker step-by-step procedure

The following is a numbered list of the tasks that must be completed to build and install a split-sole raker shore. The remainder of this section on the split-sole raker shore provides detailed instructions for constructing and installing the various elements of the shore.

1. Clear an area large enough to assemble the shore; 20 ft x 20 ft would not be too large.

2. Determine the insertion point of the raker and the length of the wall plate.

3. Prefabricate the anchor and dig the anchor hole in the soil at the appropriate position.

4. Attach the 2x4-in. top cleat to the top of the wall plate.

5. Cut the angle into the wall plate, generally 60° if inserting into the ground. Install the raker to wall plate at the base of top cleat.

6. Make sure the angle fits properly and install a 12x12-in. gusset plate.

7. Install one bottom 2x6-in. cross brace to hold the shore together.

8. Make sure the angle of the raker remains 60° at intersection of the wall plate.

9. Flip the shore over, install the other gusset plate and bottom cross brace.

10. Install the raker in position and anchor it to wall.

11. Pressurize the base of raker to accept weight of wall.

Wall plate. A wall plate typically is a 4x4. In most instances, 4x4 by 12 ft long will do the job. Make sure that the wall plate is at least 3 ft longer than the designated shore's insertion point. In many cases, the wall plate is placed above a large amount of debris, in which case, the wall plate has to be cut.

Raker. The main support element of the shore, a raker is typically a 4x4. It is cut to a specific length and has angles cut into both ends or in just one end, depending on the anchor situations.

Cleats. Cleats measuring 2x4 by 24 in. are installed on top of the raker and nailed down with (17) 16d nails each.

Gusset plates. Gusset plates composed of 12x12-in. sections of ¾-in. plywood are used to lock the top connection points together. The gussets are anchored on both sides of the shore with (13) 8d nails, using the 5-nail pattern.

Bottom braces. Bottom braces are the main difference between the solid-sole and split-sole rakers. Use two 2x6s nailed down on the bottom of the wall plate, as far down the raker as practical, and one 2x6 placed on each side. Anchor them with (5) 16d nails into the wall plate as well as into the raker.

Wedges. Wedges are usually either 2x4s or 4x4s. They are placed behind the raker and pressurized to keep the raker tight to the wall.

Diagonal braces. The last items to go on the shore, diagonal braces normally are lengths of 2x6 lumber and are nailed to the outside of the raker. They are nailed on both sides of the bottom wall plate with (5) 16d nails into the wall plate.

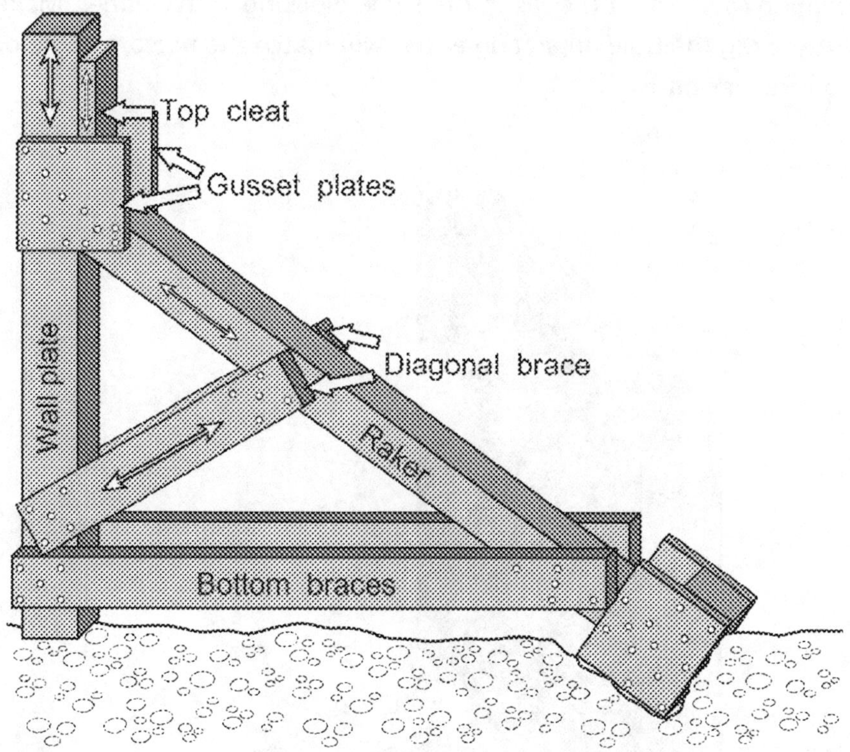

Fig. 5–53 Elements of the split-sole raker.

When constructing a split-sole raker, the first step is to determine the location of the insertion point of the shore and the angle at which it should be installed. Because the ideal angle for this type of raker is 60°, use that angle. Measure back from the wall the distance that the raker will intersect the ground and dig a hole approximately

12–15 in. deep and 2 ft wide. To determine the amount of distance back from the wall, utilize two numbers. For the 60° angle, the pitch of the raker is 7 on 12. If the insertion point of your raker shore is 9 ft, then you would dig the hole back 63 in. When you start to dig begin at the mark and use it as the center of your hole, then dig away from the wall, otherwise the hole may be too close. This should give you enough room to set the raker into some blocking with wedges. Make sure to dig the hole on an angle that will match the bottom angle of your raker's base.

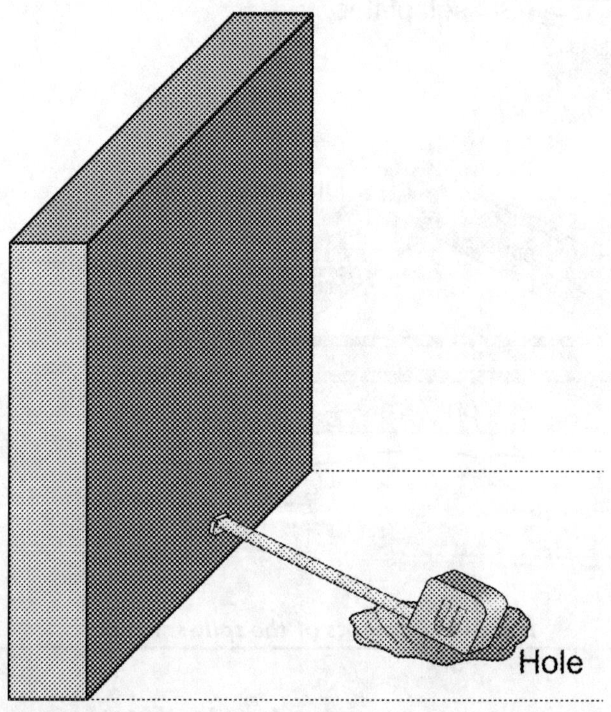

Fig. 5–54 Determine the insertion point for the raker to find where it will intersect the ground for hole placement.

If the ground is soft, you may have to dig your hole wider, for more surface contact, and use larger blocking. Place padding in the hole. It can be dimensional lumber or plywood, depending on the type of soil. When you are digging, remember to keep a watch on the unstable wall. You should keep a constant eye on the integrity of the damaged wall.

While the hole is being dug, start to assemble the shore. Determine the length of the wall plate. Since you are using the 60° angle, install a 3-ft cleat. Nail it directly on top of the wall plate at the top. This makes for easier nailing. Utilize the 5-nail pattern.

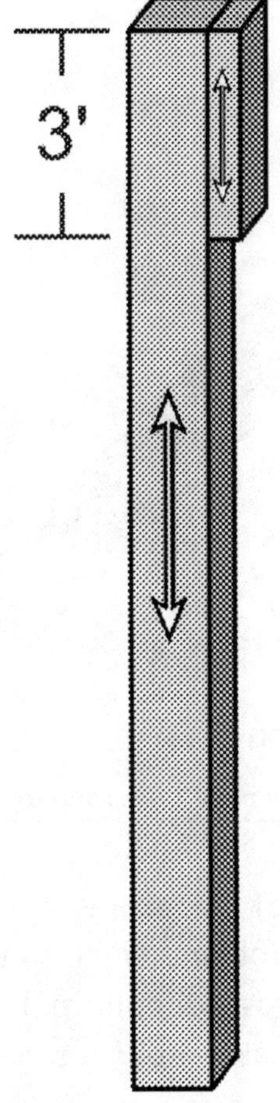

Fig. 5–55 Determine the length of the wall plate and install a 3-ft cleat.

Next, turn the shore on its side and place the raker face with the 60° cut into the plate and nail in position. Use (2) 16d nails, nailing the raker into the wall plate and following the grain on the lumber so you don't split the end of the raker. Make sure the fit is good. The base of the raker does not have to be cut; it can be left square. It will butt flush into the blocking.

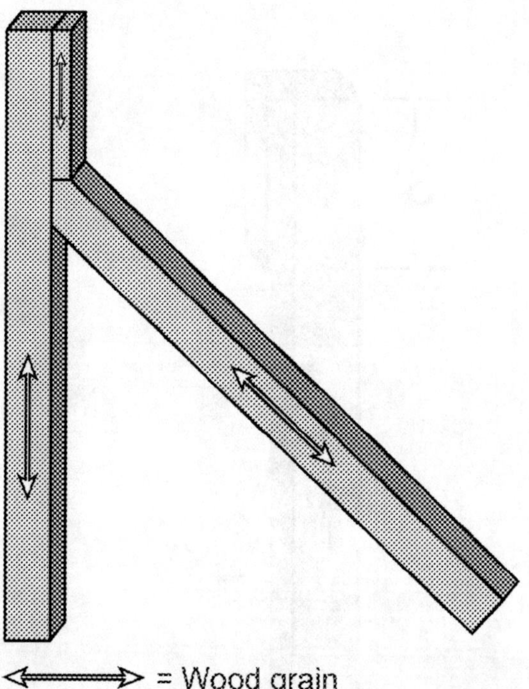

Fig. 5–56 Install the raker into the wall plate, following the grain of the lumber.

It is important that the base of the raker is in the proper position. If the raker is out of position, problems will result. If you measured down, say 6 ft, then measure across 42 in. from the wall plate to the raker. By measuring down the raker itself 6 times the number 14, you will get the proper hypotenuse of the triangle that is exactly what you are fabricating. Make a mark at that point, 84 in., and then intersect the 6 ft measurement of the wall plate to the raker mark. This will ensure that your raker is at the proper position.

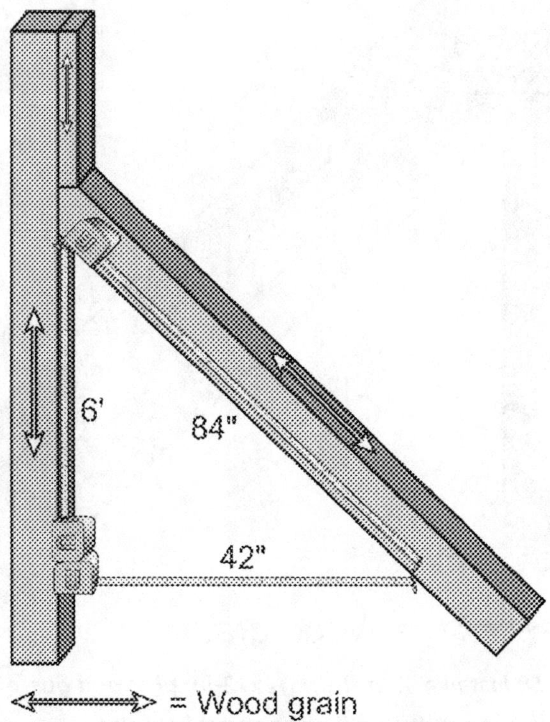

Fig. 5–57 Make sure to determine the proper hypotenuse of the triangle for what you are fabricating.

With the raker in the proper location, you can now lock the 60° raker joint into position. To do this, place a 12x12-in. gusset plate on top of the wall plate and raker joint. Nail it in place with 8d nails. On the wall plate, use eight nails; and on the raker, use five nails. For both, use the 5-nail pattern.

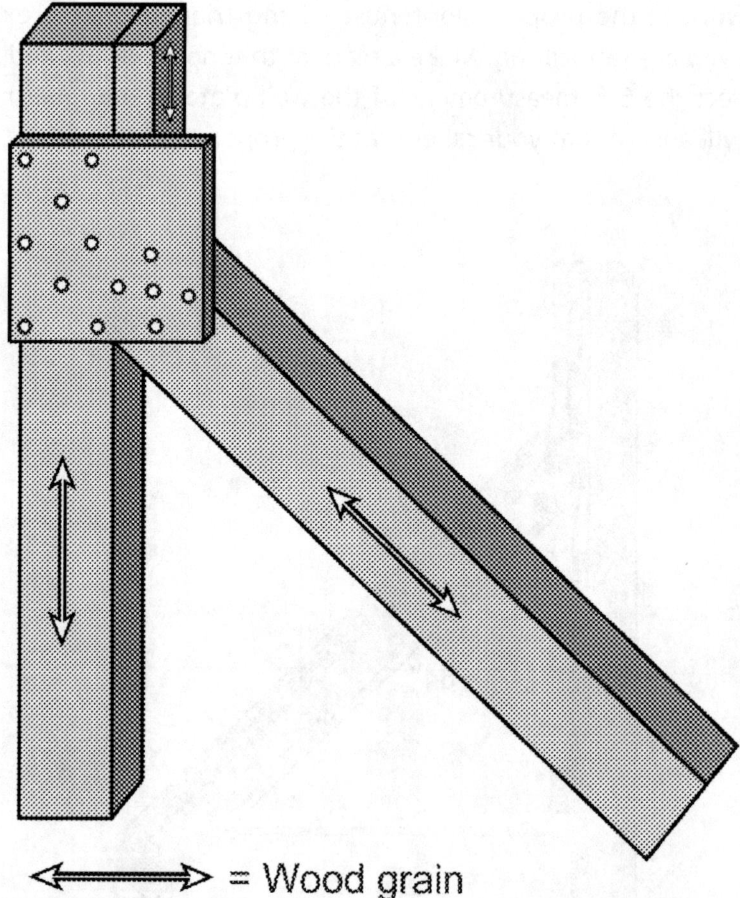

Fig. 5–58 Install a ¾ in.-thick 12x12-in. plywood gusset plate at the wall plate and raker joint.

The next step is to place the 2x6 bottom braces along both sides of the raker. Place them down low near the base of the wall plate to makes the shore's triangle as large as possible and so the shore is more efficient. At this time, square up the 2x6-in. bottom cross brace to the wall plate. Use a 2-ft framing square to do this; it is important. Nail the wall plate end flush with the face of the wall plate, using (5)16d nails in the 5-nail pattern. Keep the bottom brace approximately 18 in. up from the base of the raker; this gives you enough room to place the base of the raker in the hole without the braces interfering with the operation.

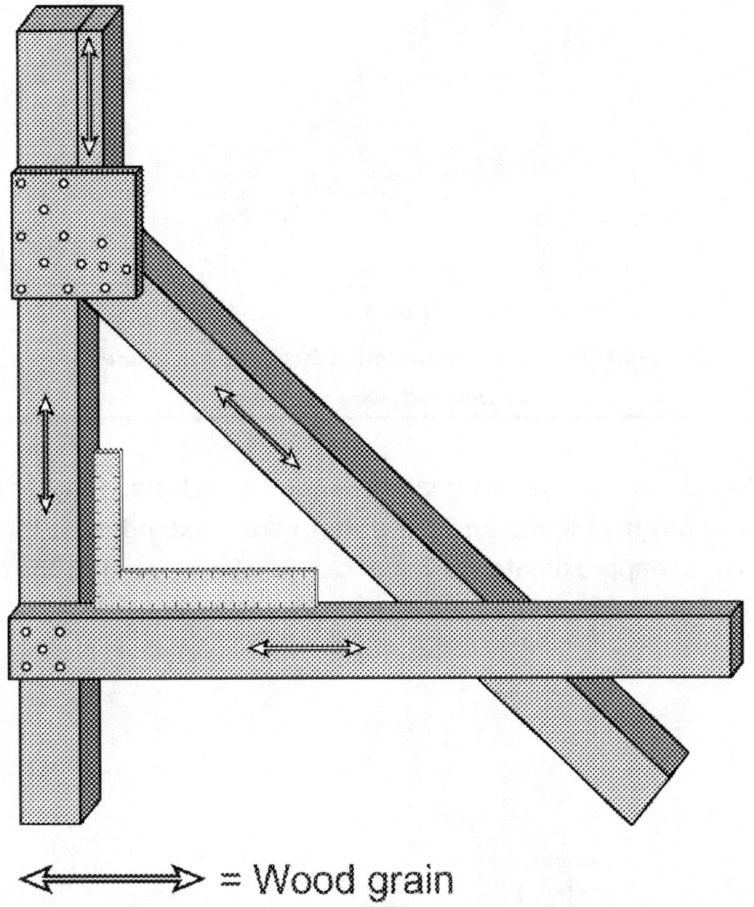

Fig. 5–59 Install the first 2x6-in. bottom brace and square it up to the wall plate.

Check your raker distance from the wall plate to make sure it is still in proper position and then anchor it down to the raker, using the same 5-nail pattern and 16d nails. The space between the inside face of the wall plate and the inside face of the raker should be 42 in.

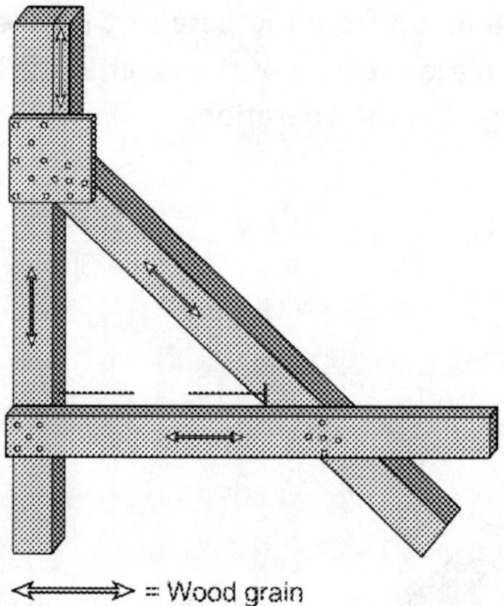

*Fig. 5–60 Check the measurement between the inside faces
of the wall plate and raker.*

Flip the shore over and nail the other gusset plate and bottom brace in position. Make sure the bottom brace extends past the end of the raker approximately 1–2 ft in case you have to adjust the raker to fit the hole. The shore is now ready for placement.

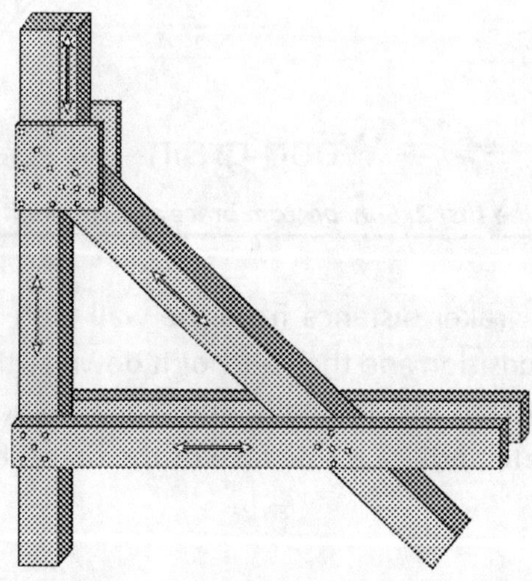

Fig. 5–61 Install the remaining gusset plate and bottom brace.

Place the shore in position and anchor it to the wall. You need at least two pins ½ in. thick in place through the wall plate to keep the wall plate from riding up the wall. Install blocking and wedges under the raker after the shore is anchored to the structure. Tighten up the wedges to pressurize the shore into the building.

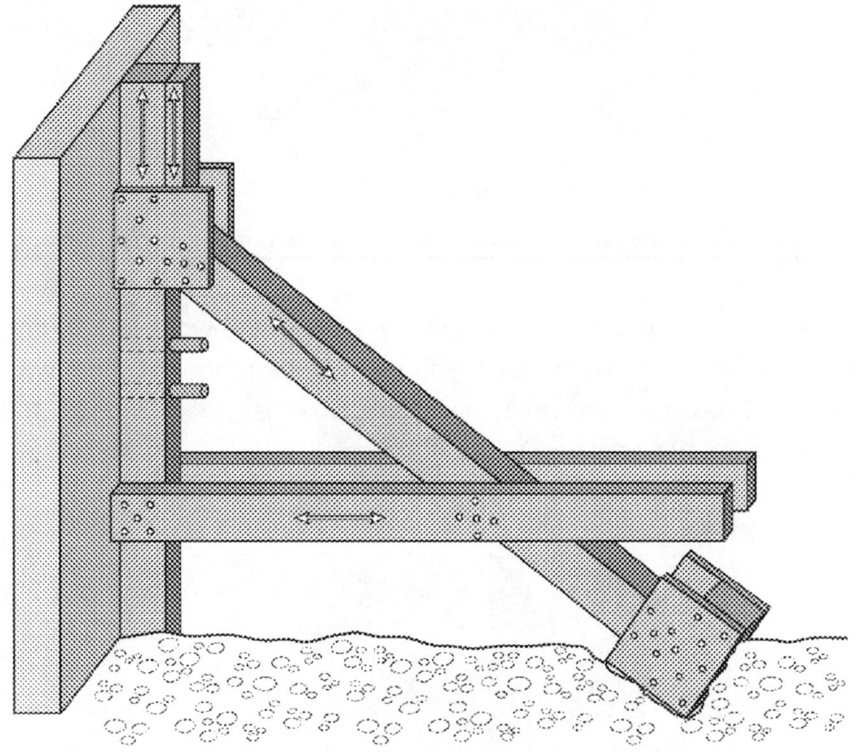

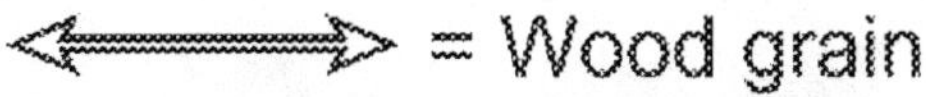

Fig. 5–62 Anchor the shore in position with two ½ in.-thick pins through the wall plate.

After the shore is placed in position and locked in place, install the two diagonal braces. Nail the 2x6s on top of the wall plate and into the raker.

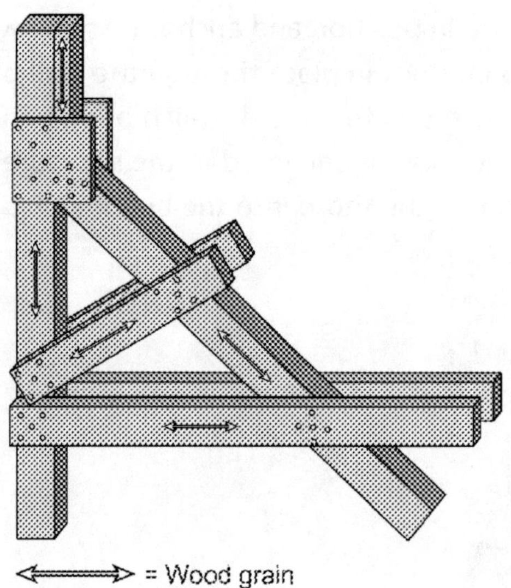

Fig. 5–63 Install the two diagonal braces after the shore is locked in place.

Install the two 2x6 diagonals on the edge of the wall plate, sitting on top of the bottom braces. If you want to be fancy, cut them with a bevel, but that is not really necessary. Now nail them with (5) 16d nails in a 5-nail pattern

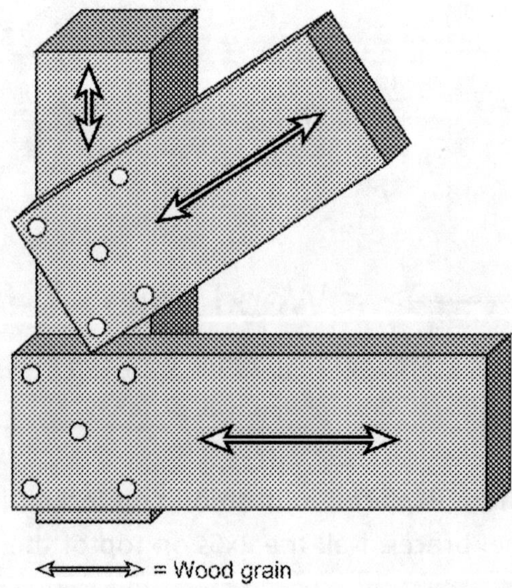

Fig. 5–64 Close-up of the 2x6 diagonal nail pattern.

The perfect instance to use this shore, whether on soil or solid surface, is when a large amount of debris is in front of the wall, blocking access to the wall face. Instead of having to remove the debris in front of an unstable wall, placing our personnel in danger, it is much easier go around it. This option not only saves precious time, it also saves quite a bit of work. In this case two things change: the height of the wall plate and the angle of the bottom braces. When you install this version of the shore, it is important that the bottom braces extend down as far as possible in order to give the raker more solid support and stabilize the shore. The wall plate is also to be cut short just above the debris as shown in Figure 5–65.

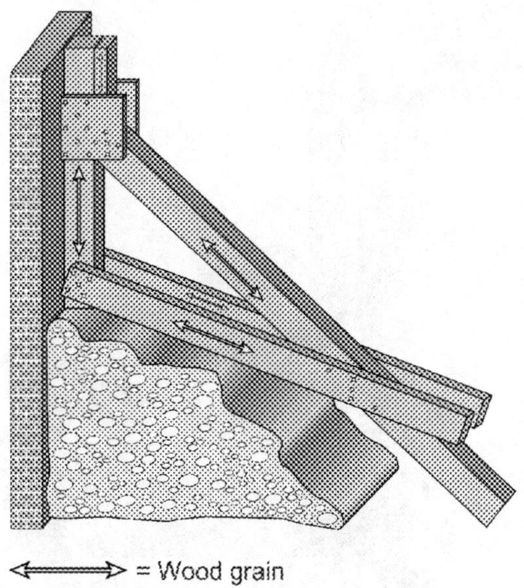

Fig. 5–65 Cut the wall plate short just above the debris.

Now you need to determine the length for the wall plate using this method. The key is to make sure that you know the raker insertion point. The rest is easy, just a little bit of adding and subtracting. For example, say the insertion point is at 10 ft, a common height encountered. The first thing you must deal with is the top cleat length. Generally with this shore into soil, it is 3 ft long; into solid surface and using a 45° angle, the cleat length is 2 ft.

The second step is to determine how high up the wall the debris is stacked. For this example, assume the debris is 5 ft high. At this time, your overall length of the wall plate is 13 ft, 10 ft to the insertion point, plus 3 ft for the cleat. You can nail the top cleat to the wall plate flush with the top of the plate. Using the same 5-nail pattern as the 24-in. cleat, you will have (26) 16d nails anchored into the cleat. Laying out the wall plate on the ground from the bottom of the top cleat, measure down 5 ft and cut the plate. The wall plate now fits at the proper insertion point and is just above the debris pile.

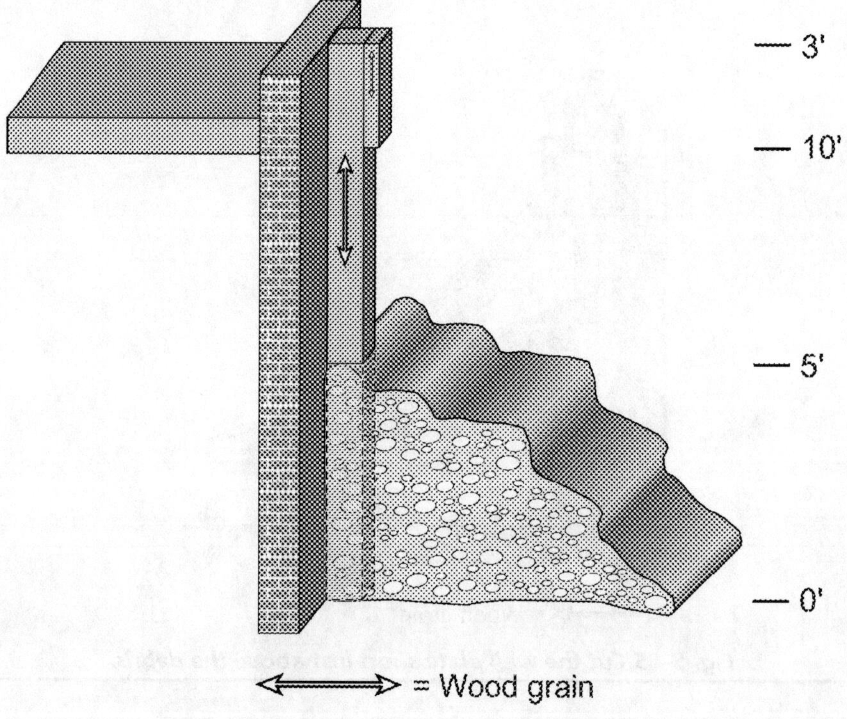

Fig. 5–66 Determine the height of the debris.

Figure 5–67 shows the split-sole raker shore constructed out of the pneumatic struts. Notice the difference in construction from the other options included in this chapter; there is no top cleat. The raker sections pin into the wall channel plate.

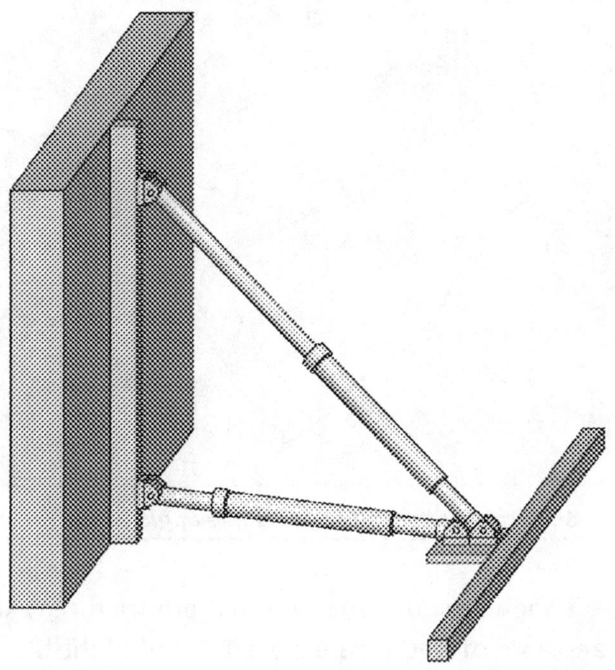

Fig. 5–67 The split-sole raker constructed out of pneumatic struts.

Split-sole raker base. There are several methods available to your team when it comes to supporting and anchoring your split-sole raker shores. The two most typically utilized ones are to use either three 2x6s or two sheets of plywood, nailed together, both at least 18x18 in. square. This generally will be sufficient in the typical soil conditions. However, you must remember, each situation has its own unique size-up concerns. In much harder ground that is difficult to dig into, smaller blocking may be adequate; and in softer soils or sand-like conditions, a much larger surface area must be established for properly anchoring the shore.

A popular way to secure the shore is to dig your hole down at least 1 ft deep. **Note: The base of the raker must be below the surface of the ground.** Excavate the soil on the same angle as the base of the raker so that the blocking falls inline with the bottom of the raker. Always place a set of wedges between the raker and any blocking. This has to be done in order to pressurize the shore (Fig 5–68).

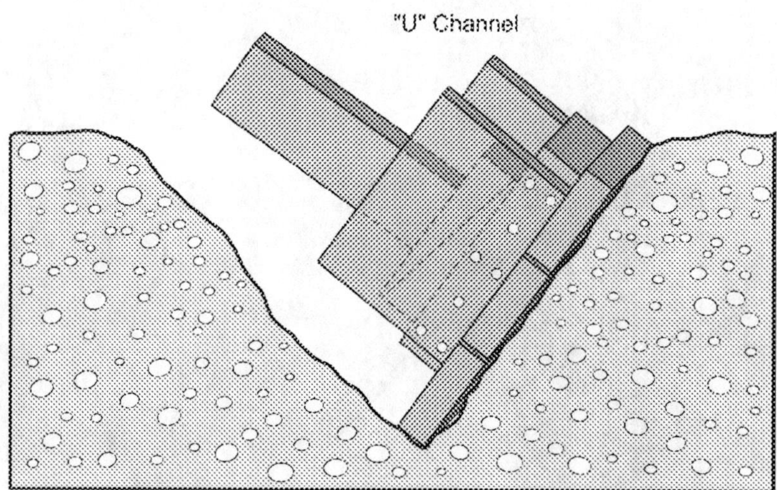

Fig. 5–68 Excavate the soil on the same angle as the raker.

Figure 5–69 shows a situation in which there are three 2x6s placed in the hole, and a set of wedges are placed on top of the blocks. Make sure the wedges are at right angles to the blocking. The 2x6s must be 18 in. long. If they are any shorter, the shore may be unable to support the proper load. Make sure the 2x6s are at right angles to the wedges. If you place the wedges parallel to the blocking, you would only be pressurizing one 2x6. That is not enough to support the load.

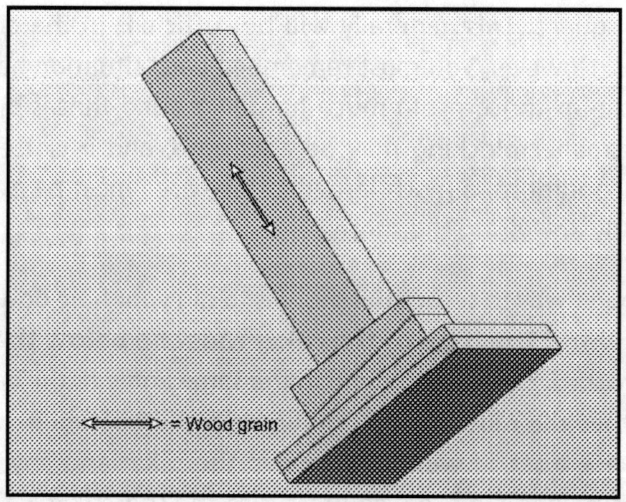

Fig. 5–69 A raker base option using three 18 in.-long 2x6s.

In the option depicted in Figure 5–70, the same basic blocking principle described in the previous paragraph is used; however, this option uses two sections of plywood. These plywood sections must be 18x18s nailed together. When digging your holes, make sure that the hole is flat and that the full face of the plate contacts soil.

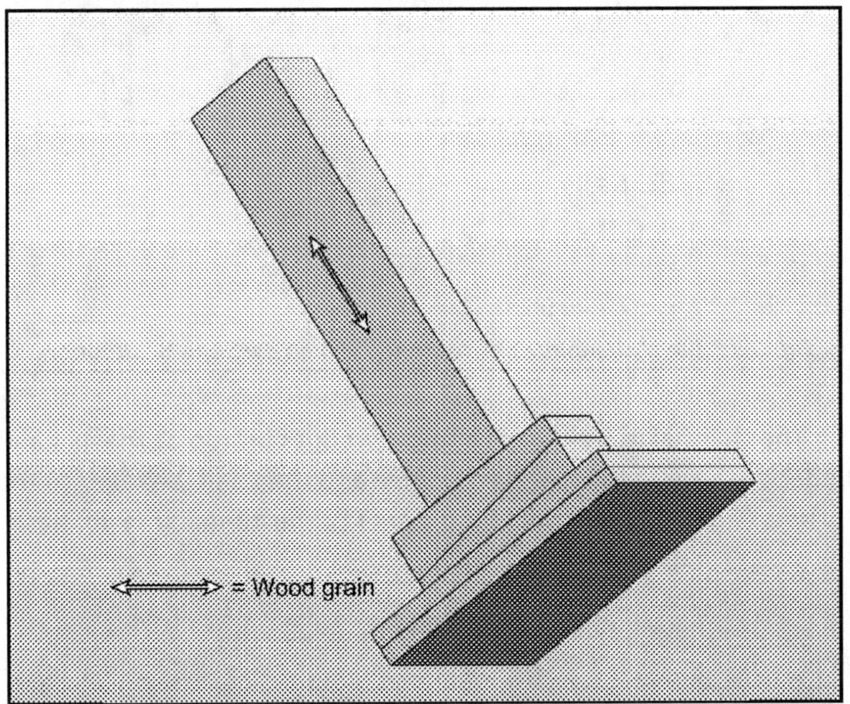

Fig. 5–70 A raker base option using two 18x18-in. sections of plywood.

If you want, you can use the option pictured in Figure 5–71 any time; but it should be your first choice in an earthquake situation. Nail the raker into the plywood gusset. The *U* channel is two pieces of ¾-in. plywood, 12x12 in., and anchored to an 18-in. 4x4 with (8) 8d nails anchored in a 5-nail pattern.

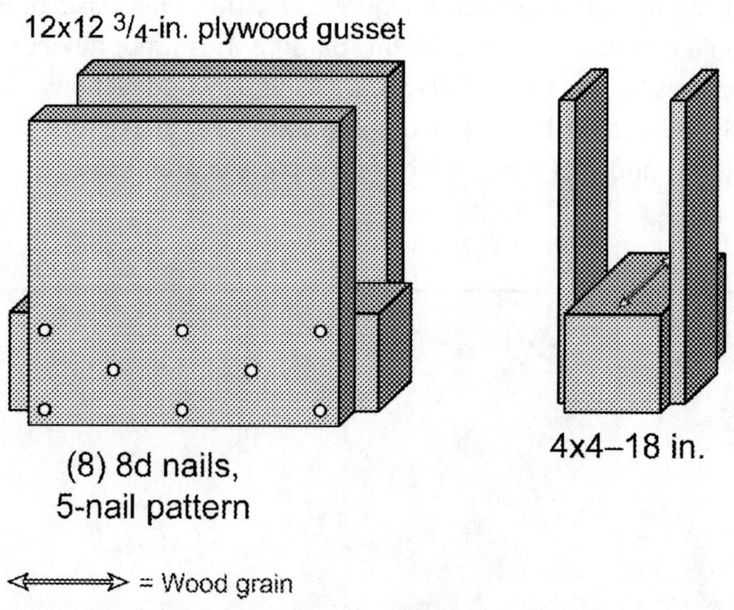

Fig. 5–71 The U-channel, your first choice in an earthquake situation.

The channel in use, place a set of wedges on top of the 4x4 and under the base of the raker. Pressurize and then nail the raker to the gussets (Fig. 5–72). Anchor the channel to the 18-in. by 18-in. base.

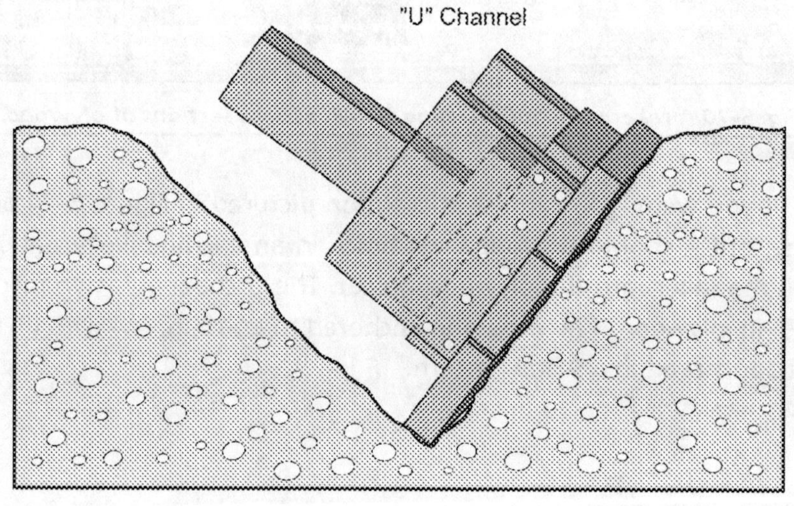

Fig. 5–72 The U-channel in use.

Another option open to you is the use of a trough, which is two 2x6s 3–4 in. long, nailed to a 2x4. Anchor it down on manufactured surfaces with bolts or pins. Nail the 2x6s into the 2x4, placing the 16d nails alternating top and bottom at 3-in. intervals.

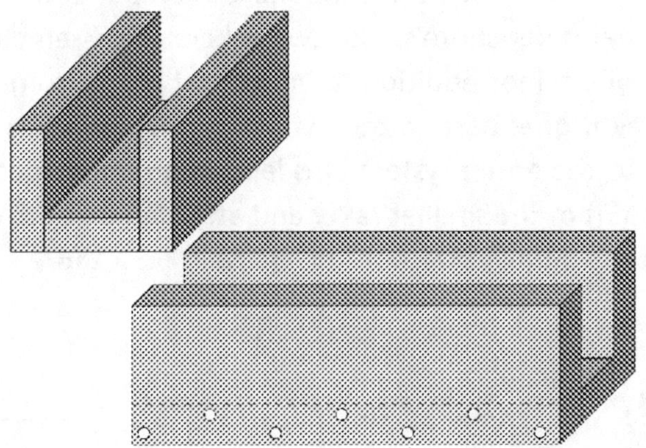

Fig. 5–73 The trough, another raker base option.

Figure 5–74 shows an expanded view of the blocking. This must be anchored to the ground in the same manner as the anchors of the solid-sole rakers. Place a 24-in. 2x4 block behind the raker against the return cut and nail it down with a 5-nail pattern of 17 nails.

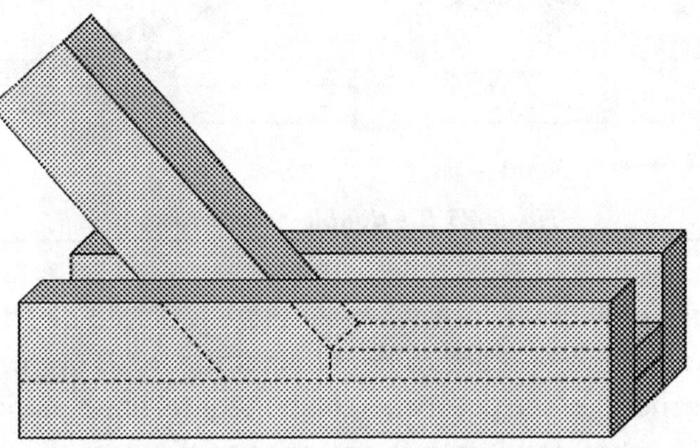

Fig. 5–74 Expanded view of trough blocking.

Double-raker shore

The main reason to install a double-raker shore is for the following two situations. The first is to shore a wall that has a bad bulge or crack in it. The other reason is to shore more than one floor, typically the second and third floors. All the shore's components are the same as the normal raker shore's. The two rakers are assembled as usual except for one minor addition: a horizontal brace from the face of the small raker goes back to the main raker. This brace is installed as a stiffener to the entire system. It is lapped on both sides of the top insertion point of the smaller raker and attached on both sides of the shore with the 5-nail pattern using 16d nails on a 2x6.

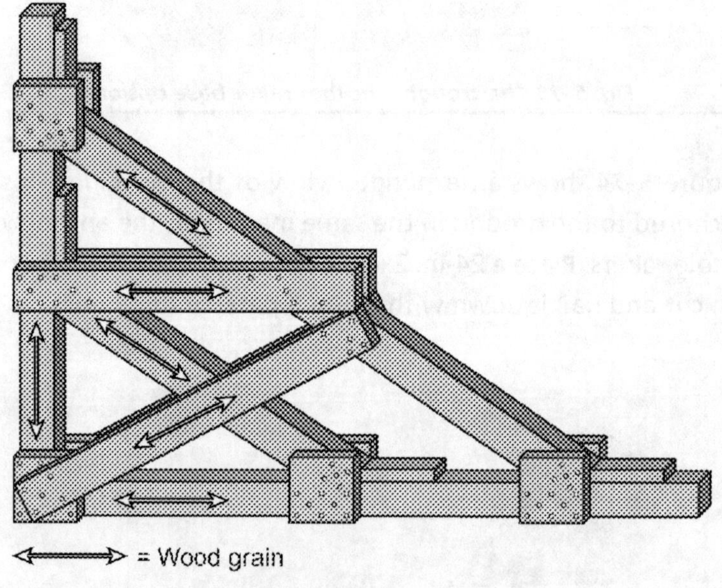

Fig. 5–75 The double-raker shore.

The diagonal brace for the whole shore is placed directly underneath this horizontal brace to help funnel the loads evenly through the whole shore. When you install this horizontal brace along side the joint of the raker, nail it into the raker as well as into the wall plate to eliminate the need for a gusset plate.

Double-raker shore step-by-step procedure.

The following is a numbered list of the tasks that must be completed to build and install a double-raker shore. The remainder of this section on the double-raker shore provides detailed instructions for constructing and installing the various elements of the shore.

1. Clear an area large enough to preassemble this shore; 20 ft x 20 ft would not be too big an area.

2. Determine the angle of your shore and the insertion point to use.

3. Place the wall plate and sole plate at right angles to each other, forming an *L,* and nail together.

4. Square up the plates and mark the raker insertion point.

5. Anchor the corner gusset plate in position, keeping the plates square.

6. Install the TOP raker in the designated position and nail it down.

7. Install top and bottom 2x4-in. cleats and nail them down properly.

8. Install top and bottom gusset plates, nailing them down with proper nail patterns.

9. Flip the shore over and nail down the top corner and the bottom gusset plates on this side of the shore.

10. Install shore in position, pressurizing against anchor.

11. Install the lower raker in position and nail it in place.

12. Place cleats at the top and bottom of the lower raker and nail it in place.

13. Place the center brace where the face of the wall plate intersects the lower raker face. Do this on both sides.

14. Pressurize the raker into the building; driving both sets of wedges into position at the top and the lower raker.

15. Fasten the bottom gusset plates to lower raker.

16. Pressurize the raker to the wall, anchor it to wall, and install the center diagonal braces.

Wall plate. Typically the wall plate is a 4x6 or even a 6x6. In most instances, 4x6 by 16–24 ft long is sufficient. Make sure that the wall plate is at least 2½ ft longer than the designated shore's insertion point.

Sole plate. Another section of 4x6 and generally 16–20 ft long, a sole plate is nailed into the wall plate at the base and behind the wall plate.

Rakers. Rakers are the main support elements of the shore. They typically are made of 4x6s or on occasion, of 6x6s. A raker is cut to a specific length and has angles cut into both ends. For the most part, the angles are 60°. The shore's lumber will not be as big as a 45° angle raker and can still support a larger load.

Cleats. Cleats are 2x6 by 36 in. and are installed on top of the raker. At that base of the raker, use 24-in. cleats. Nail both in place using the 5-nail pattern on each.

Gusset plates. To lock all connection points together (except at the top of the lower raker), install 12x12-in. sections of ¾-in. plywood. Anchor the gussets on both sides of the shore with (13) 8d nails using the 5-nail pattern.

Wedges. Usually 4x6s, wedges are placed behind the rakers and pressurized to keep the shore tight to the wall in question.

Cross braces. Cross braces are two 2x6-in. braces from the wall plate to the outside of the top raker, one on each side. Anchor them on the face of the lower raker at roughly the midpoint of the top raker.

Diagonal braces. The last items to go on the shore, diagonal braces normally are lengths of 2x6-in. lumber and are nailed to the outside of the raker. They are nailed on both sides of the bottom corner gusset plates with (5) 16d nails into the wall plate and sole plate.

Assemble the raker out of the collapse area. You can assemble the shore with both rakers if you have the manpower. These rakers have a tendency to weigh quite a bit more than the conventional raker. One option is to assemble the top raker only and to install the smaller raker when the shore is in position. It is generally a good idea to use the 60° angle for this shore. Doing so results in a smaller shore that weighs less than a 45° angle shore.

Tape out and place marks for both rakers, and use the 3, 4, 5 method to square up the shore.

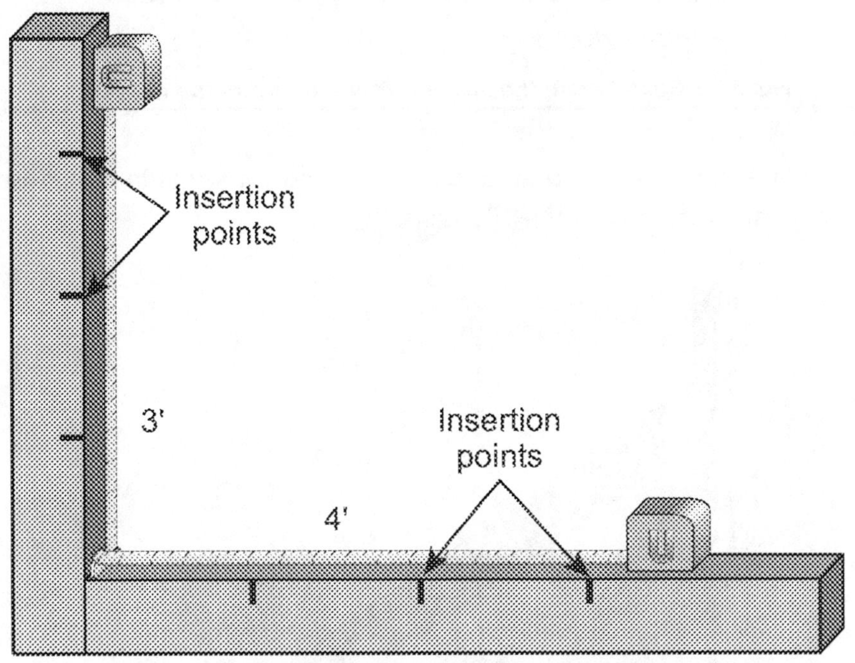

Fig. 5–76 Measure out both insertion points.

Gusset plate the inside corner when it's squared up and install the top raker, making sure the raker fits right and the shore is square.

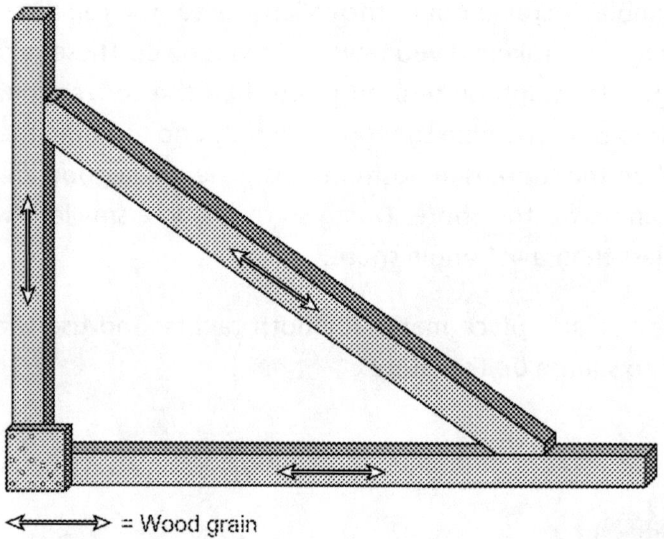

Fig. 5–77 Gusset plate the inside corner and install the top raker.

Place the 36-in. top cleat and the 24-in. bottom cleat in place; use the 5-nail pattern with the 16d nails.

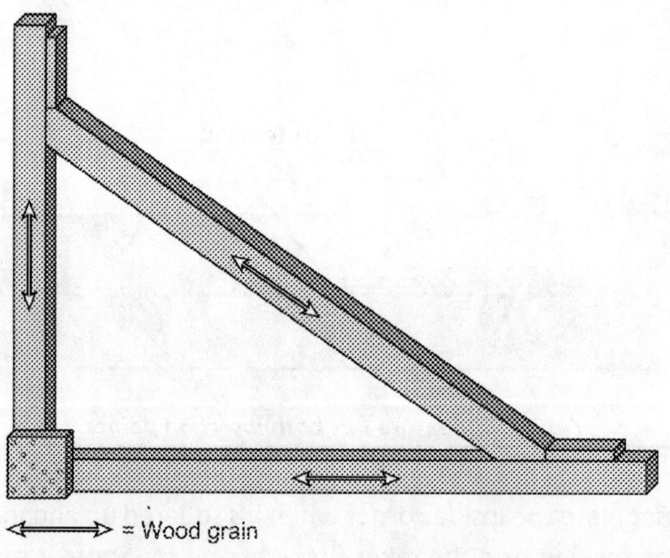

Fig. 5–78 Install the 36-in. top cleat and 24-in. bottom cleat in place.

Place the gusset plates on the shore as normal and nail them down with the regular nail pattern in the usual places.

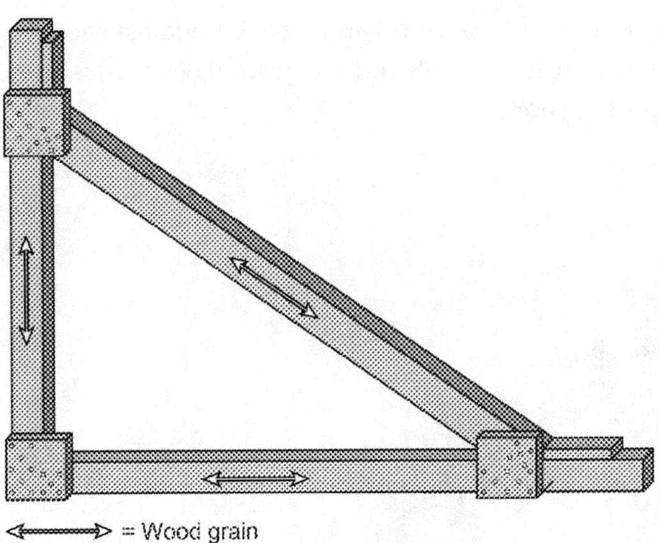

Fig. 5–79 Install the usual gusset plates.

The next step is to flip the shore over and nail the gussets to this side. Because the shore is probably big and heavy, use several men to flip the shore.

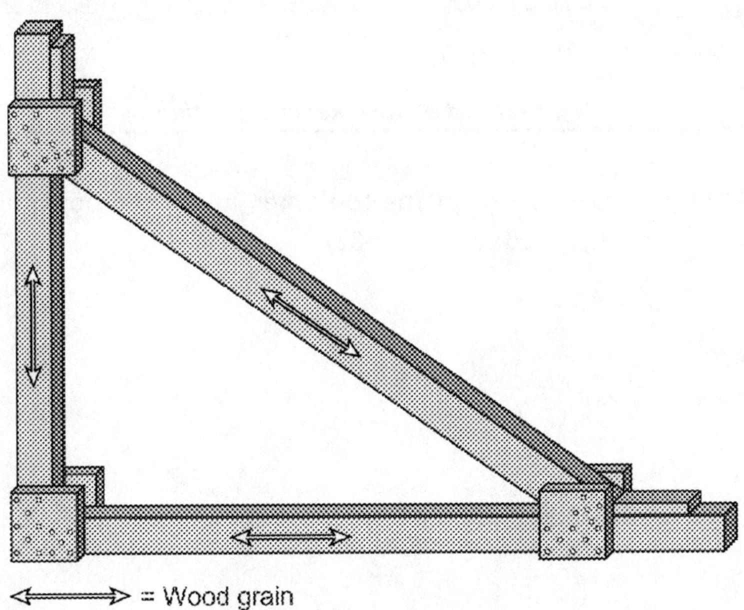

Fig. 5–80 Flip the shore over and install gusset plates on the remaining side.

Lift the raker into position and install it against the wall, making sure to anchor it to the wall and the ground. Pressurize the top raker and secure the shore.

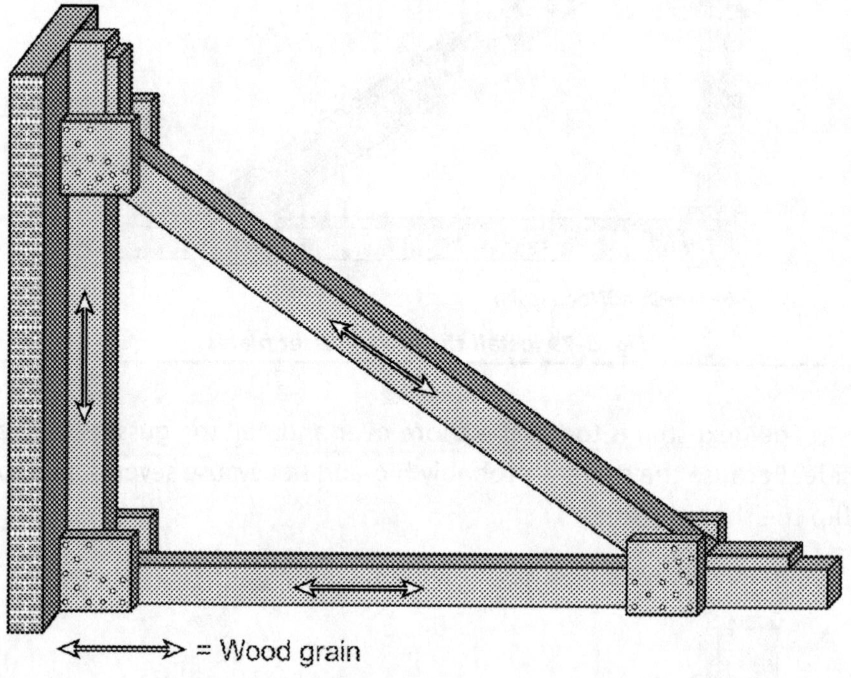

Fig. 5–81 Install the raker into position.

After you have safely set the top raker, install the bottom one. Cleat it and set the wedges (Fig. 5–82).

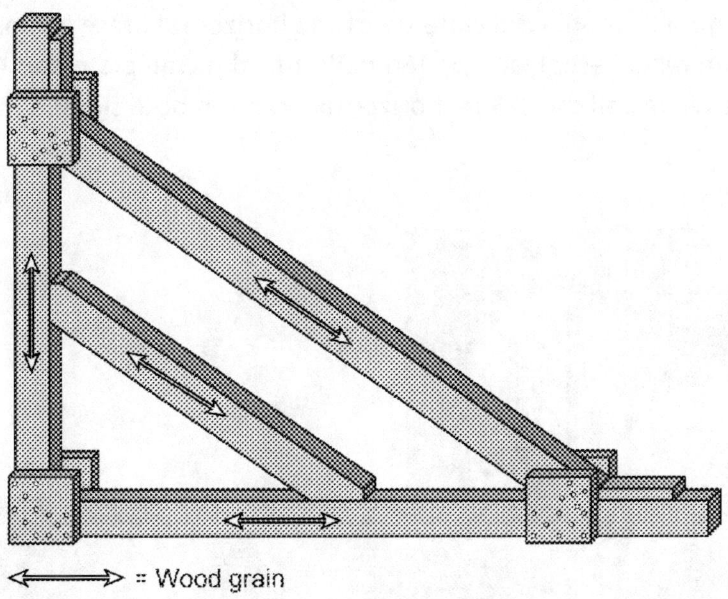

Fig. 5–82 Install the bottom raker after the top raker is safely set.

At this time, install the 2x6-in. horizontal cross braces and gusset plate the bottom of the inside raker. Nail it into position.

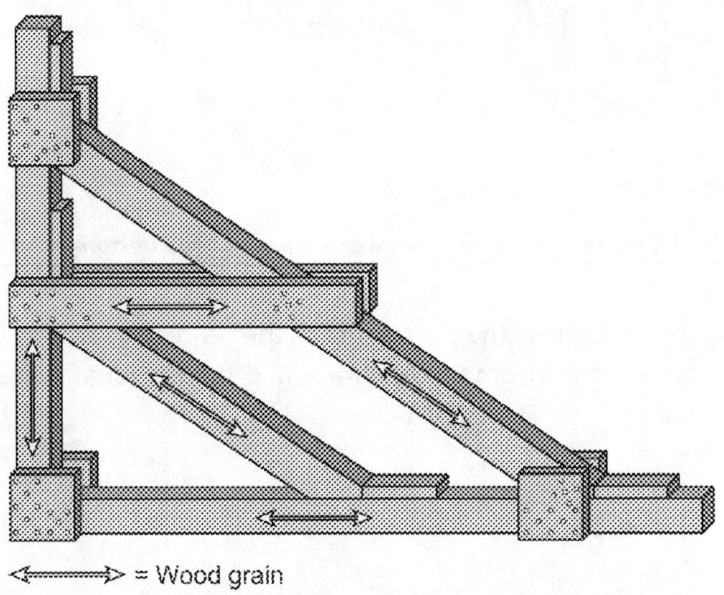

Fig. 5–83 Install the 2x6-in. horizontal cross braces.

Figure 5–84 shows a close-up of the horizontal brace against the bottom raker face. Place (5) 16d nails into the wall plate and 4 into the raker. Install the 2x6-in. horizontal brace on both sides.

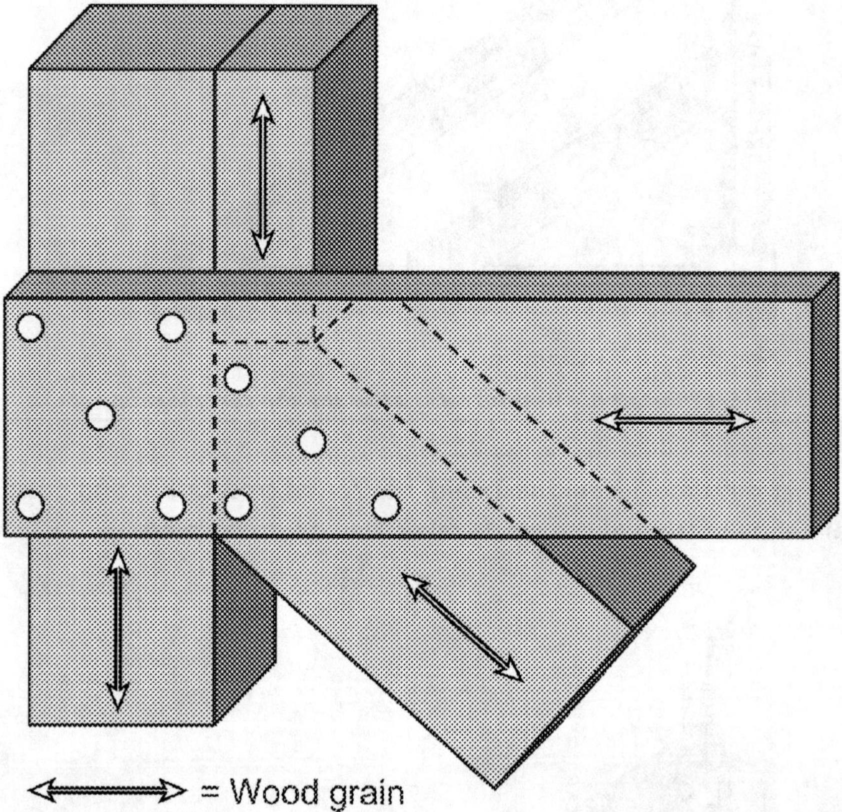

Fig. 5–84 Close-up of the horizontal brace against the bottom raker face.

Figure 5–85 shows the proper fit for the braces against the raker. Secure both the horizontal cross brace and the diagonal brace with (5) 16d nails.

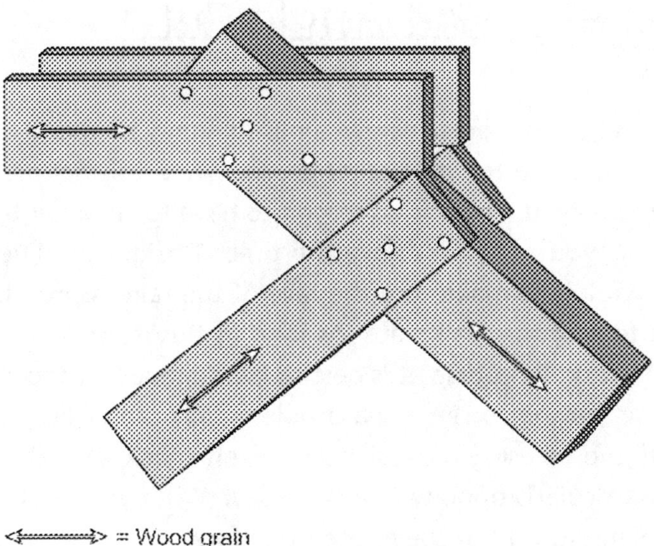

Fig. 5–85 Close-up of the proper fit for the braces against the raker.

Diagonally brace the shore on both sides as usual. Drive (5) 16d nails into each raker. Make sure the diagonal brace is installed directly under the horizontal cross braces as shown in Figure 5–86.

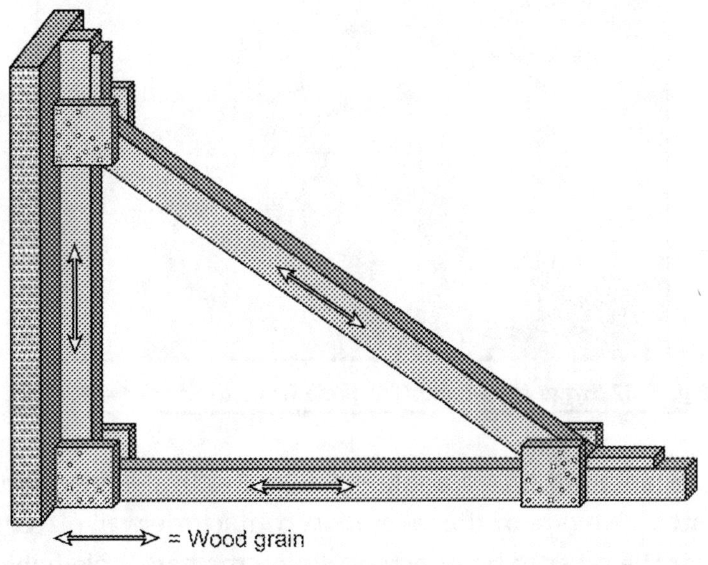

Fig. 5–86 Install the braces as shown.

Bad Angle Cut

On occasion, of course rarely, a mistake may be made in cutting the specified angle for your raker shore. When this happens, it can be easily remedied. There is no reason to have to cut an entirely new raker piece; you just don't have the time. The bottom line for the shore to work is to make sure the face of the raker is in full contact with the face of the wall plate. As long as this is accomplished, the raker will work. Of course, it is best to have a good fit the first time, rather than having to remedy a problem. You can achieve a proper angle cut and fit using a circular saw, a mitre box, or a chain saw if you are particularly good with a chain saw. What you must strive for is a complete flush fit of the entire raker face.

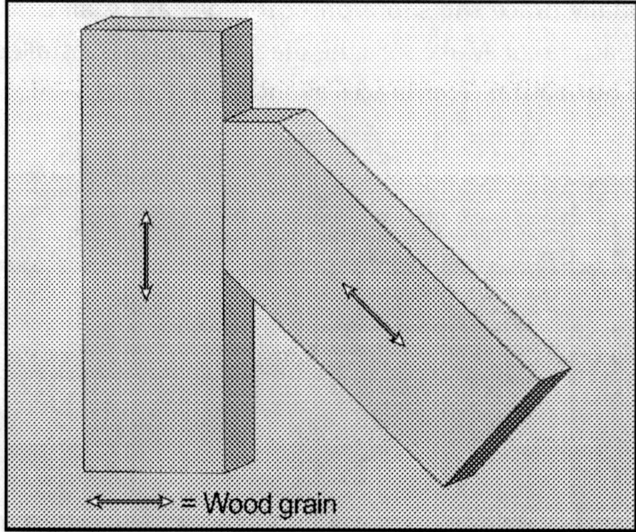

Fig. 5–87 Strive for a complete flush fit of the entire raker face.

In instances of a bad angle cut (Fig. 5–88), the space must be eliminated. The face of the raker must contact the wall plate fully in order for the raker to be effective. Hiding the bad angle behind the gusset plates doesn't do it!

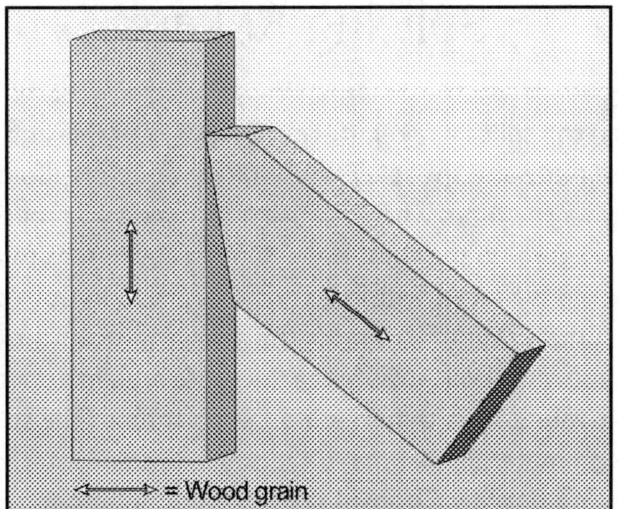

Fig. 5–88 The gap must be eliminated in a bad angle cut.

If the raker cannot be adjusted up or down for some reason, which may happen, one option is to fill the gap. Figure 5–89 shows a shim filling the gap. Nail the shim to the wall plate to secure it, using 8d nails. Be careful not to split the shim.

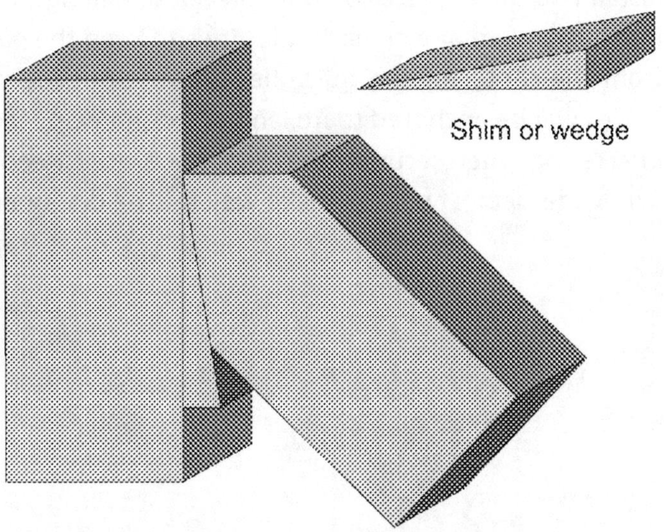

Fig. 5–89 Fill the gap with a shim.

Splicing Rakers

Yes, you can splice a raker; it is perfectly acceptable and so is splicing the wall plate and the sole plate, which would be spliced the same way. There may be times when this becomes necessary especially when you have to shore up the third floor. The big key here is to make sure that the splice is braced within 2 ft of the splice, in both directions. However, some firefighters prefer the splice right on the joint. The one thing you must do is cross brace the splice area in both directions. Keep this in mind when determining the position of the splice. If you have enough material, just make the splice in the center where you will be bracing anyway. This will save time and material.

Splicing a raker: step-by-step procedure

1. Make sure the ends of both raker sections are square; double-check this.

2. Toenail the sections together.

3. Install two 36-in. pieces of ¾-in. plywood, nailing (8) 8d nails into each raker, one on each side. Try and keep the nails away from the center of the splice because the outside diagonal braces will be anchored there, and you don't want these nails interfering when securing the diagonal braces. Keep a space 6 in. wide clear of nails to make way for the diagonals.

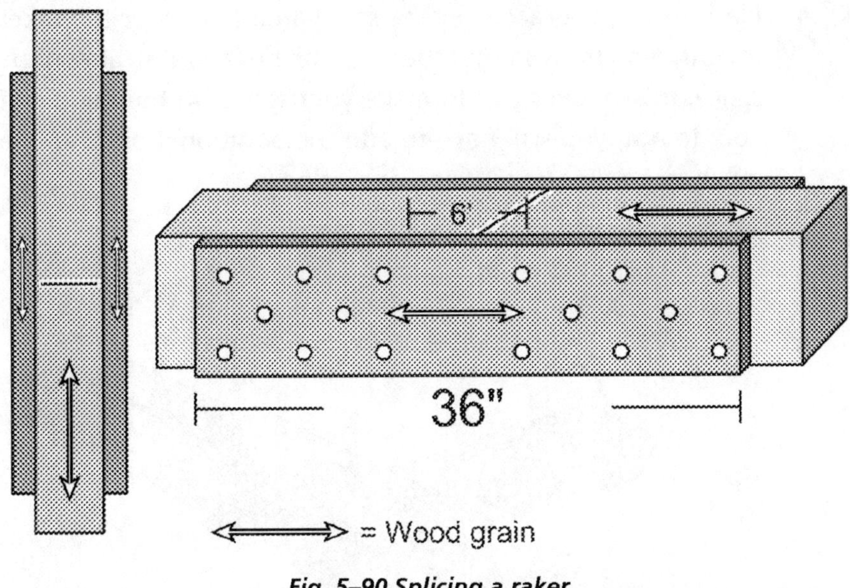

Fig. 5–90 Splicing a raker.

Figure 5–91 shows the top view of the splice when finished with diagonal braces of 2x6s going from the base of the raker shore on both sides.

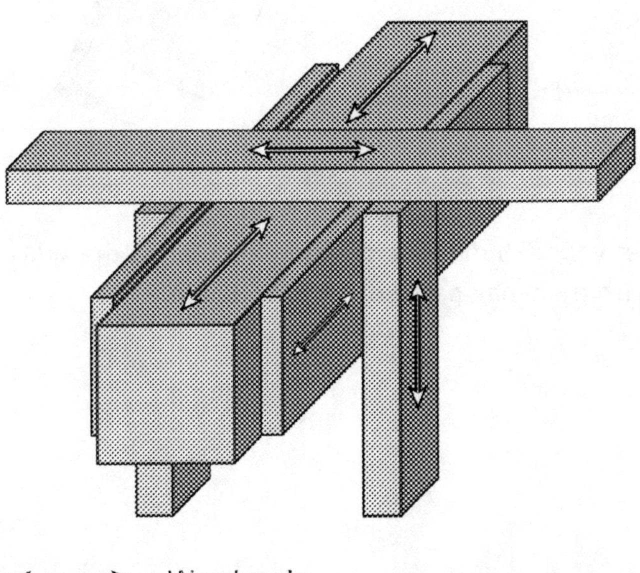

Fig. 5–91 Top view of the spliced raker with 2x6 diagonals installed.

4. Place the cross brace over the same area to anchor the raker
    in both directions so that there can be no lateral strain on the
    splice. It behooves you to make your splices in the middle. If
    you do not, you will have to add the additional bracing in a
    much more time-consuming operation.

    Figure 5–92 shows a view from the bottom.

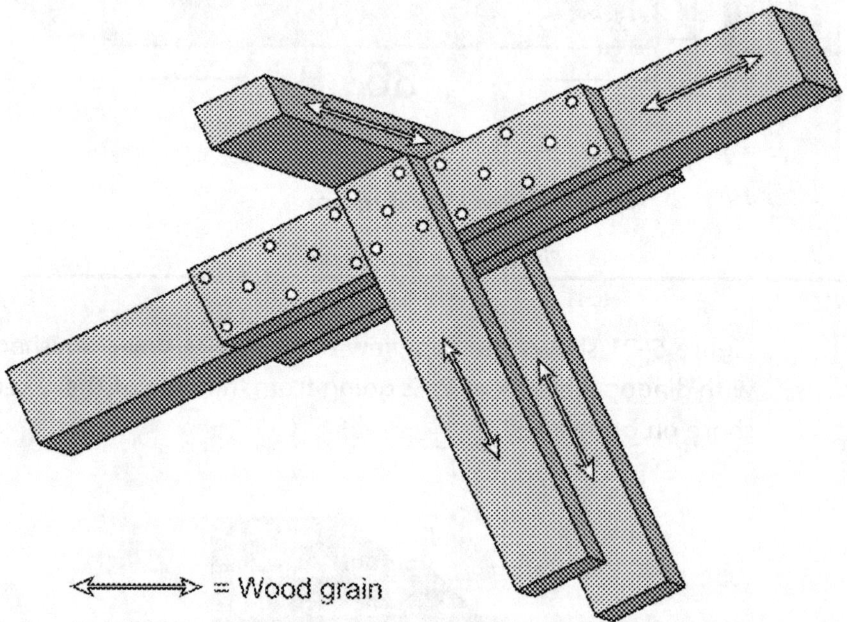

Fig. 5–92 Bottom view of the spliced raker with 2x6 diagonals installed.

5. Set your 2x6 diagonals into the center of the splice and nail
    with the 5-nail pattern, using 16d nails.

Raker Bracing Systems

Whenever installing raker shores, you must install at least two of them. One raker shore, no matter the type, is not sufficient. A raker is laterally unstable and would fall under a minimum load. Therefore, install at least two rakers all the time. When erecting two shores, always connect them together in order to unite them as one system. They now are able to withstand a lateral load. To connect them together, both the split-sole and solid-sole type, lay essentially what is a truss on top of the shores. The truss acts in tension and compression on a lateral plane, keeping all the rakers from shifting.

To accomplish this effectively, you want all the elements of the bracing system on top of the shores nailed together in the same plane. Placing bracing on top of and underneath the rakers defeats the purpose of the truss because all the elements are not tied together; therefore, the bracing system is weakened. For proper bracing, use 2x6s and anchor them with 16d nails. If your team doesn't have 2x6s available, not to worry, you can use 2x4s. All you have to do is double up on the 2x4s, lay them one along side the other, and nail them both down with (3) 16d nails each.

To complete the bracing system properly, you must install a series of *X*s on each end of the system. First make sure that the first bay and the last bay both have diagonal *X*s in order to keep the system from shifting when lateral pressure is applied to the shores. These *X*s are 2x6s nailed down with the 5-nail pattern using 16d nails.

The horizontal bracing system is used to support and collect the rakers as a group. The top brace, middle brace, and bottom brace are made of 2x6 lumber.

When the raker wall insertion point is more than 8 ft in height, install a middle horizontal brace directly over and in the center of the raker. Doing this provides lateral stability in both directions against the shores. For a raker with an insertion point more than 8 ft in height, use three horizontal braces (Fig. 5–93).

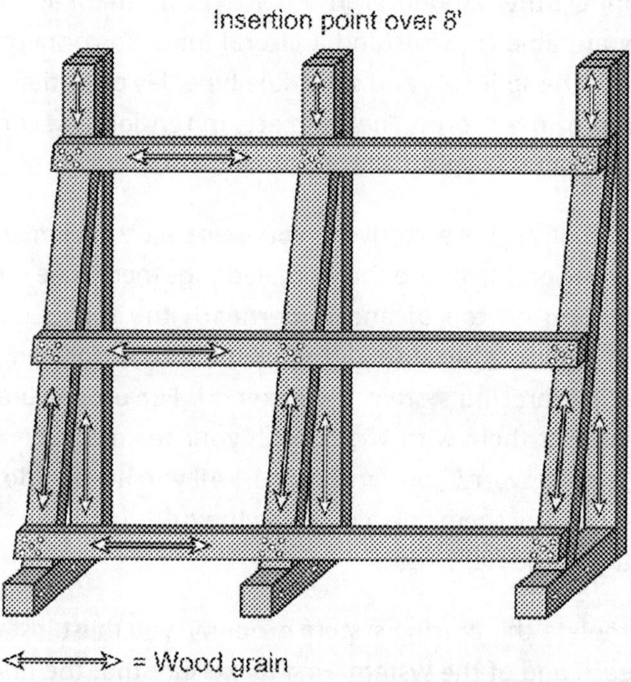

Fig. 5–93 Use three horizontal braces for rakers using an insertion point greater than 8 ft.

When the wall insertion point is 8 ft or less, there is no need for a center horizontal brace on the raker shore because of the short span of the raker. Therefore, you need to place only two braces: a top and bottom horizontal brace (5–94).

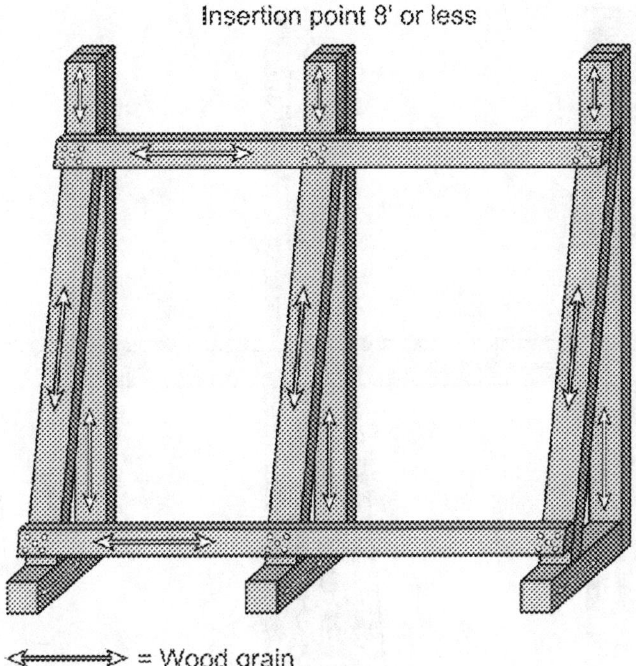

Fig. 5–94 Use two horizontal braces for rakers using an insertion point less than 8 ft high.

Raker bracing step-by-step procedure

To install the horizontal bracing system, as with most shoring items, start in the middle (Fig. 5–95).

1. Place the middle horizontal brace as close to the center of the raker as possible, directly on top of the raker. It's important that you do this so that the diagonal braces are not all different sizes.

2. Anchor the brace directly to each raker with the 5-nail pattern, using (5) 16d nails.

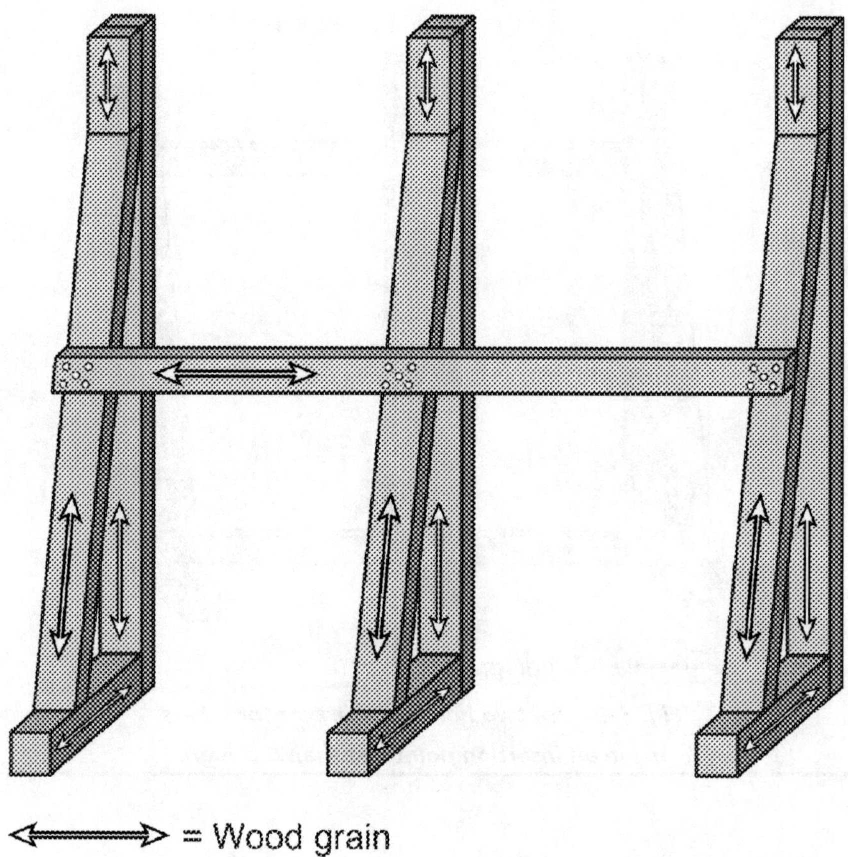

Fig. 5–95 Install the middle brace first.

3. Install the bottom horizontal brace by anchoring it directly on top of the raker and nail with the 5-nail pattern (Fig. 5–96). Keep the spacing the same from the bottom brace to this middle brace. The squarer you keep the horizontal braces, the better the diagonal braces fits.

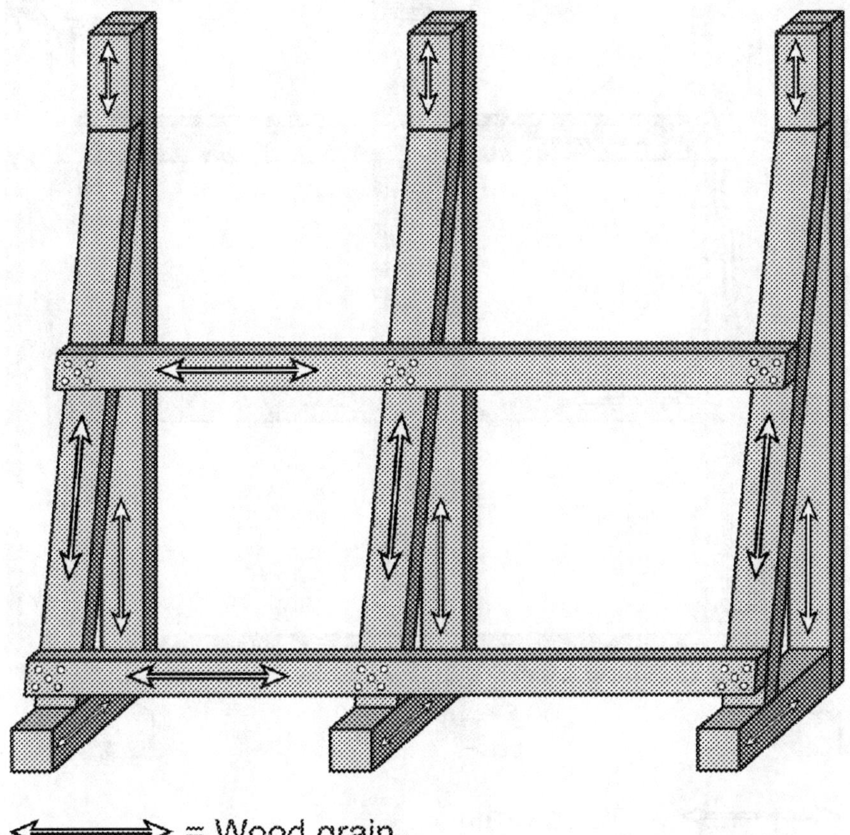

Fig. 5–96 Install the bottom brace next.

4. Install the top, horizontal brace, placing it up as far as possible. The wider the bracing system, the better it works for the shores.

5. Anchor the top, horizontal brace on top of the raker with the same nail pattern used to anchor the other horizontal braces (Fig. 5–97).

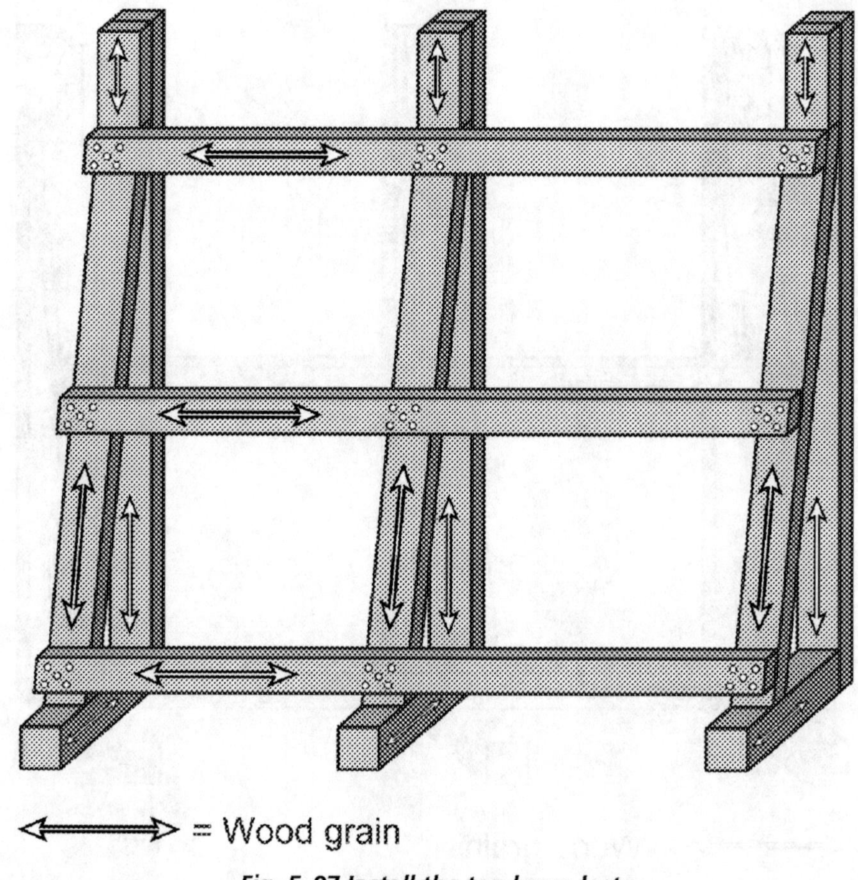

Fig. 5–97 Install the top brace last.

Figure 5–98 shows a close-up of the proper positioning of the brace: directly contacting the top gusset plates. The (5) 16d nails are to be evenly spaced across the 2x6 brace.

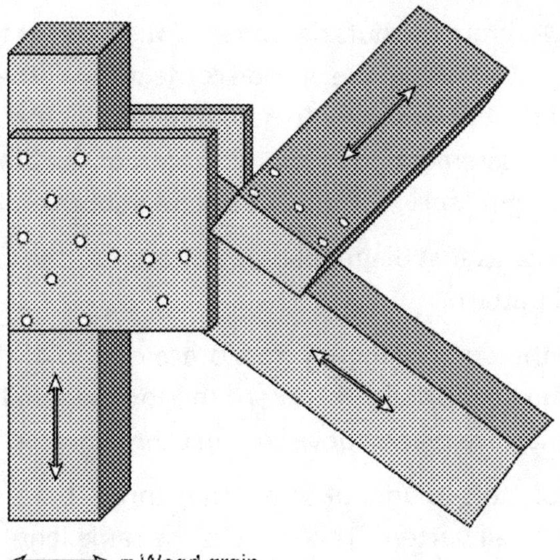

Fig. 5–98 Close-up of the proper positioning of the top braces—
directly contacting the top gusset plates.

Figure 5–99 is a close-up of the positioning of the bottom brace, showing the proper nail pattern. The brace should just touch the bottom gusset plates.

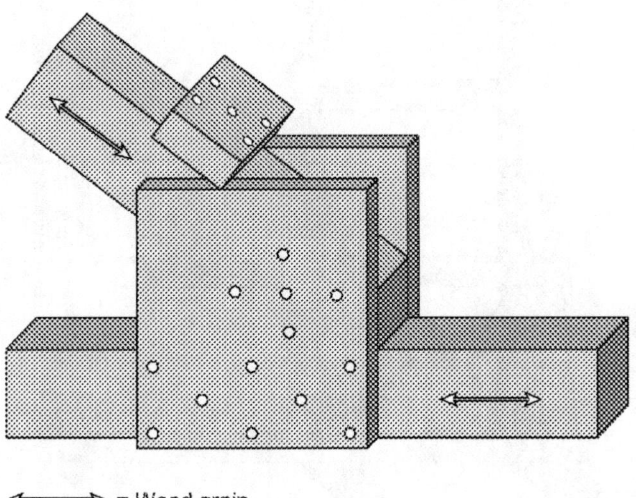

Fig. 5–99 Close-up of the proper positioning of the bottom brace—
touching the bottom gusset plates.

6. Measure from the outside corner of the raker at the bottom of the middle brace to the outside corner of the other raker at the top of the bottom brace to determine the length of the braces. This measurement is the overall length of the brace. You don't have to get fancy; cut the 2x6 square at both ends.

7. Anchor this first diagonal brace directly to the raker with the 5-nail pattern.

8. Place the other diagonal cross brace on *top* of the first brace and the horizontals. This keeps the top diagonal brace on the same surface plane above the first three braces.

9. Anchor this diagonal into the three braces with your (5) 16d-nail pattern. This brace is the same length as the first. **NOTE**: Do not place this brace end on top of the horizontal where it is nailed to the raker. You do not want to nail over nails (Fig. 5–100).

10. Install the top section diagonal brace set up. The brace lengths are the same if the middle horizontal brace is dead center of the raker.

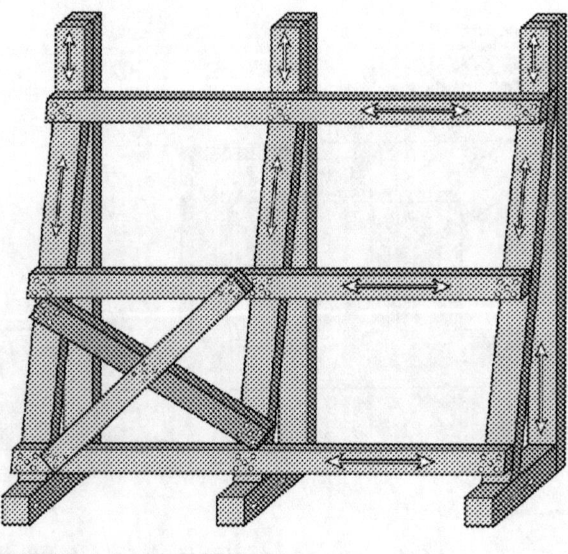

Fig. 5–100 Install the diagonal cross braces, making sure to not nail over nails.

11. Install the upper set exactly the same way as you did the bottom set. The braces must run the same directions! Notice that the first brace on both sections runs the same direction, a condition that must exist, otherwise the braces will not fit properly.

12. Anchor this set the same way as the bottom set of diagonal braces and with the same number of nails (Fig. 5–101).

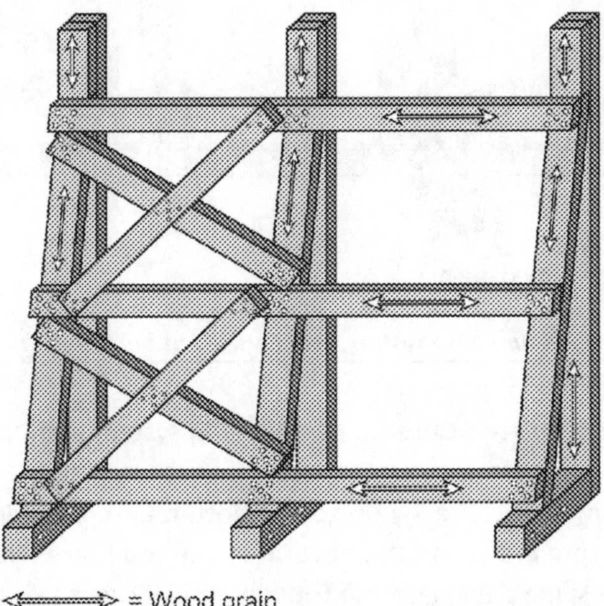

Fig. 5–101 Install the second set of diagonal cross braces the same way as the bottom set.

Figure 5–102 shows a 4-shore system. With the raker shores placed 8 ft on center, it makes it easy for us to run 16-ft 2x6s as horizontal bracing members. When running these horizontal braces, make sure you alternate the joints on the different rakers. It is important that you cross brace both ends for stability. You do not have to cross brace every bay (the space between individual rakers). You can do each end and every fourth bay. That will be enough to sustain the load.

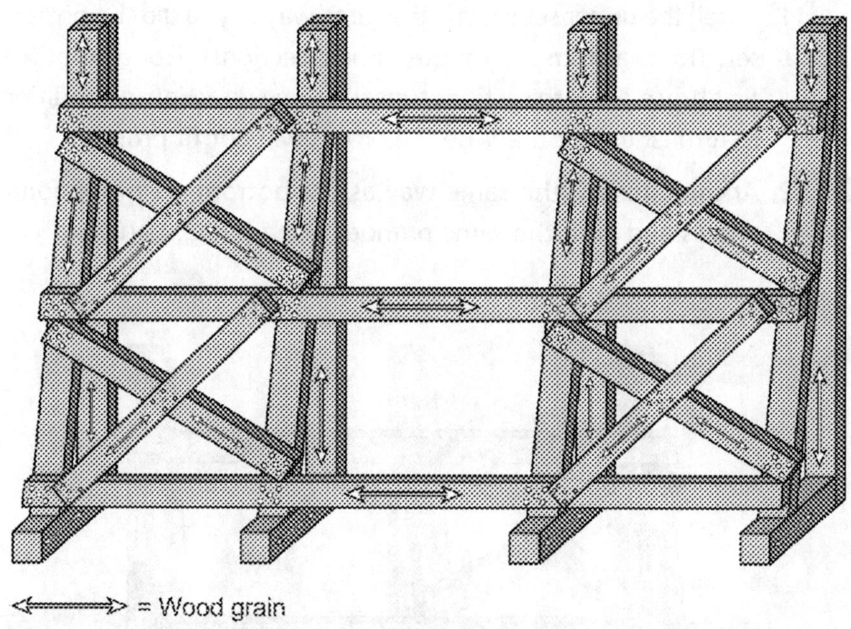

Fig. 5–102 A four-shore system, cross braced at both ends for stability.

Figure 5–103 is a close-up of a bracing system with less than an 8-ft wall height properly assembled and anchored using cross braces. The bottom one is directly on top of both rakers, touching the top of the bottom brace and the bottom of the top brace. These braces can be the same size. Place the top diagonal inside the nail patterns of the horizontal braces. DO NOT anchor nails over nails. By keeping the top diagonal in tighter angle, you can use the same size piece of 2x6 as the bottom diagonal. Nail down the center of the X where the braces cross.

Fig. 5–103 Close-up of a bracing system with less than an 8-ft wall height.

In Figure 5–104, there are no 2x6s available, so 2x4s were used. You must double up on the 2x4s as shown. Anchor each one down with (3) 16d nails. Assemble this setup the same way as you assembled the 2x6s. As a rule of thumb, nail all dimensional lumber with 16d nails and place one less nail into the material than it is wide. For example, place five nails in a 2x6 with a 5-nail pattern; place 3 nails in a 2x4 in a diagonal line.

Fig. 5–104 Using 2x4s for cross-bracing when 2x6s are not available.

Figure 5–105 shows a lateral view of two rakers with an insertion point less than 8 ft in height that is anchored and braced properly.

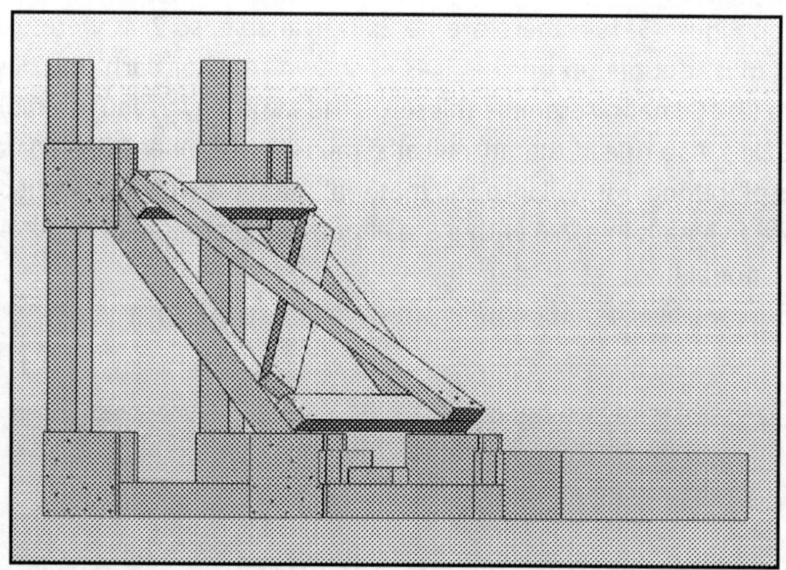

Fig. 5–105 Lateral view of two rakers
with an insertion point less than 8 ft in height.

As another rule of thumb, if you have a freestanding wall that must be braced, make the insertion point roughly at the top ⅓ of the wall if there are no major cracks, buckling, or visible deformations.

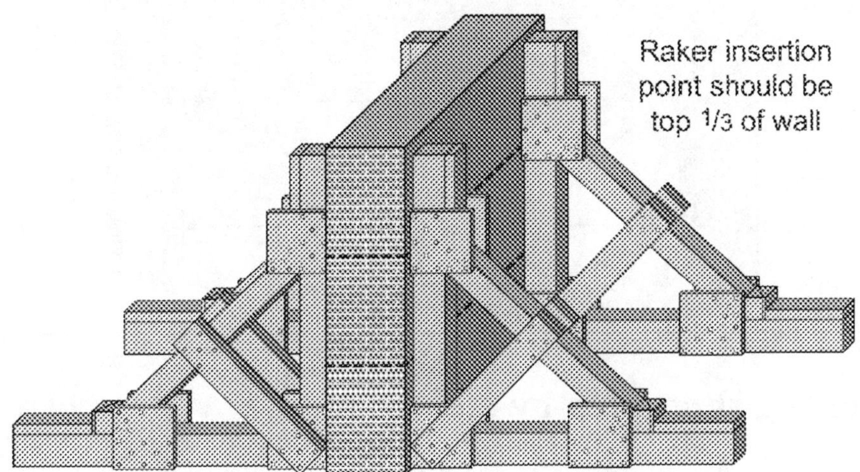

Fig. 5–106 Make sure the insertion point is at roughly the top third of the wall.

The Flying Shore

Although a flying shore may look very complicated, it really isn't. This shore is basically an exterior horizontal shore used to support an exterior wall of one structure with another. It can be used in shaft ways or alleyways. When pressuring the system, you must make entirely certain that the wall you are going to use as a support can actually take the additional pressure from the damaged structure. To erect a multistory flying shore, place one on top of the other. Have the wall plates just continue up the side of the building and place struts at each floor level.

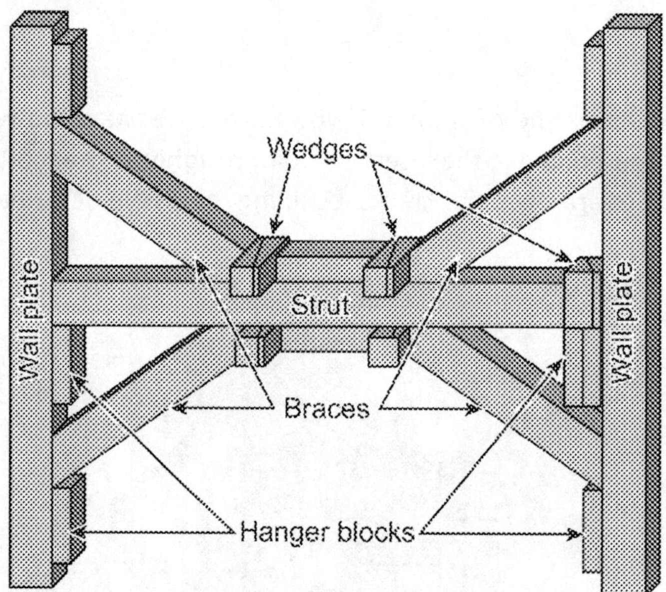

Fig. 5–107 Elements of the flying shore.

Flying shore step-by-step procedure

The following is numbered list of the tasks that must be completed to build and install a flying shore. The remainder of this section on the flying shore provides detailed instructions for constructing and installing the various elements of the shore.

1. Clear an area large enough for your team to safely construct the shore because flying shores have to be erected in position. The size of the area varies with each situation you encounter.

2. Determine the size material you will need for the shore; 4x4s, 6x6s, and larger have been utilized in the past.

3. Place the wall plates up against both walls, keeping them plumb, and install hanger cleats. Anchor all in place.

4. Measure the opening for the strut, making sure you leave room for your wedges.

5. Install the strut and wedges, keeping them level and then pressurize.

6. Determine the angle and length of the diagonal braces. These vary with each situation.

7. Install the bottom section braces first and cleat them at the bottom.

8. Install the top section braces and cleat them at the top.

9. Install the bottom center cleat and wedges and snug up wedges but do not fully tighten the wedges.

10. Install the top center cleat and wedges; tighten up the wedges at this time.

11. Retighten the bottom wedges and check the top wedges.

12. Gusset plate the shore if necessary.

Wall plates. There are two wall plates, one each side, that must be at least 8 ft long and on occasion may have to be 12 ft long. Use 4x4s with a 4x4 strut, 4x6s with a 6x6 strut; and for larger struts, use the 4-in. width of the rectangle lumber and the same dimension as the strut, (8x8, use 4x8s as the wall plates). For the cleats, use 2-in. dimensional lumber the same width as the wall plates. They must be at least 12 in. long; but in many cases, 18 in. works better.

Strut. The main support element of the shore, struts are used with a set of wedges to apply pressure between two wall plates. Struts are always square 4x4s, 6x6s, or 8x8s, depending on the length and the amount of support that is needed.

Diagonal braces. Diagonal braces are usually made of 4-in. dimensional lumber (4x4 or 4x6) placed on an angle. This angle can be 45° or 60°, again determined by the situation. Use four diagonal braces, two top and two bottom, spanning from wall plate to strut.

Horizontal cleats. Two cleats that sit on top and underneath the center of the strut, horizontal cleats are constructed of 2-in. dimensional material. Place them between the two diagonal braces and anchor to the center of the strut.

Wedges. A flying shore has five sets of wedges. The first set pressurizes the strut to the walls. The other four sets pressurize the diagonal braces to the strut and the wall plates.

Gusset plates. Any vibration or an earthquake situation warrants the use of gusset plates where the joints of the strut intersect the wall plate and the diagonals intersect the wall plate and strut.

Determine the position of the shore and the location of the wall plates. Place the plates against both walls; shim and plumb if necessary. If the wall being supported is buckled or deformed, it is a good idea to install the wall plates first and then measure. Determine where to place the horizontal strut and measure between the two plates. Don't forget to deduct the width of your wedges from the measurement. Place two cleats in position so you can rest the strut on top of them. These hanger cleats should be 2x4s at least 1 ft long. On the side where the wedges will be installed between the strut and the wall plate, install two cleats on which you can rest the strut similar to the method used on the interior horizontal shore.

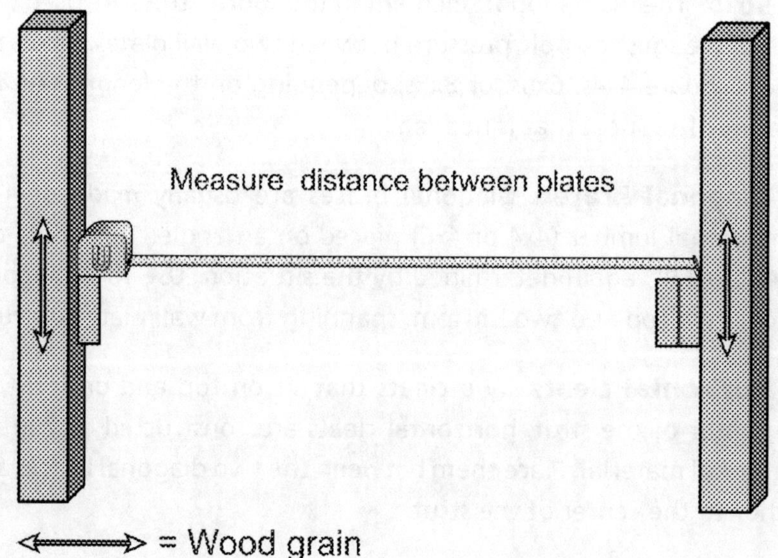

Fig. 5–108 Determine where to place the horizontal strut and measure between the two plates.

Construction of the shore pictured in Figure 5–109 begins the same way as the horizontal shore. Place the wall plates and the center horizontal strut in position just like you did with the interior horizontal shore. Remember to keep all the shore's elements as plumb and level as possible. Pressurize the wedges to secure the strut. Toenail the strut to the wall plate and the cleats on the wedge side. Place the wedges on the side of the good wall for safety reasons.

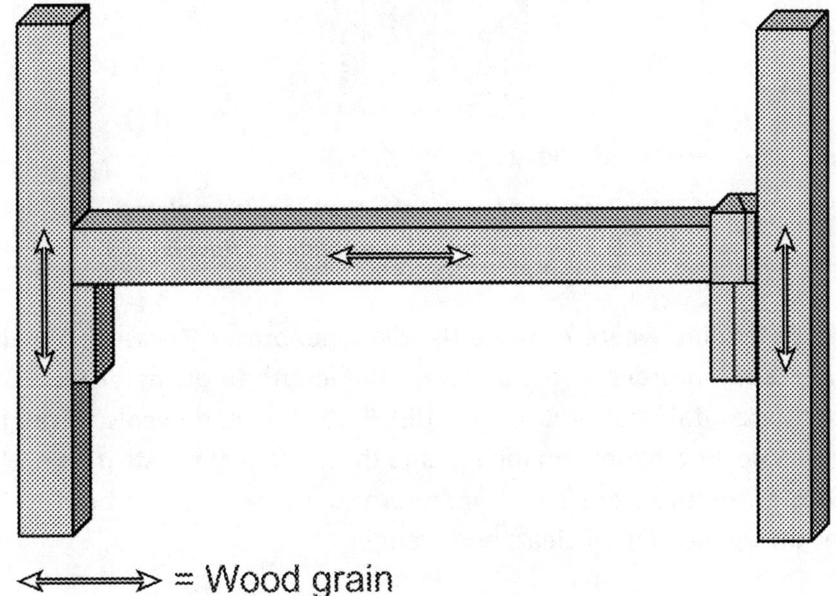

Fig. 5–109 Place the wall plates and the center horizontal strut in position.

Figure 5–110 is a close-up of where the strut should be placed, directly in line and in front of the building's floor beams. Placing the strut in the middle of the wall does not support anything. In fact, it could cause a collapse, especially in unreinforced masonry construction.

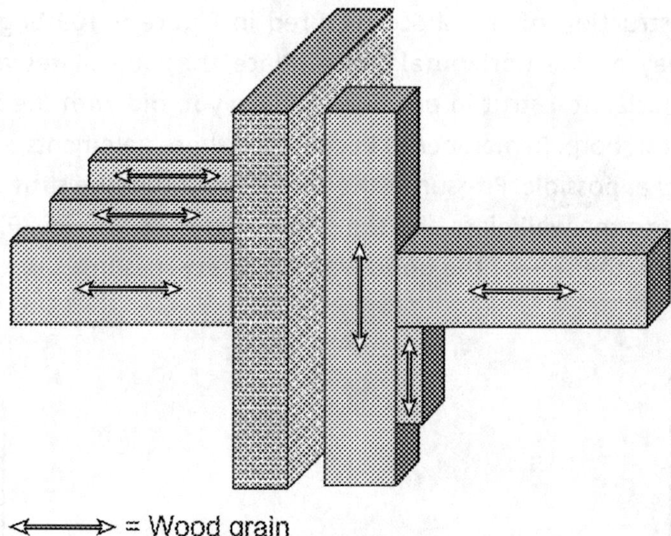

*Fig. 5–110 Close-up of where the strut should be placed—
directly in line and in front of the building's floor beams.*

Determine where to place the diagonal braces. These diagonals are placed in order to cut down on the length-to-diameter ratio of the horizontal strut and to help distribute the load evenly through the shore. Install them on the top and the bottom of the strut on both sides. Make sure there is a 1½-in. return cut in the braces on both ends for the application of cleats and wedges.

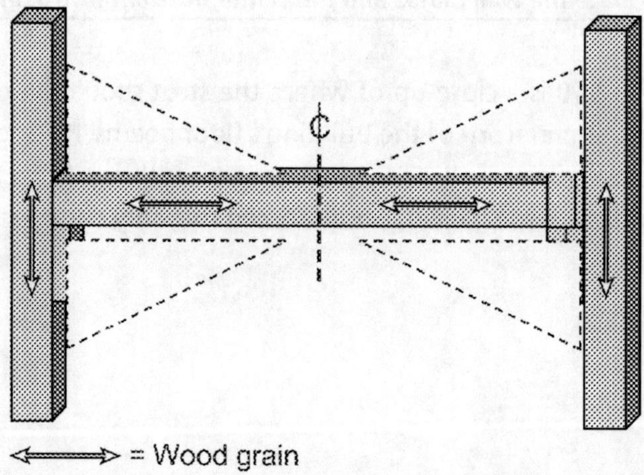

Fig. 5–111 Determine where to place the diagonal braces.

Install the bottom set of braces first, lining them up for a good fit. Toenail them into the wall plate and strut on the outside face of the brace only.

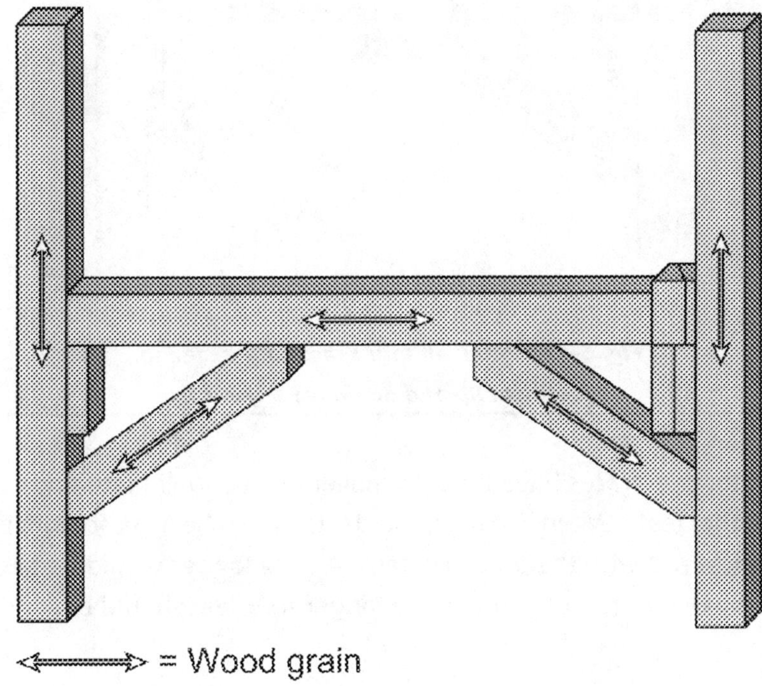

Fig. 5–112 Install the bottom set of braces first.

When all four of the braces are installed, slide them up and down to achieve a good fit. These braces are typically constructed of material the same thickness as the strut. The typical angle is 45°, but it can be 30° or 60° if necessary for a long strut. If the strut is 6x6 lumber, the braces can be 4x6 and constructed on the flat, that is, with the 6-in. section parallel to the 6x6 timber.

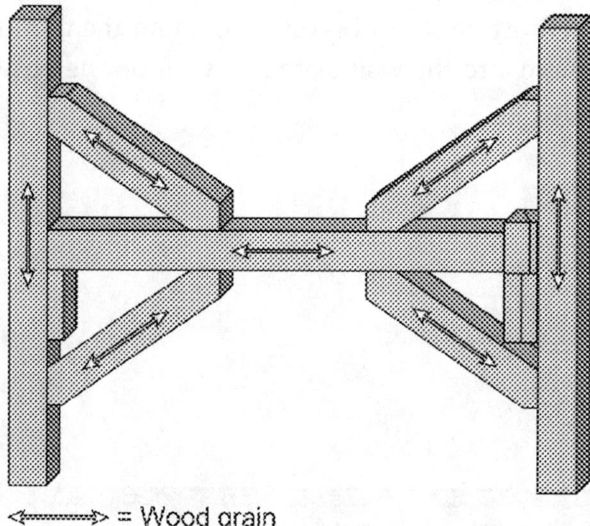

**Fig. 5–113 When all four braces are installed,
slide them up and down for a good fit.**

After the braces have been toenailed to the wall plate and strut, install the cleats. When installed, the cleats should be at least 18 in. long and anchored with the 5-nail pattern. A little longer would be better. Basically the larger the lumber, the longer the cleat should be.

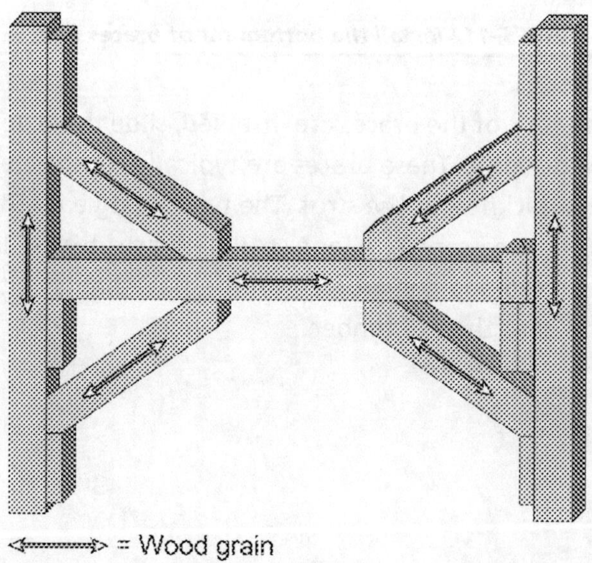

**Fig. 5–114 Install the cleats after the braces have been toenailed
to the wall plate and strut.**

Place a 2-in. dimensional piece of lumber the same width as the strut on the bottom of the horizontal strut. Leave room for wedges on each side. Install a set of wedges on both ends of the cleat and snug them up. Don't pressurize the braces at this time; wait until the top braces are installed.

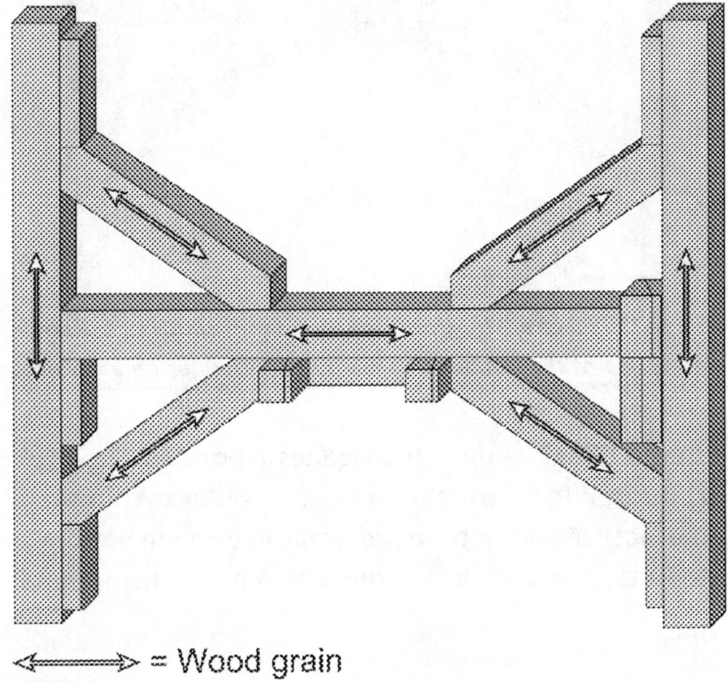

Fig. 5–115 Place a 2 in.-thick cleat the same with as the strut on the bottom of the horizontal strut, install wedges on each side.

Install the other horizontal cleat on top of the strut. Place a set of wedges in position at the base of each brace and tighten them equally. Do one side at a time. Tighten up the bottom first just enough to pressurize the brace. Do not tighten more than necessary or you could deflect the strut upwards, bending it. Next, tighten up the top wedges to lock the plate into the wall securely.

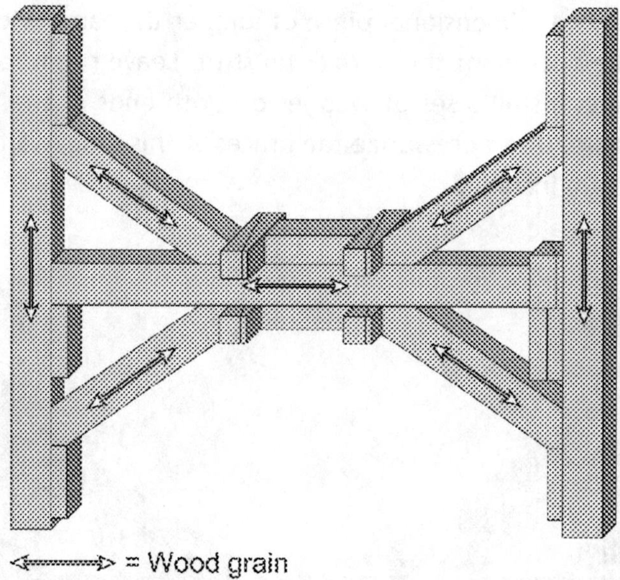

Fig. 5–116 Place a 2 in.-thick cleat the same width as the strut on the top of the horizontal strut, install wedges on each side.

Figure 5–117 is close-up of the wedges in position. The 2x blocking and the bottom 2-in. thrust block keep the wedges in position should they come loose. For long-term use, you can trim the bottom wedges and install a cleat under them in order to prevent them from falling.

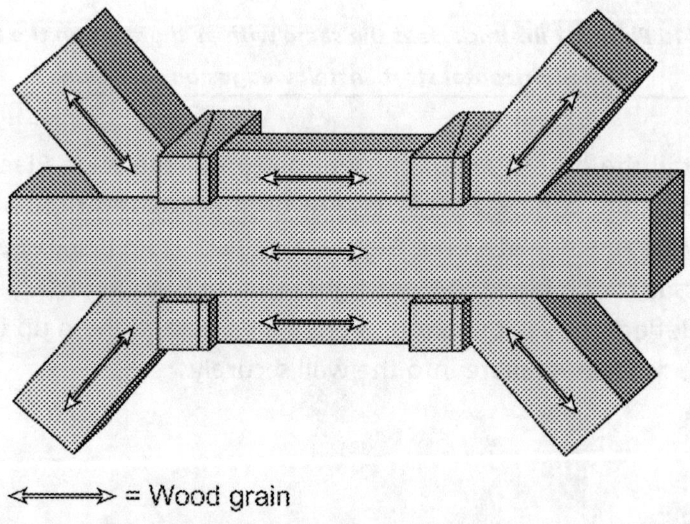

Fig. 5–117 Close-up of the wedges in position.

Figure 5–118 shows the flying shore in position with cleats and gussets installed. Make sure all the joints are locked in position. It is a good idea to install the gussets on both sides of the shore in or to lock the joint tighter for an earthquake or major vibration situation.

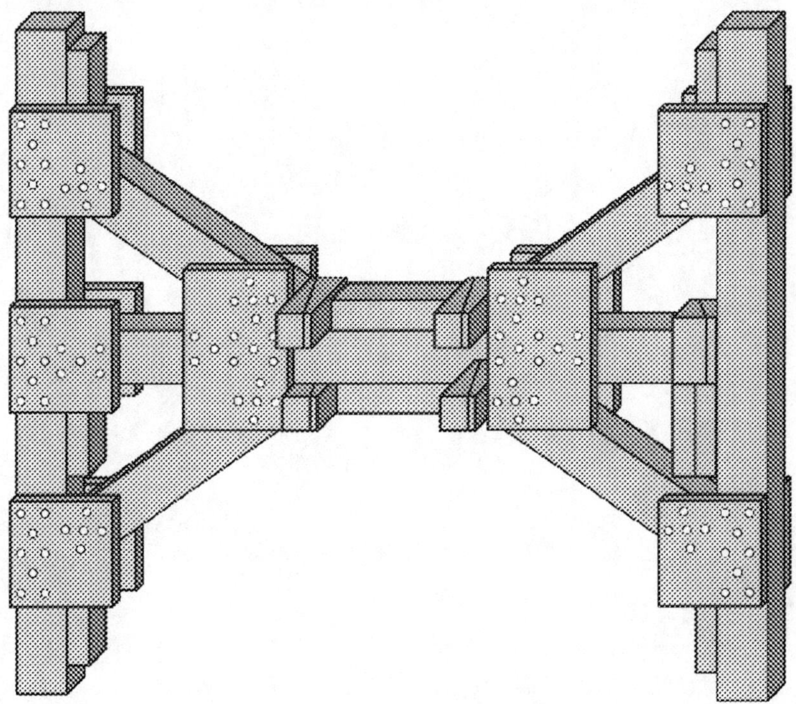

Fig. 5–118 The flying shore in position with cleats and gussets installed

The Shoring and Cribbing of Sloped Surfaces

The Sloped-floor Shore

The sloped-floor shore is used for large sections of flooring that may be leaning, propped up, or otherwise lying at a drastic angle to the ground. The drastic sloping of the floor has created a void where the floor has moved from its original position. If the floor has risen several feet (more than 3 ft), the use of other types of shoring may be more efficient than box cribbing. Box cribbing is a series of short pieces of lumber stacked on top of one another. It is utilized for support of buildings, building materials, or collapse debris. However, it is not suited for working with severely sloped surfaces. Instead of erecting box cribbing, the sloped-floor shore could possibly fit the bill.

There are times when you may be searching inside a collapsed structure that has drastically sloped floors—or even walls for that matter. In many cases where these structures are heavily damaged and deemed unsafe to enter without shoring, some stabilization may have to be accomplished in order to complete your objective. Normally, box cribbing would be your first choice for smaller openings. However in many situations, this may not always be your best option. When the opening you are attempting to shore is greater than 3 ft high, it may be easier and require far less material if you utilize what we call a *slope-floor shore system*.

Another situation will also present itself when the use of slope-floor shores is indicated. This is the situation in which the slope of the floor to be stabilized has an angle greater than 15° or 30% slope. At this angle, the load applied against a crib many times cannot be funneled through the center of the crib. This causes the crib to become unbalanced and possibly fail. Since the art of collapse shoring is not exact, try to give several different types of shoring options that would be available to the rescue specialists. In this situation, explore a couple of different possible solutions to the problem.

Whenever your team is determining the type of shore to erect and the best location to place it, the one thing that must be considered is the generation of physical forces. When your team knows where the forces are being applied, it is much easier to design the shores and know exactly where to place them for optimum efficiency. This concept is particularly important in the support of the sloped-floor situation.

Size-up

A couple of items must be looked at before you attempt to erect this type of shore. The size-up includes the following:

- The identification of the unstable load situation

- The integrity of the floor area you will be shoring

- The support potential of the floor area on which the shore will sit

- The weight to be supported with the erection of the shoring system

- The accessibility to the area to be shored

- The space available for the rescuers' movement within the collapse area

Identification. When sizing-up a collapse area, take a look at the situation and determine the type of shoring that would work the best given that specific damage. One of the biggest issues is the height of the void area; less than 3 ft high suggests box cribbing. A height greater than 3 ft suggests either a small vertical shore or a sloped-floor shore. In some cases if there is enough room, a small vertical shore may be the best option.

Floor integrity. When you enter a structure with a sloped floor, you must identify the type of floor system you face such as wood or concrete floors or steel and concrete floors. The different types have unique problems associated with them. Wooden floors tend to bend and deflect quite a bit, thereby necessitating a variety of angles in the shore, which can give you some trouble when assembling and anchoring the shore. Concrete floors may be very heavy and require large size lumber. If the slabs of concrete are cracked too badly, it may be ineffective to place shoring against them as the entire integrity of the floor may be compromised. As a result, the condition of the floors must be thoroughly examined before placing the shores.

Support potential. The area the sloped-floor shore is to sit on and to which it transfers the overload is a high priority in your size-up. The area must have the potential to absorb the load or be able to help transfer the additional load to the ground. You must be able to examine underneath where you are going to place your shore and determine that possibility. If you determine that the area cannot support the additional load, do not bother to erect the shore. Other stabilization procedures or debris removal may have to be considered as an alternative.

Weight issue. The issue of how much weight is to be supported is determined with little effort. Most masonry debris weighs roughly 125 lb per cu ft. Estimate the cubic footage around the area to be shored and calculate the approximate weight to be supported. Once you have done that, you can determine how many shores you need or if you need to increase the size of the shoring material from 4x4 to 6x6.

Shore spacing and access. When you assemble shoring in any void situation, you must always consider team access and egress. There are times when quite a bit of debris removal must occur to continue the rescue operation. Make sure when you install any shoring that the installation does not block access to other rescue areas. This may take some planning on the part of the team to determine the best location for the shores. Don't box yourself in or cut off your only means of egress. There are two types of sloped-floor shores: perpendicular and friction.

Sloped-floor shore – perpendicular

The perpendicular sloped-floor shore is just that, perpendicular to the forces being applied to the sloped floor. This type of sloped-floor shore is erected under a slab that is pinned or fixed in position and will not slide on you. This shore generally is the first choice, especially if you know that the lower base of the sloped floor is anchored and not able to move. The type of floor and the amount of debris to be supported dictates the size lumber to utilize; generally speaking, use either 4x4s or 6x6s.

Your first step in safely erecting this shore in position is to install some sort of temporary safety shoring. Your two main choices normally are either manufactured adjustable aluminum rescue struts (such as the Paratech rescue strut system) or the standard T-shore fabricated from wood. Whichever type you choose, use several of them in order to cover the necessary area.

When you erect the sloped shores, assemble and place them in twos, just like the raker shore systems. Do this in order to increase the stability of the shoring system. This is why the necessary number of safety shores must be installed.

Step-by-step procedure

1. Determine the area to be shored and clear it of debris.

2. Place sole plate down and install the header above.

3. Determine the two locations for the posts and the angle of the cut.

4. Measure for both posts.

5. Install the posts with the bevel side on the sole plate and nail them to header.

6. Anchor down the sole plate.

7. Cleat and wedge behind both posts to pressurize them to the floor above.

8. Install a second shore within 8 ft of the first one. Usually, this second shore is installed no more than 4 ft from the first one.

9. Cross brace both shore sections inside and outside.

10. Cross brace both shores together at right angles to the shores.

Figure 6–1 shows the sloped-floor shore perpendicular method type 2 style (with a solid sole plate).

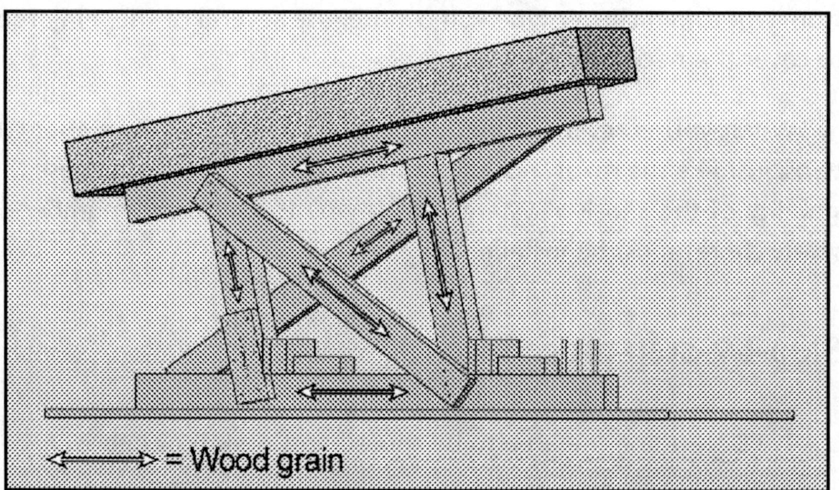

Fig. 6–1 The perpendicular sloped-floor shore, type 2.

Header. The header of a sloped-floor shore is usually a 4x4 or can be a 4x6 for larger shores. Anchor it to the damaged floor.

Sole plate. A sole plate is also usually a 4x4 or a 4x6 for a larger shore. It is anchored after the posts have been placed in position. Use three 1-in. steel pins or equivalent to hole the sole plate in place. Both the header and soleplate must be approximately 2 ft longer than your post spacing to be effective.

Posts. Generally 4x4s, but in some instances 6x6s, posts usually come in pairs in a slope-floor shore although on rare occasions more than two can be used. Keep them both within 12 in. of the ends of the header and soleplate. One end stays square, and the other end has the specific angle cut into it.

Cleats. Cleats are generally 2x4s or 2x6s and at least 18 in. long; but in many cases, a 24-in. length might be preferred, especially if the angle of the post is steep.

Wedges. There is one set of wedges behind each post. Each set is pressurized against the cleats to tighten up the shore against the damaged floor.

Cross braces. The last items in a sloped-floor shore can be a 2x4 in a small shore but are generally 2x6s. Cross brace each shore section, and then cross brace the two sections together, joining the two legs of the shore and making the elements one shore.

Determine the position of the header and the size material to use. When this is done, place the header in position and anchor it to the floor slab. Several methods can be used to install the header to the floor. You can drive a bolt through the header and into the floor or anchor the header with plates. Either metal or wood anchor is another option. The same methods utilized to anchor rakers to structures can be implemented in this situation. Just remember, before you continue with the rest of the shoring operation, you must pin the header into place.

After the header is installed and anchored, the next item to be addressed is the installation of the sole plate. Place the sole plate in position directly underneath the header; make sure it is in line with the header. Try to keep it on a level plane and in vertical alignment with the header so that it will be much easier to install the posts when the time comes. Don't anchor the sole plate to the ground just yet. You may have to fine tune and adjust the shore.

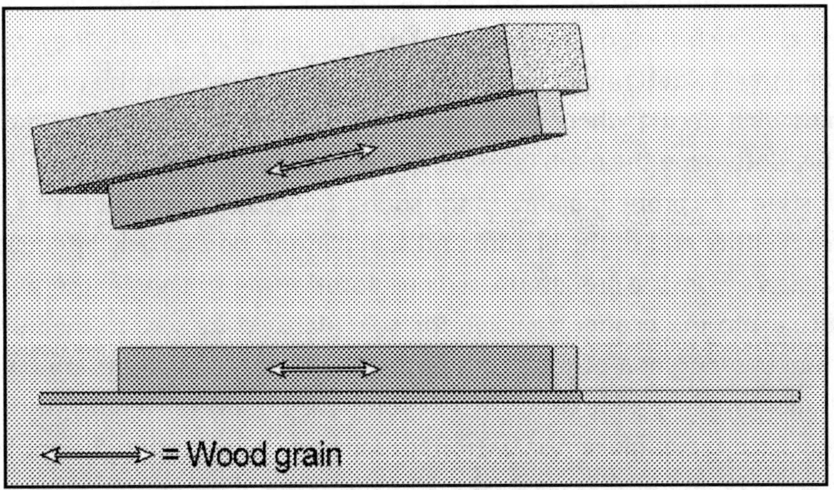

Fig. 6–2 Place the sole plate in position directly underneath the header.

The next step is to determine the angle of the slope of the floor to be shored. There are several methods available to accomplish this. You can use an angle finder, a computer-type level that gives you the angle, or pitch, on a digital readout. Or you can use a tape measure and a torpedo level. Then there is the old standby method: scribe the post to the angle. In Figure 6–3 an angle finder was used.

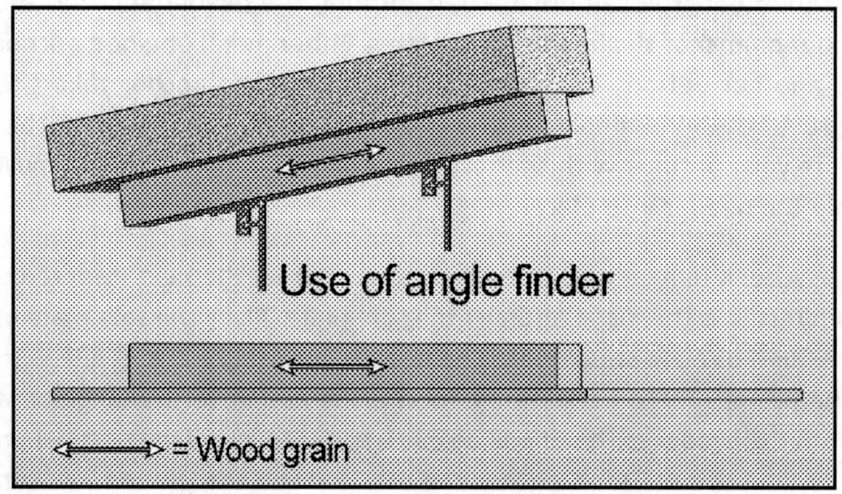

Fig. 6–3 Determine the angle of the slope on the floor.

Once your shoring team has determined the angle for the two posts, which is normally the number you generally install, they can then start to determine the length of each post. To accomplish this, you must first decide where the two posts should be best situated. To do this, determine the main pressure areas and the best placement of the posts for the proper balance of the shore. The posts must be at least 1 ft from either end of the header and sole plate to be properly effective. Measure from the predetermined post location at right angles (perpendicular) to the header down to the sole plate. Make sure to measure from the long end of the post's position in order to make fitting post angles properly a lot easier. When you have the overall length of the posts, cut the angle on the bottom of the post. After cutting the angle, place a 1½-in. return into the back of the angle, if necessary.

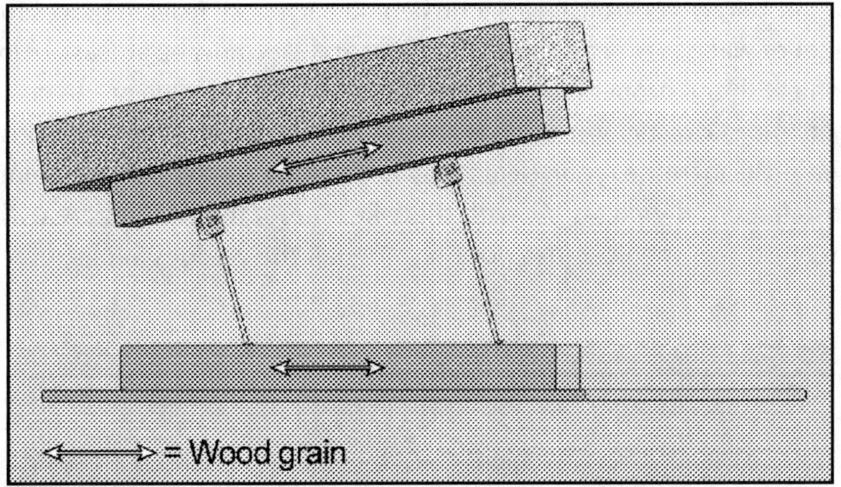

Fig. 6–4 Determine the length of each post.

At this time, install the posts and set them into position, toenailing each post into the sole plate and header. Make sure the sides of the posts are aligned with the face of the sole plate and that they are flush with the header and sole plate. Anchor down the sole plate in at least three places. Use three 1-in. pins or some substantial anchoring setup that is equivalent.

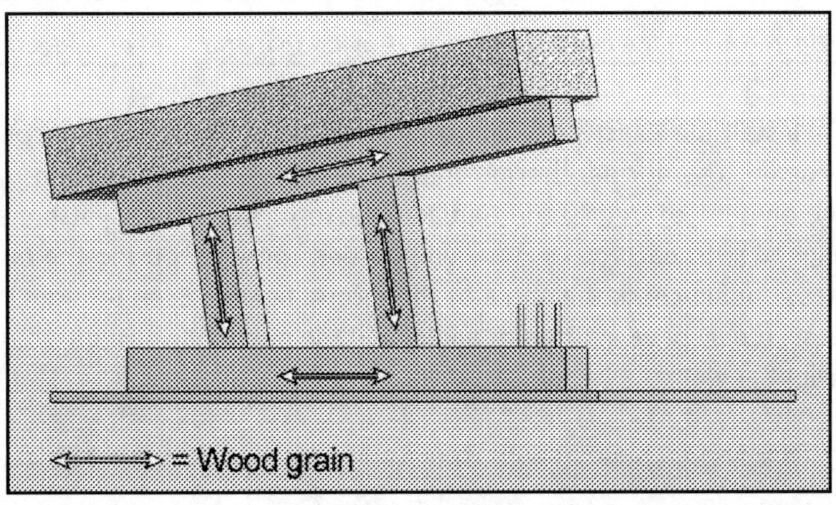

Fig. 6–5 Install the posts and set them into position.

Install two bottom cleats, one behind each post, leaving spaces for a set of wedges for each post. The bottom cleats must be at least 18 in. long with the proper nail patterns; however, as the angle of the posts gets more drastic and the possibility that the lateral forces against the posts increase, your cleats may have to be longer. At this point, install a set of wedges. As with the raker shore, the wedges are used to pressurize the post into position. Toenail the wedges in place.

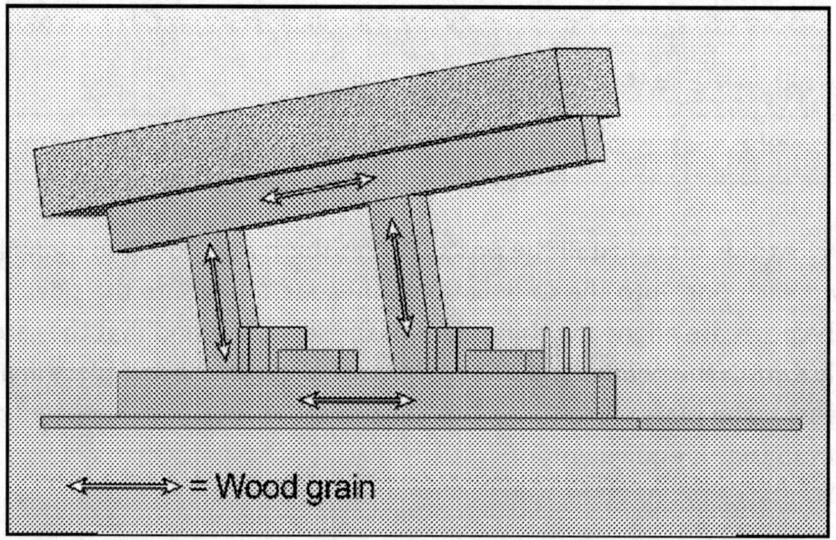

Fig. 6–6 Install the two bottom cleats, leaving space for a set of wedges.

Install the exact same shore alongside the one previously constructed. The normal shoring spacing can be 4–8 ft, depending on the amount of debris and damage above it. Keep the two shores aligned with each other; they will be tied together during the last step. Double-check the angle of the second header. Even though it's close, it may have a different angle, and the posts may be a different height as well.

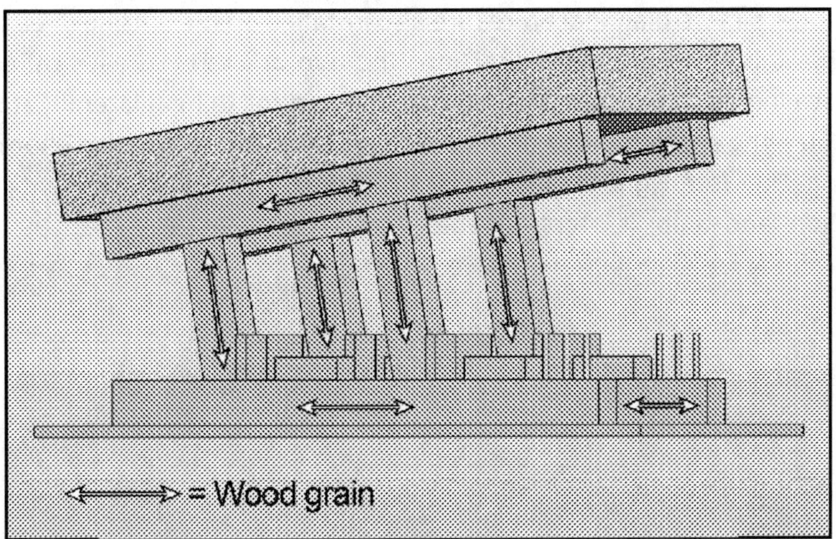

Fig. 6–7 Install a duplicate shore alongside the previously constructed one.

The next items to go in are the diagonal cross braces. The positioning of these braces is very important. They must be placed at the top of one post and at the bottom of the other post. They should be placed on the inside of each set of shores and in the same direction.

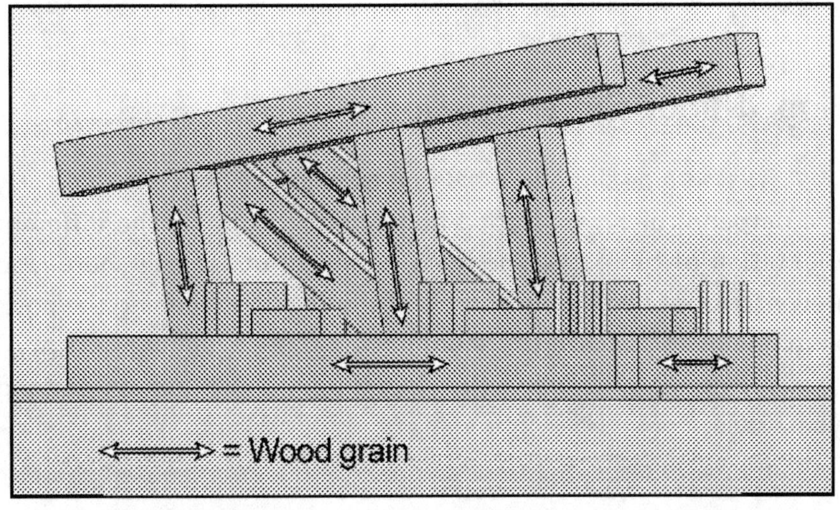

Fig. 6–8 Install the diagonal cross braces on the inside of each shore.

Two more diagonal braces are also installed on the outside of each shore, both in the same direction. The two braces cross each other, forming an *X*, which is what you are looking for to help laterally stabilize the shore in both directions. Next, cross brace the two shores together (see Fig. 6–32 later in this chapter). This boxes in the entire shore and keeps it laterally stable in all directions.

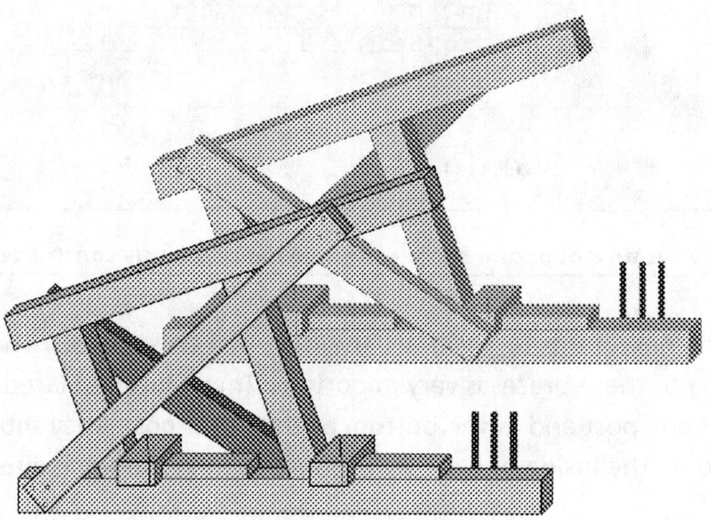

Fig. 6–9 Install two more diagonal braces on the outside of each shore.

Sloped-floor shore—friction

A friction type sloped-floor shore is generally built the same way as a perpendicular type sloped-floor shore except for the different orientation of the posts. The main reason to utilize this shore is to prevent a loose floor or roof section from shifting. Keeping the posts plumb helps to stabilize the shore in case there is any movement of the slab. Figure 6–10 shows the sloped-floor shore friction method type 2 style.

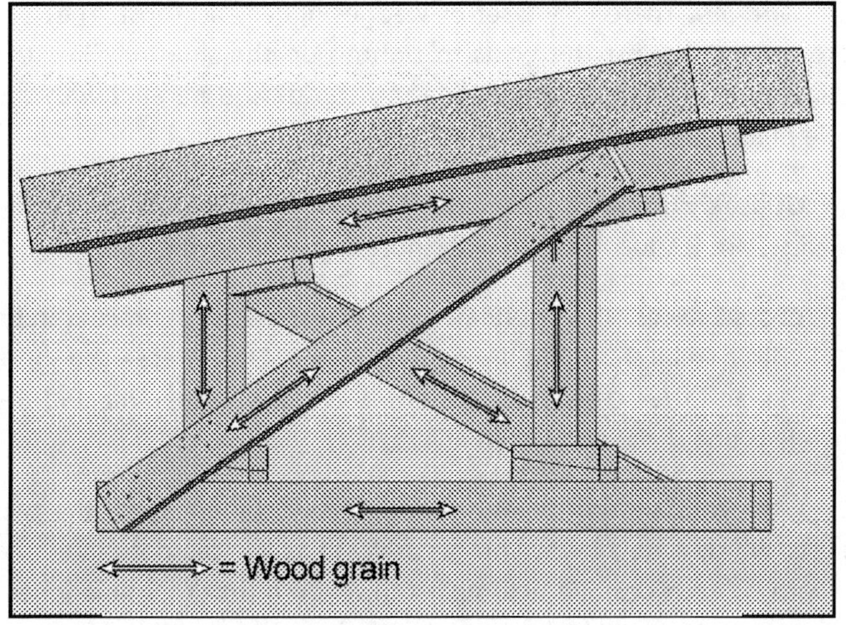

Fig. 6–10 The friction slope-floor shore, type 2.

Header. The header of a sloped–floor, friction shore is usually a 4x4 or a 4x6 for larger shores. It is anchored to a damaged floor.

Sole plate. Also usually a 4x4 or a 4x6 for a larger shore, the sole plate is anchored down after the posts are placed in position. Use two 1-in. steel pins or equivalent. Both the header and sole plate must be approximately 2 ft longer than the post spacing to be effective.

Posts. Generally made of 4x4s (but in some instances made of 6x6s) posts are usually installed in pairs in the system although on rare occasions there can be more than two. Keep them both within 12 in. of the ends of the header and sole plate. One end stays square and the other end has the specific angle cut placed into it. With this friction type shore, the angle cut will be placed at the header.

Cleats. Cleats are generally made of 2x4s or 2x6s and at least 12 in. long In many cases, they can be 18 in. long—especially if the angle of the posts is steep. Cleats are nailed behind the posts directly to the header with a 16d 5-nail pattern.

Wedges. There is one set of wedges under each post. They are pressurized against the posts to tighten up the shore against the damage floor. Keep these wedges on top of and parallel with the sole plate.

Gusset plates. Gusset plates placed on both posts beside the wedges are frequently used with shores after an earthquake.

Cross braces. The last items to go in a shore, cross braces can be 2x4 with a small shore but are generally 2x6s. Cross brace each shore section, and then cross brace the two sections together, making the two legs one shore and stable.

Once your shoring team has determined the angle for the two posts, which is how many you generally install, it can determine the length of each post. To accomplish this, you must determine where the two posts would be best situated. This will be done by examining the main pressure areas and the best placement of the posts for the proper balance of the shore. The posts must be at least 1 ft from either end of the header to be properly effective.

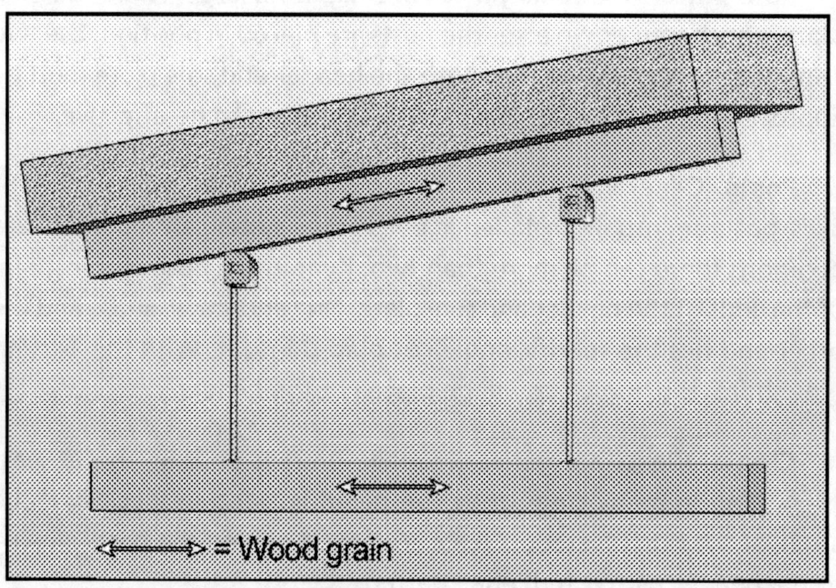

Fig. 6–11 Determine the length of each post.

The next task to be done is to measure from the predetermined post location at right angles (perpendicular) from the sole plate to the header. Measure on a plumb line from the header to the sole plate. Make sure to measure from the long end of the post's position in order to make fitting the post angles properly a lot easier. Because you will install a set of wedges under the post, don't forget to deduct the width of the wedges from the overall post length. When you have the overall length of the posts, you can then cut the angle on the top of the post. Now place a 1½-in. return into the back of the angle, if necessary.

Install the lower post first for ease of installation. Place a set of wedges under the post, toenail them into the header, then snug up the wedges. Don't tighten the wedges too much, or you could push the post up the header and loosen up the toenails. Make sure all the edges of the header, sole plate, post, and wedges are flush with one another.

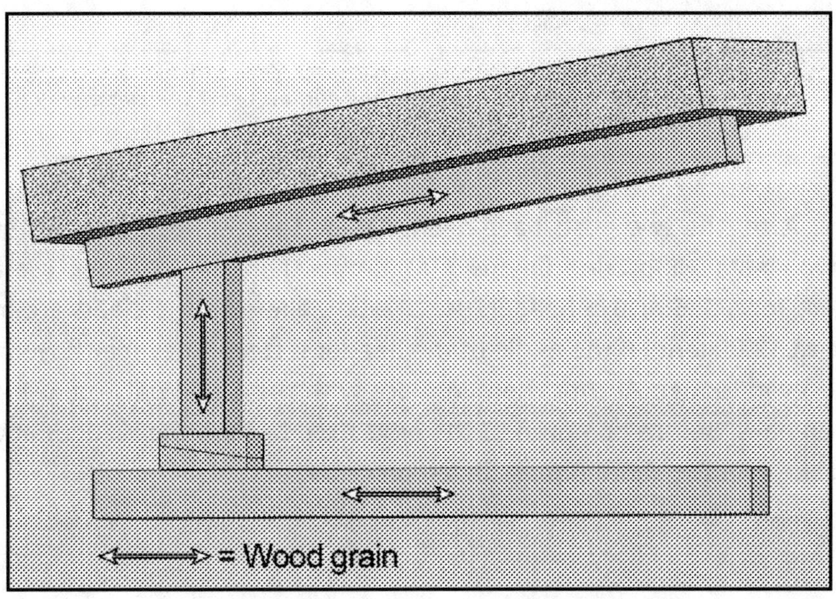

Fig. 6–12 Install the lower post first.

Install the second post, anchor and adjust it the same way as the first post, making sure the angles fit properly. A poor fit will make the shore much less effective.

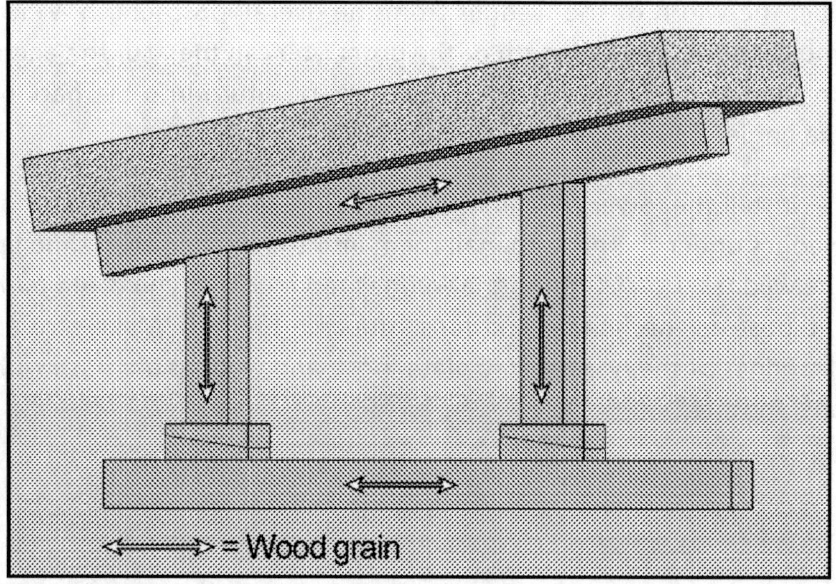

Fig. 6–13 Install the second post, anchor and adjust it the same way as the first post.

After the two posts have been installed and snugged up, install two cleats using 2-in. dimensional lumber. For a 4x4 header, use a 2x4; for a 4x6 or 6x6 header, use a 2x6. Make sure the cleats are at least 12 in. long. If the angle of the header is steep, they may have to be even longer. Anchor them in place with the 5-nail pattern, using 16d nails. Before going on to the next step, tighten the wedges and make sure the shore is pressurized properly.

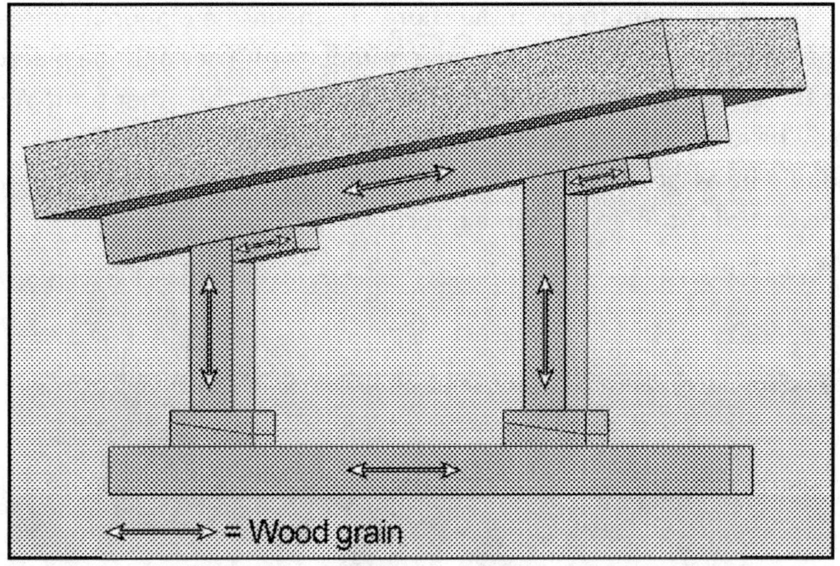

Fig. 6–14 Install two cleats using 2-in. dimensional lumber.

The next task to be completed is to construct the same shore adjacent to the one you just erected. Depending on conditions and the amount of damage and debris above, this shore is normally 4–8 ft from the first shore. Keep the two sections in line with each other (parallel) and check the angle of the new header. It may be different even though it's close by.

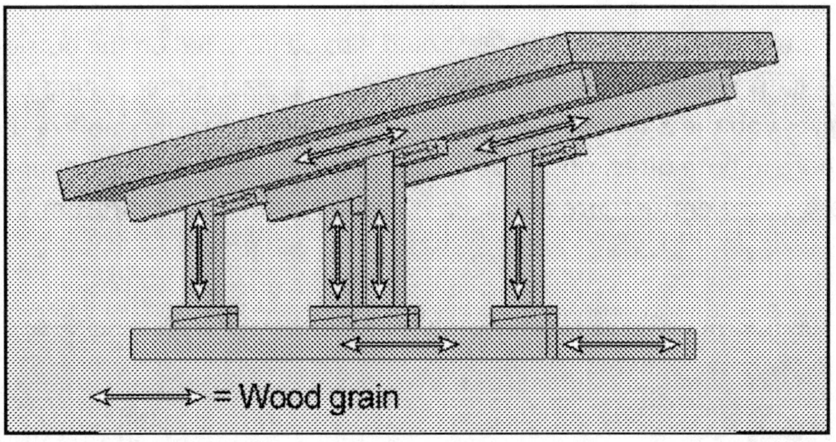

Fig. 6–15 Install a duplicate shore 4–8 ft from the previously constructed one.

The next items to go in the shore are the inside diagonal braces. The positioning of these braces is very important. They must be placed at the top of the one post and at the bottom of the other post. You may have to scribe the top of the brace and cut it to fit. The brace should be as close to the post as possible and placed on the inside of each set of shores in the same direction.

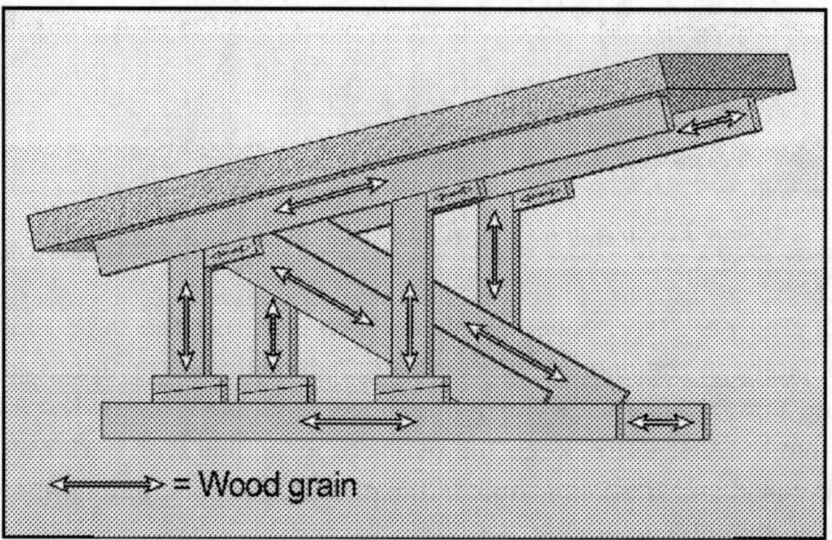

Fig. 6–16 Install the diagonal cross braces on the inside of each shore.

Two more diagonal braces should also be installed on the outside of each shore, also in the same direction as each other. Doing this gives you the X that you are looking for to help laterally stabilize the shore in both directions. Next, cross brace the two shores together to box in the entire shore and keep it laterally stable in all directions.

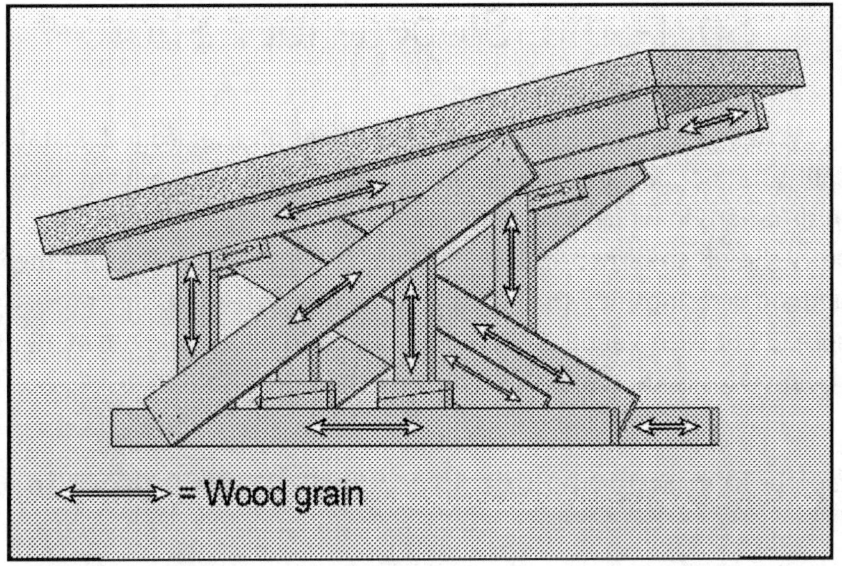

Fig. 6–17 Install two more diagonal braces on the outside of each shore.

In an earthquake situation, you need to place gusset plates along the bottom where the wedges are so that they will not dislodge during an after shock.

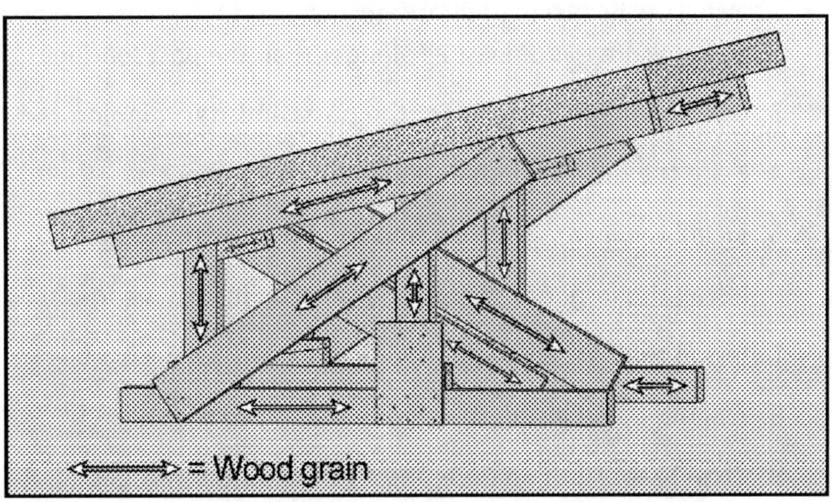

Fig. 6–18 Install gusset plates along the bottom wedges in an earthquake situation.

Split-sole, Sloped-floor Shore

When you need to have your shore bear on the ground, install a split-sole, sloped-floor shore. This shore is particularly advantageous in situations in which a building leans and collapses or if a soft second story slid off the structure. When using this method, make sure there is good bearing into the soil and that the post loads are properly spread out with sleepers. The sleepers should be a minimum of 18 in. square. This is known as the type 1 method (into soil).

Split-sole, sloped-floor shore step-by-step procedure

1. Install some sort of initial temporary shoring to safely erect this shore in position. When the sloped shores are erected, they must be assembled and placed in twos, just like the raker shore systems and the other slope-floor shores. Clear an area that is at least 24 in. wider than the length and width of the total shore. Make sure that you have cleared the debris down to good ground.

2. Install the header in position and anchor it to the damaged area, making sure it is at least 2 ft longer than the post spacing.

3. Determine the location for the posts and start digging.

4. Place the ground pads and measure for the posts.

5. Install the lower post first and snug it up with wedges.

6. Install the larger post next and snug up it with wedges.

7. Install the two, split-sole, 2x6 bottom braces.

8. Assemble and install the second section of the shore.

9. Cross brace both sections and tighten up the wedges.

10. Cross brace the two sections together.

Figure 6–19 shows the sloped-floor shore, also known as type 1.

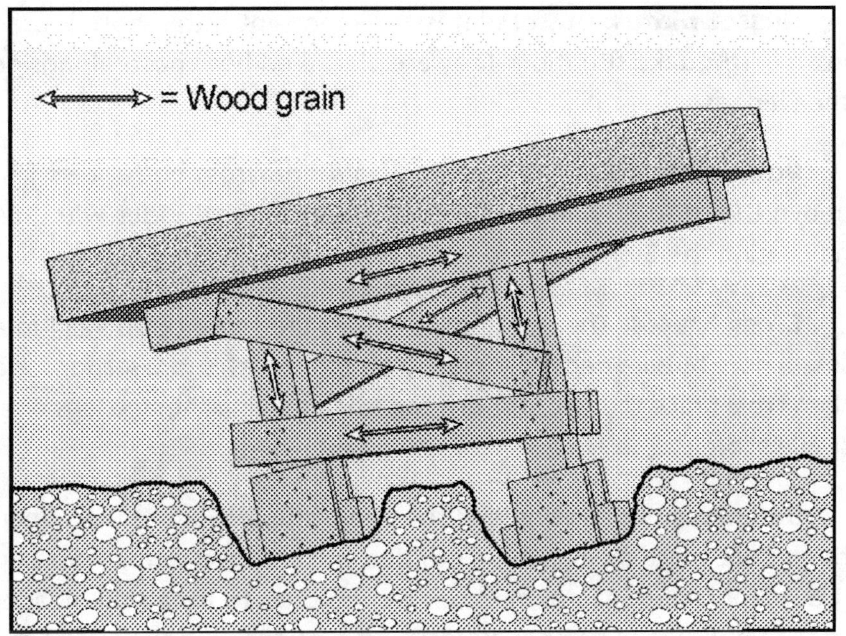

Fig. 6–19 The split-sole, sloped-floor shore.

Header. Usually a 4x4 or a 4x6 for larger shores, anchor the header to the damaged floor. The header must be at least 2 ft longer than your post spacing.

Posts. Posts are generally 4x4s. There are usually two in the system although, on rare occasions, there can be more than two. Keep the posts both within 12 in. of the ends of the header. Both ends of the posts must stay square.

Ground pads. Sections of plywood or 2x6s, ground pads are a minimum of 18x18 but may be larger if the ground is soft. Make certain they are placed below the surface at least 12 in.

Wedges. There is one set of wedges under each post. They are pressurized against the posts to tighten up the shore against the damaged floor. Keep these wedges on top of the sole plate and parallel with that plate.

Gusset plates. Gusset plates are important if the shore is used in an earthquake situation. They are placed on both posts alongside the wedges.

Bottom braces. With two 2x6s placed across both posts just above the ground, nail the bottom braces on both sides with 16d nails using the 5-nail pattern.

Cross braces. The last items to go in the shore can be 2x4 in a small shore but are generally 2x6s. Cross brace each shore section, and then cross brace the two sections together, making the two legs one, stable shore.

Determine the position and the size of the header to use. When this is done place the header in position and anchor it to the floor slab if at all possible. Several methods can be used to install the header to the floor. Just remember, before you continue with the rest of the shoring operation, it will be much easier if you can pin the header into place before continuing with the rest of the procedure. Determine the location for the two posts to be installed, using the information obtained during the size-up of the damage floor area.

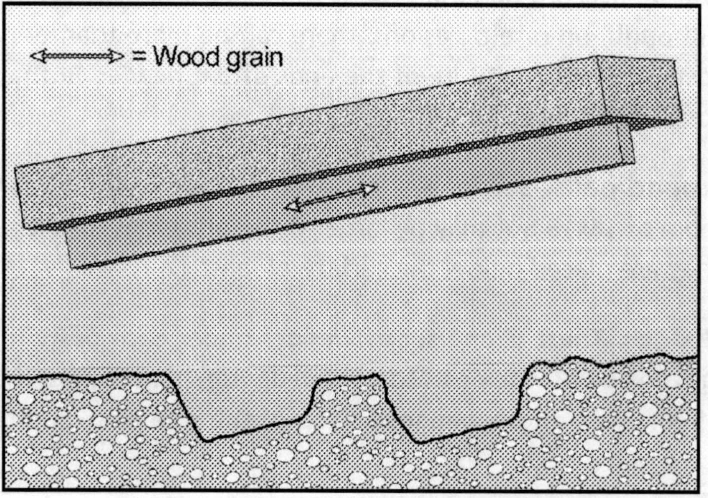

Fig. 6–20 Place the header in position and anchor to the floor slab.

The next step is to excavate the holes necessary for the pads and wedges. It is important that the holes are dug parallel to the slope of the slab to be stabilized in order to keep both ends of the posts square. You won't have to worry about any angle cuts in this situation. These holes must be deep enough for the bottoms of the posts to be below the surface of the ground. The holes can be roughly 24x24 in good soil conditions; poorer soil conditions may dictate larger or wider holes in order to get more bearing for proper stability. Softer soil will dictate larger ground pads, necessitating larger holes.

Place ground pads, or sleepers, on the ground in the holes just dug. Make sure that the 4x4 pads have full contact with the ground and the sleepers. There cannot be any space between the sleepers and the soil or the shore will move. Measure from the 4x4 pad to the header, remembering to deduct the width of your wedges. Measure for both the posts and pay attention to the angle of your tape. It has to be at right angles to the header.

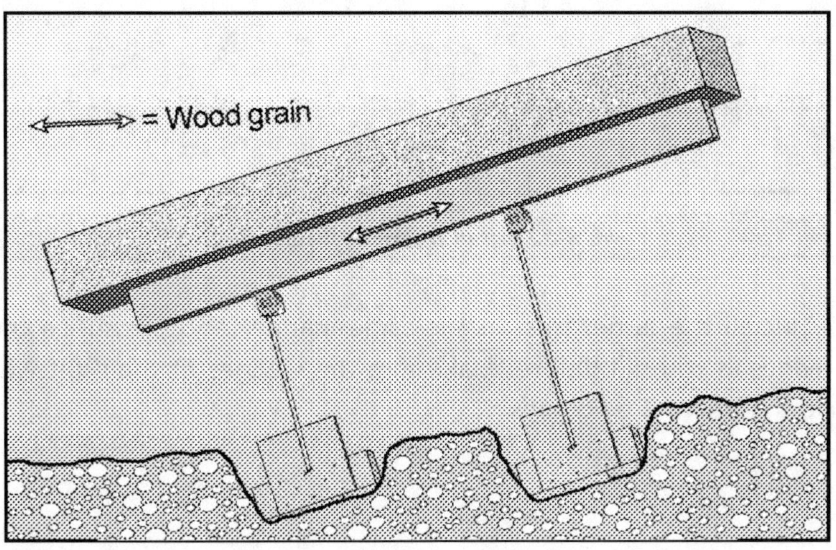

Fig. 6–21 Measure from the pads to the header, remembering to deduct the width of the wedges.

Install the smaller post first so that it is a lot easier to erect the shore. To install the post, slide it into the channel block on top of the ground pads and toenail it to the header. Place the wedges in position in the trough under the post and pressurize them. Make sure there are no spaces between the channel and the soil then tighten up the wedges until you have full bearing.

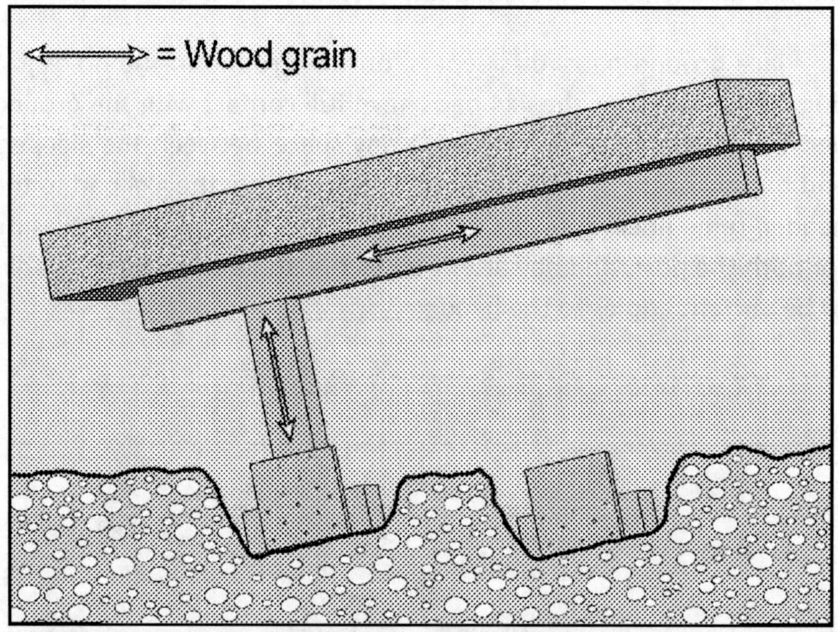

Fig. 6–22 Install the smaller post first.

Install the larger post in the same manner and pressurize the wedges as you did before. Make sure the wedges fit the right way and that there are no gaps.

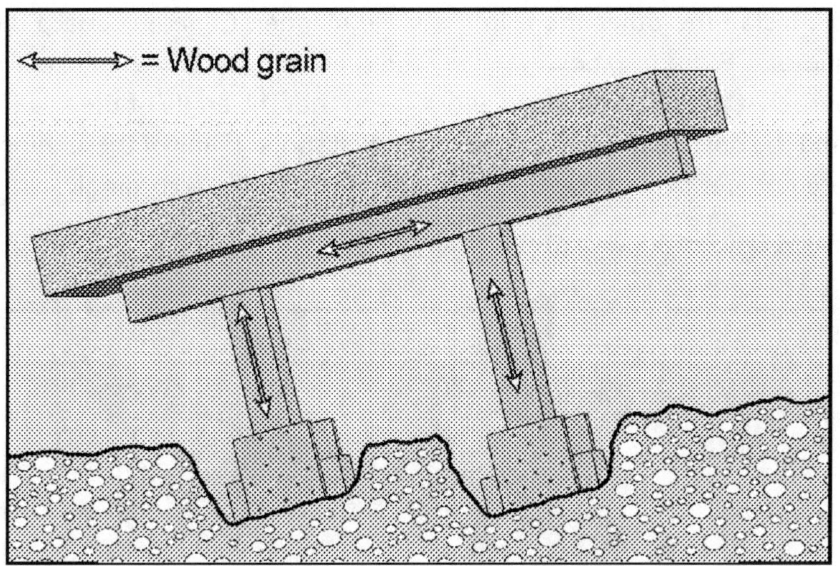

Fig. 6–23 Install the larger post in the same manner.

At the base of the posts close to the ground, place a 2x6 horizontally on both sides of the posts. Anchor in place with the 5-nail pattern using 16d nails. Keep the braces in line with each other and keep the overhang to a minimum (6 in. is fine).

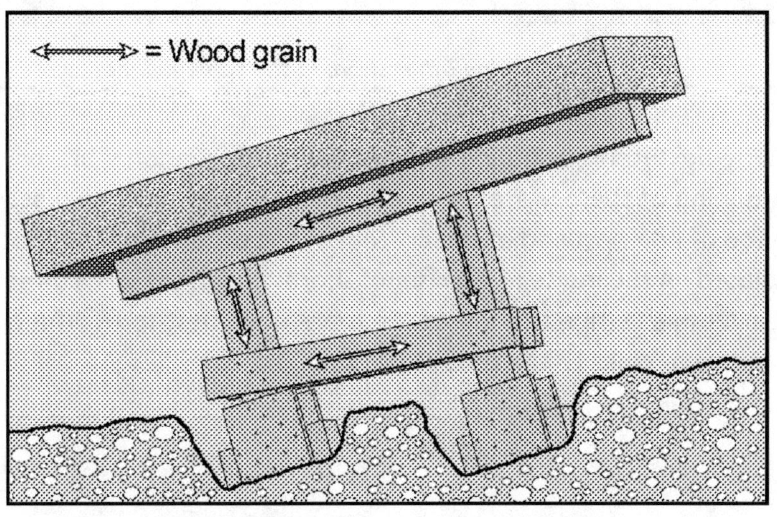

Fig. 6–24 Place a 2x6 horizontally on both sides of the posts.

Install the other shore in the same manner. Generally, space the shores 4–8 ft apart; however, vary the space if necessary. Again, make sure that the holes are deep enough and that the angle of the header and the post height is the same as the first shore. If any of these things are not identical to the first post, adjust the pieces accordingly.

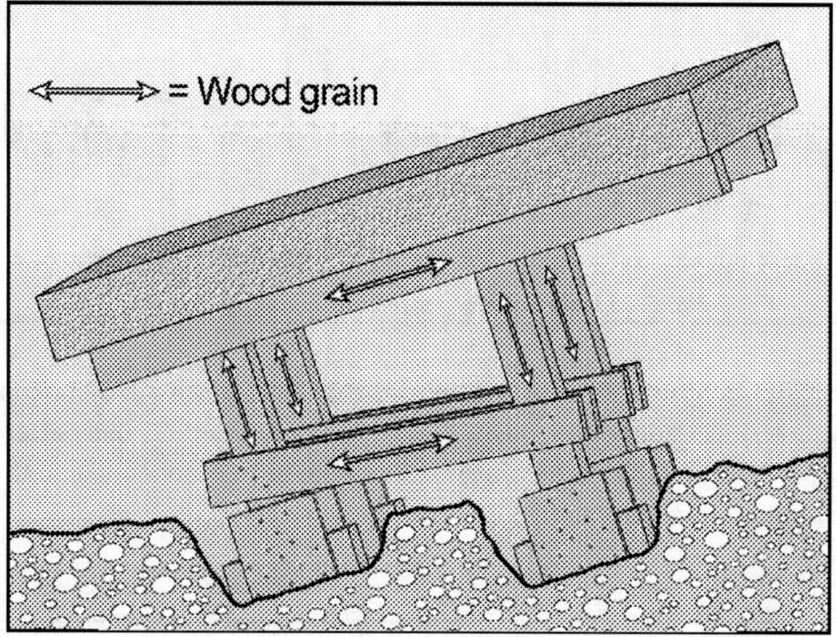

Fig. 6–25 Install the other shore in the same manner.

The next items to go in place are the inside diagonal braces. The positioning of these braces is very important. They must be placed at the top of one post and at the bottom of the other post. They should be placed on the inside of each set of shores and in the same direction as each other. Scribe the top of the brace and cut it to fit. The brace should be as close to the post as possible.

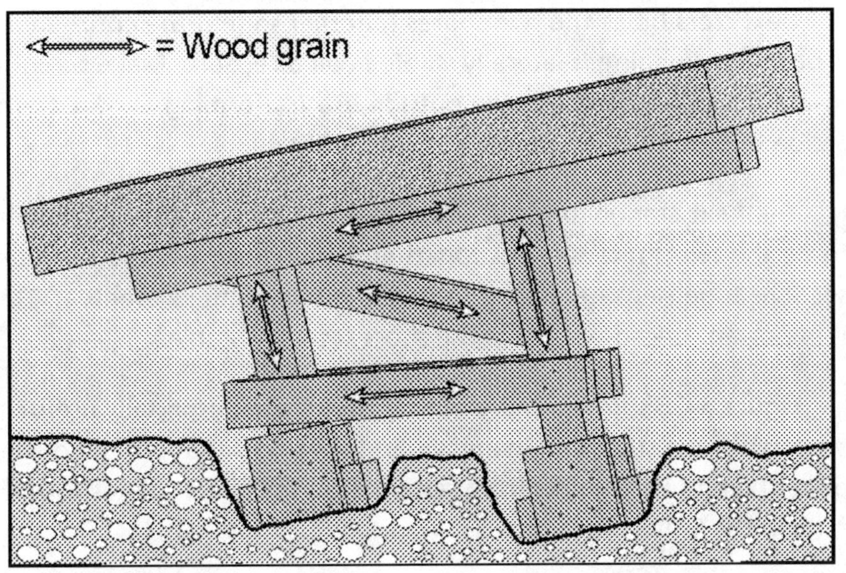

Fig. 6–26 Install the inside diagonal braces.

Two more diagonal braces are to be installed on the outside of each shore, also in the same direction as each other. This installation results in the X that helps to laterally stabilize the shore in both directions.

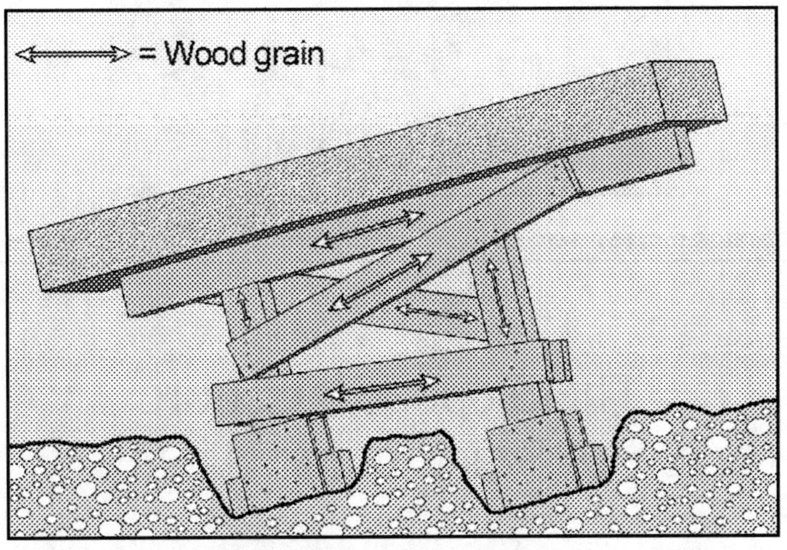

Fig. 6–27 Install the outside diagonal braces.

Next cross brace the two shores together so that the entire shore is boxed in, keeping it laterally stable in all directions. Make sure the inside braces are opposite the outside braces, forming an *X* pattern. Scribe the top of these braces and cut to fit.

The next step is to place another set of cross braces in position to tie the two shore sections together. The first set ties the two post sections together; the second set ties the two sections together. This is important because it keeps the shore laterally stable.

Figure 6–28 shows a shore with gusset plates, especially useful in an earthquake situation. In any violent shaking situation, the gusset plates help to keep joints from separating. Nail gusset plates with 8d nails in the 5-nail pattern.

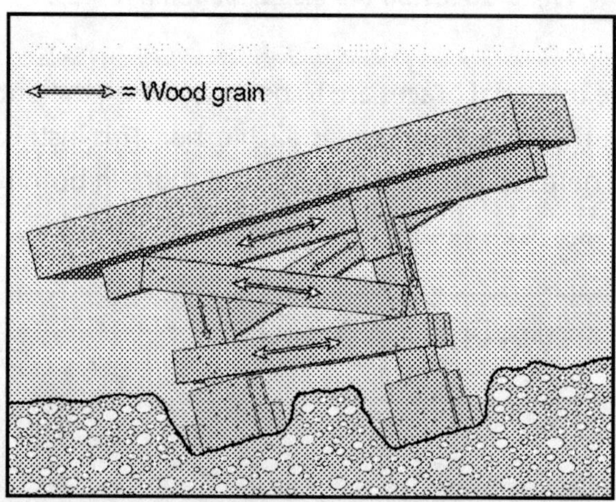

Fig. 6–28 Gusset plates are especially useful in an earthquake situation.

The Slope-floor Shore
Cross Bracing and Angles

Figure 6–29 shows the two sections placed next to each other ready to be cross braced together to form one complete two-section system.

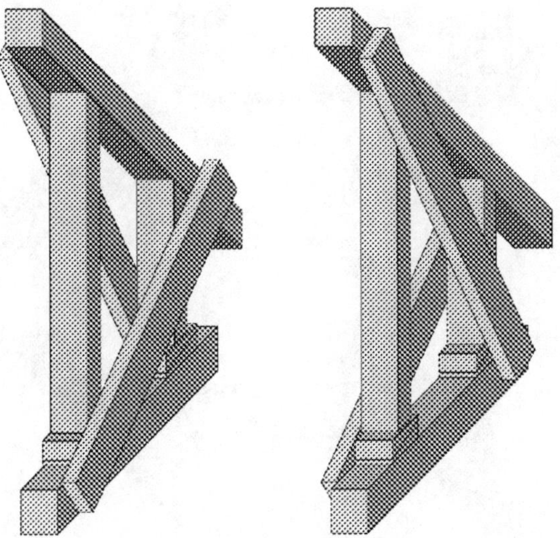

Fig. 6–29 Two sections ready to be cross braced together.

The big difference between bracing the shores is basically how far apart and how small the shore sections are. The length for the spacing between shores is normally 4–8 ft. This spacing predicts the number of diagonals installed. For shores spaced 5 ft or less from each other, use only one; greater than 5 ft in length, then make the X with two diagonals. For taller shores with more room, you will be able to do the same horizontals and Xs in the front as well. In some cases, the shores are too small for that, in which case, use the options that are suggested in Figures 6-33 and 6-34.

When cross bracing the two slope shores, it doesn't matter which type you install. All are braced in this similar pattern. The two shores must be laced together so they are stable enough to hold anything if there is lateral movement in the area.

The first step is to place the two 2x6 horizontals in position. Place the top horizontal just under the two headers and in contact with them. Nail it to the two back posts with (5) 16d nails in the 5-nail pattern as usual. Next place the bottom horizontal in position directly on top of the two sole plates or, in the case of the split sole, on top of the two bottom braces. Anchor it the same way as the top horizontal.

Fig. 6–30 Install the two horizontal braces.

If your shores are 5 ft apart or less, all you need is one 2x6 diagonal brace laced in under the two horizontals. The direction you use doesn't really matter. Just make sure the diagonal is up tight against the bottom of the top horizontal and the top of the bottom one. Anchor it with the 5-nail pattern, using the 16d nails.

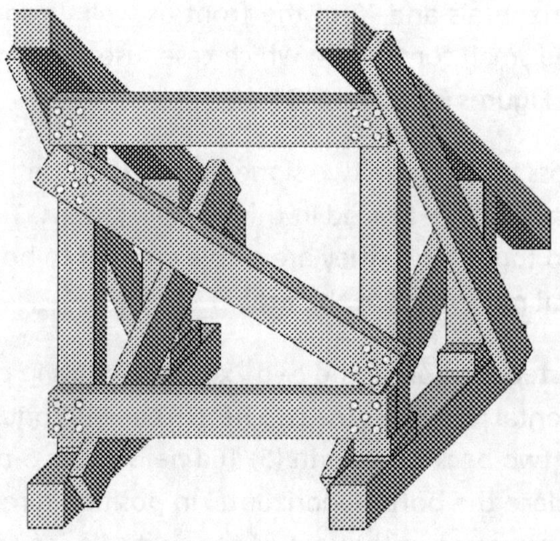

Fig. 6–31 Only one 2x6-in. diagonal brace is needed
in shores 5 ft apart or less.

If your shore is more than 5 ft from the other one, place two diagonal braces into the bracing system. Form the *X* as shown in Figure 6–32 and anchor it with the same nail patterns as before. Make sure to nail the two diagonals where they intersect each other, using the 5-nail pattern with (5) 16d nails.

Fig. 6–32 Place two diagonal braces in shores more than 5 ft apart.

If the shores are too short just nail the two 2x6s horizontally across the front posts—this works fine.

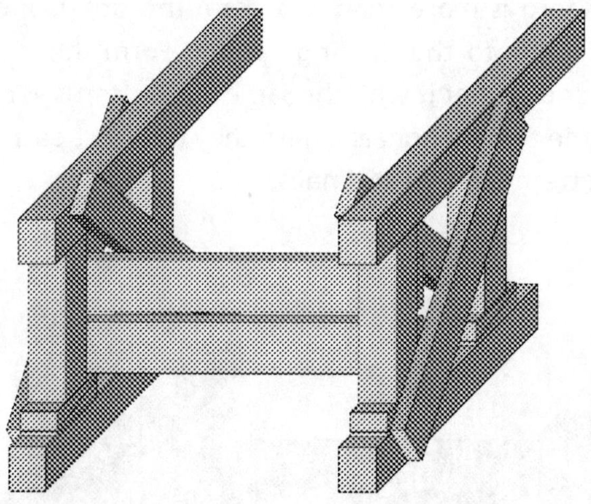

Fig. 6–33 If the shores are too short,
two 2x6s placed horizontally across the front posts work fine.

Another option is to use a section of plywood if the shores are spaced 5 ft or less apart. Use 8d nails for the plywood. If your plywood is 12 in. wide, use eight nails with the 5-nail pattern.

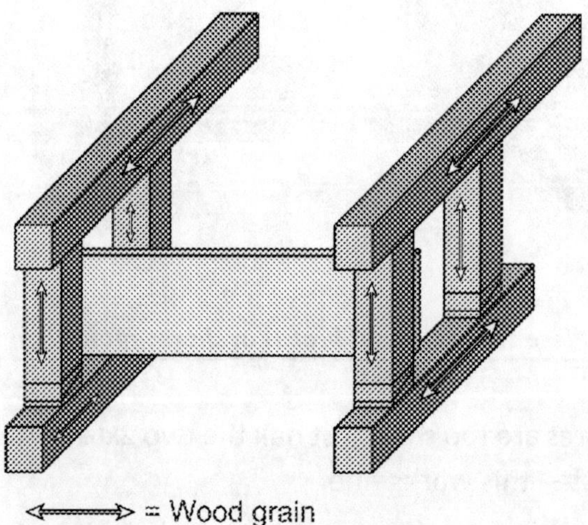

Fig. 6–34 A section of plywood is an option
for bracing shores spaced 5 ft or less apart.

Figure 6–35 is a close-up of how the diagonal braces must look with the angle cuts. It is necessary to install diagonal braces on all of the slope-floor shores because of the angle of the header versus the depth of the shore. The brace must be as close to the top of the post as possible to work the most efficiently.

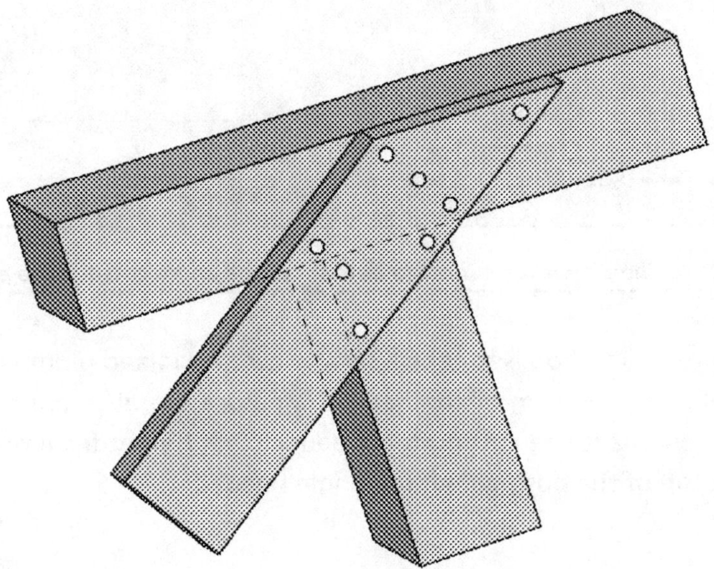

Fig. 6–35 Close-up of the diagonal brace with the angle cut.

Determining the angle cuts for the posts. There are numerous ways of determining the angle of the header. The following describes three of the simpler ones. There are two others utilizing the framing square, but it is more complex. The angle cut into the posts has to be accurate in order for the shores to work to their best efficiency.

Figure 6–36 shows the simplest method. Place the base of an angle finder against the header, and it will read the angle. You can then tell the cutting team the angle to cut in the posts. The team can use the speed square and mark the angle rather quickly.

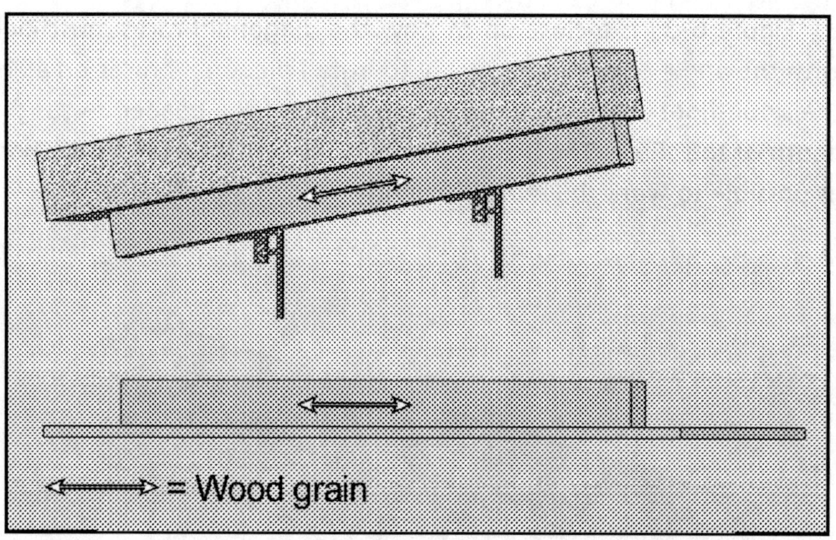

Fig. 6–36 The simplest method to determine the angle cuts for the posts.

Another method is to take a scrap piece of 4x4 and plumb it with a level to the bottom of the header. Measure the difference in the height and have the cutting team deduct that height from one end of the top of the post, giving the angle needed.

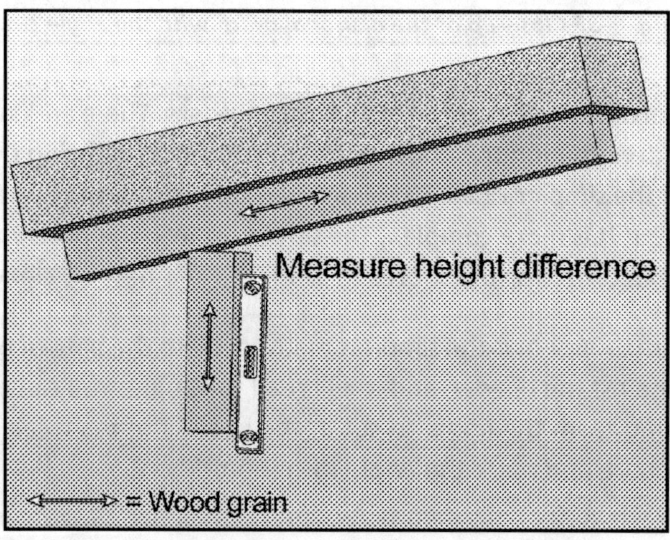

Fig. 6–37 Determining the angle cuts for the posts using a scrap piece of 4x4.

The easiest way to scribe the line next to the header is to plumb a piece of material with the post, using a level. You can use the post itself or a scrap of wood. The only drawback of this method is the fact that the piece you just marked has to be given to the cutting team. Most of the time the cutting station is not in proximity to the shoring operations. Therefore, this method could lead to a large delay in erecting your shoring.

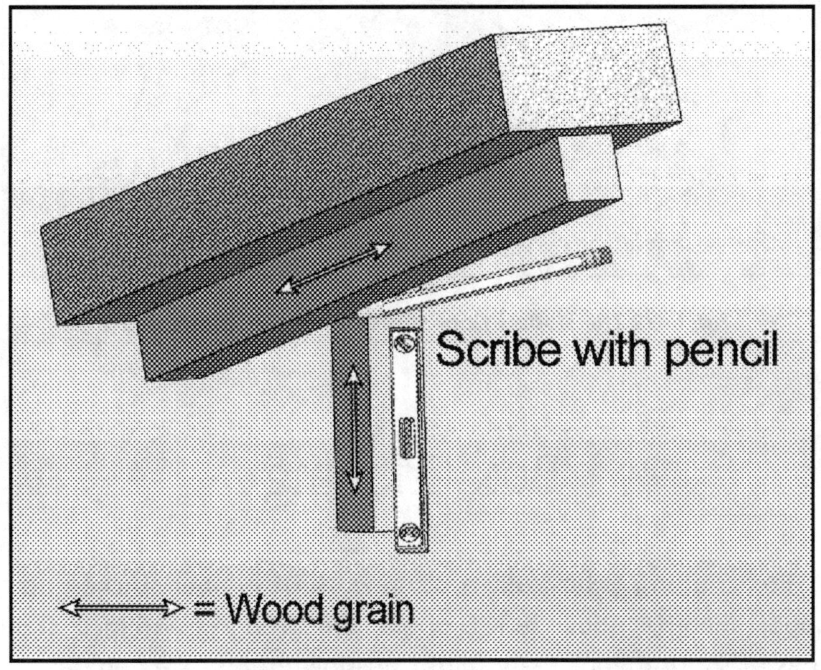

Fig. 6–38 The easiest way to scribe the line next to the header.

Figure 6–39 shows a slope-floor shore with two pneumatic struts instead of wood.

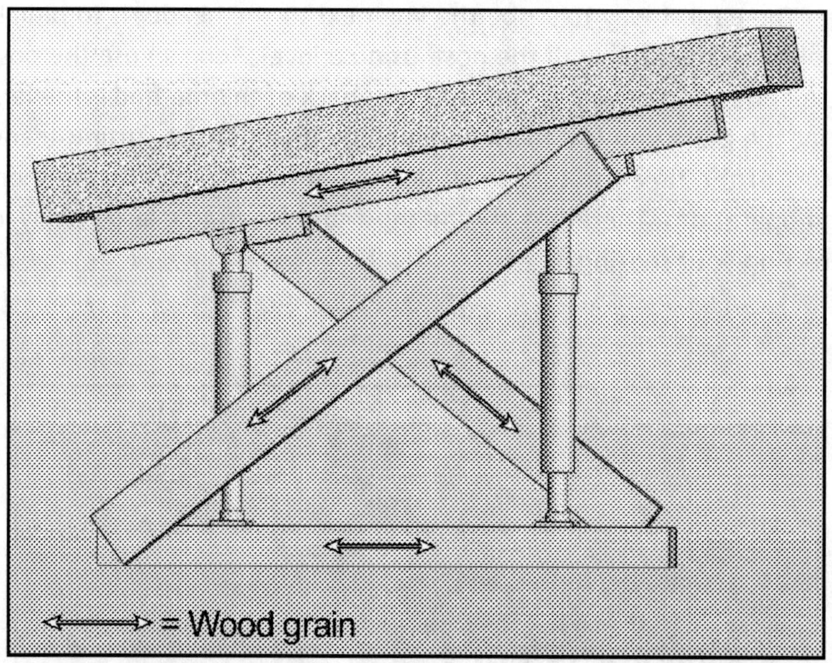

Fig. 6–39 The sloped-floor shore using pneumatic struts.

Box Cribbing

Box cribbing is basically a framework of lumber stacked up to support material above it. It is one of the strongest and most stable types of shoring your team might possibly erect. As a general rule, if the height of a void opening is 3 ft or less, box cribbing is the most efficient shoring method. On the other hand if the opening is more than 3 ft high, you have the option of utilizing other types of shoring—vertical shores or slope-floor shores to name a few. If you are using 4x4s for cribbing (which is the most common size) and if you decide to place a 42-in. high crib (not really that high), you would need 12 tiers (layers) of cribbing. Also suppose that you are using a 4-*X* crib (4 pieces per tier), your team would have to bring into the area 48 pieces of cribbing! (A bit much, don't you think?)

A slope-floor shore uses lots less lumber. For a shore that is 3 ft and under, cribbing is the way to go. In most voids encountered between floors in typical URM or lighter wood-framed building collapses, 4x4s more than suffice as box cribbing material. If necessary, for instance, in much larger building and in structures with more substantial structural elements, you may want to use something larger, possibly 6x6s. For collapse voids in structural steel or concrete buildings, you may need larger timbers, although the 6x6s usually suffice.

Box cribbing step-by-step procedure

1. Determine the location for your box cribbing. Try to determine the location of the existing floor beams and support at least two of them if possible. Transferring the load directly to those beams is far more efficient than just using the floor sections themselves to support the load. In larger voids with not much debris, it may be possible to support more than two beams with a single crib without interfering with the void access and egress. By supporting the box crib with at least two beams, the crib is more stable and transfers additional impact loads more evenly through the flooring. If the crib is supported by only one beam, it may not be able to absorb the load and could possibly fail. When spanning the two beams, the contact points of your crib must cover the entire surface of both beams to actually be effective. The edges of the crib cannot be the items to cover the beams; they may not be efficient enough to transfer the load properly.

2. Clear the area where the box cribbing is to be erected. The area should be at least 18 in. wider than the width of the crib to keep debris from falling back in your way as you work. This will also enable you to slide the shore back and forth several inches if necessary to get a good fit above. There must be enough room to shift or even rotate the crib.

3. Place the first tier into position directly under the area you want to support. For the best stability and the most efficient use of the box crib, lay it at right angles to the floor beams in order to support as much weight as possible. As you assemble the shore, you should bring out the edges of the second tier past the outside face of the bottom tier in order to maximize the support of the cribbing material. Basically as a rule of thumb, overlap the cribbing the width of the material used. If your team is erecting cribbing with 4x4s, overlap the ends of the lumber from 3½ in. to 4 in. The main reason for this is safety. If your crib starts to become overloaded for some reason, the material fails at the ends first. The ends split and check. If they are directly on top of the bottom tier, the strength of the crib is compromised. However with the overlap, as the cribbing material starts to split, your team has time to add more material or start an additional crib before the original one fails. If your team decides to utilize 6x6s, overlap the ends the width of the material from 5½ in. to 6 in.

4. Nail the lumber in the tiers of the crib together to make it easier to shift the entire set-up when and if necessary. Toenail tier to tier at the contact point where the material crosses each other. Use a Paslode or a pneumatic nailer so as not to disturb the crib and to keep vibrations to a minimum. Toenailing stabilizes the crib; and if done properly, it will not affect the strength of the crib. Be careful of splitting the material. Utilize 16d nails and keep the angle of the nail steep enough to get proper penetration into the lower element.

   As you get the majority of the crib set-up close to where it ultimately will be installed, nail it together. Toenail the crib whenever possible, especially when aftershocks are a possibility or when the crib is being erected on or is to support a sloped floor. This will lock the crib, keeping it stable in case any movement occurs that may loosen or dislodge it. At this time, do not nail the bottom tier to the floor. You want to be able to slide the crib from side to side slightly so you can easily insert the wedges and the last pieces of material.

5. Place wedges or thin strips of material under the beams when supporting a wooden beam floor. To correctly transfer the loads from above, you must have direct and full bearing with the contact points of the crib. You must insert a sufficient number of wedges and shims for this purpose. Shifting the crib slightly may help you fit the wedges properly.

6. Nail the box crib to the floor and the ceiling when you are satisfied that the crib has enough bearing and is properly transferring the load. Constantly monitor the crib during the operation. Try to make sure the crib doesn't loosen. If the crib gets hit or if the debris from above shifts, retighten the crib. Be careful; you must constantly monitor stability of the box crib for your safety.

Safety considerations

Before your team starts erecting box cribbing, it must determine the route rescuers will use to remove victims they may come upon in the collapsed voids. Your team must ensure that the crib will not close off that route or hinder victim extrication. If they are to install multiple cribs, one of their primary objectives is to constantly check that each crib remains tight. Remember throughout the operation, collapse debris may shift and settle. If the debris is being removed from above the void area, its weight is taken off the crib, possibly resulting in the loosening of the crib. That affects its stability.

Always size up the item you are going to crib before constructing the crib. Make sure to use the proper size lumber and the necessary number of pieces per tier. When in doubt, always put in a little more material. Play it safe and do not be afraid to over shore. Remember, there is only one way to know if your shoring is adequate. That is when a secondary collapse attacks the crib and it remains standing when the dust clears. You do not get a second chance!

When erecting cribbing, shore from a safe area into an unsafe area, giving yourself a stable and safe position in which to work. Always work from the good area into the bad area. Protect yourself at all times and always be prepared for problems. If you are prepared, then you can deal with problems much more efficiently. It is imperative that you do not put any part of your body, hands, fingers, arms, and so forth between the box crib and the items you are shoring. If debris shifts or a secondary collapse occurs, you could be pinned and become one of the victims. Use a piece of lumber or a wedge to slide elements of the crib into place. This will be particularly important as the crib reaches its load point and you put the final pieces into position.

Box cribbing strengths

The following are some of the acceptable weights that box cribbing can support. These numbers are round figures derived after extensive testing by the engineering field. For Douglas fir and southern pine, these numbers work fine. The strength of the lumber is determined by the square-inch capacity of the material in question. Typically for good grain 4x4s, your team can expect at each contact point a support of roughly 6000 lb. This is determined by the following calculation: true width of the material used multiplied by the typical compressive weight of the material used equals the weight the crib can support in pounds per contact point.

For example, the true width of a 4x4, which is 3.5 in. x 3.5 in. = 12.5 sq in. Multiplying 12.5 sq in. by the typical compressive strength of Douglas fir, which is 500 lb per sq in. (psi), results in an answer of (rounded to) 6125 lb per contact point.

For a 6x6, the calculation is 5.5 x 5.5 = 30.25 multiplied by 500 lb. This equals 15,125 lb, rounded to 15,000 lb per contact point.

Box cribbing size-up

As with any type of shoring scenario, there has to be some type of initial evaluation of the unstable area. Your rescue shoring team must size up the area and the predicament that it faces at any given time. The following information describes a few items to look for when installing box cribbing. However, because every situation is different even in the same collapse incident, the size-up may include more than these listed.

Item to support. One of the first items to address, obviously, is the item or items to resupport or stabilize in position. Several factors come into play: the size of the object, the location of the object and its relationship to a debris pile, and the weight of the object to be shored.

Support base. As with any type of shoring situation, it is critical to determine on what object the cribbing will be bearing. Without a solid base for the cribs, there will not be sufficient support, making for an obvious unstable situation. It is important that any shoring your team installs works effectively. Without the proper support for your shores, this will not happen.

Access and egress. Even though you need to place the cribbing in areas where it will be most effective, that location can pose an access/egress problem. Some creative thinking may be required, for example, setting the cribs at angles or placing a header across two cribs for additional supports.

Slope of the crib. In collapse rescue operations, many things change on a regular basis, requiring continual size-up; therefore, the shoring operation is dynamic from its inception. This situation exists for the duration of the incident. As a result, any time your rescue team decides to install box cribbing, several items must be looked at before proceeding. There are two major points to look at for your box cribbing size-up. The first and generally the most important is the slope of the floor or ceiling in which you want to place your crib system for support. The generally accepted rule of thumb that prohibits the installation of cribbing is not to install cribbing if point of slope is 30%, or roughly 15°. On slopes greater than 30%, there is a major chance of your crib being unbalanced and unable to support a possible impact load. The cribbing has the possibility of being forced out of position and failing, something you cannot allow to happen. One easy way to determine the slope is to examine the area and determine if in 10 ft of the floor there is no more than a 3-ft difference in height from one end to the other. Another rule of thumb is if the height of the area is greater than 3 ft, use other shoring options available to you, most commonly sloped-floor shores.

Figure 6–40 shows a typical box crib. When installing the crib, be sure to overlap the tiers the width of the material, 3–4 in. for 4x4s and 5–6 in. for 6x6s. Each layer of the crib is known as a *tier*. The crib is identified by the number of pieces per tier. The correct way to describe the crib shown in Figure 6–40 is 7 tier 3 by (3x) crib; where *3 by* means three pieces per tier and *4 by* means four pieces per tier.

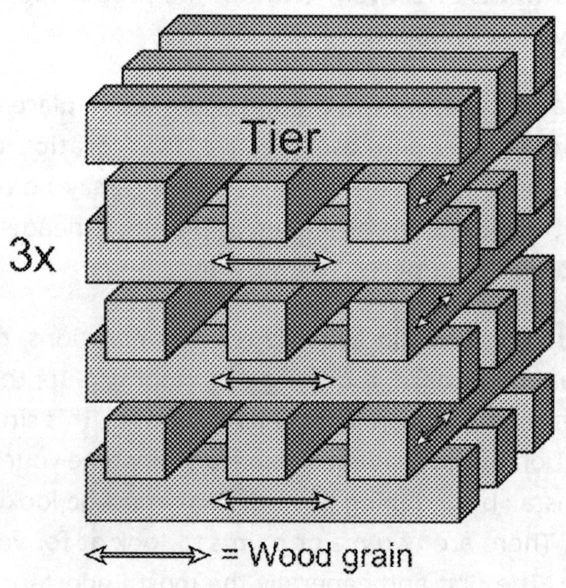

Fig. 6–40 A typical box crib.

The following shows the general rule of thumb for crib balance and height.

The following shows the option to use when cribbing is not appropriate. Cribbing is unstable when a slope is 30%.

Figure 6–41 shows a quick rule of thumb to determine the 30% slope, or 15° angle, when you are operating in a collapsed structure. For slopes any steeper than this, you cannot utilize cribbing. Switch to another type of shore.

Rules of Thumb

- *Overlap ends the width of the material 3 times the height to the width*
- *Angled cribs use 1½ times the height to the width*
- *Over 3 ft high can use other shoring methods*

Sloped Floor vs Box Crib

- *Sloped floor shores should take precedence over box cribbing when the slope of the floor in question is greater than 30%*

–or–

- *When the opening in question is greater than 3 ft high*

On another note, the solid crib is the strongest and most stable of the cribs, but it requires a substantial amount of material and takes considerably longer to erect than the 2x and 3x cribs.

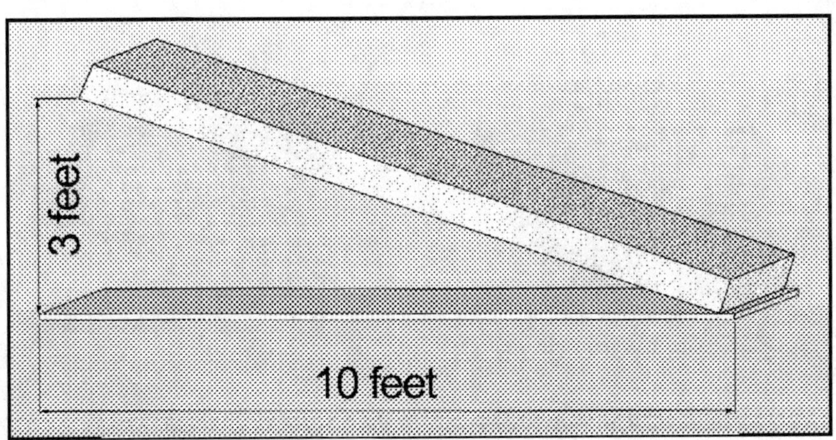

Fig. 6–41 A quick way to determine the 30% slope (15° angle).

Figure 6–42 shows the most common types of cribbing ensembles. They are as follows (clockwise from the top):

- A 2x crib with four contact (bearing) points

- A 3x crib with nine contact points

- A 4x crib with 16 contact points

- A solid crib, which has nearly full contact

The amount of contact of a solid crib of 4x4s 24 in. long has a full contact surface of rough 17.5 in., giving you a support of nearly 105,000 lb!

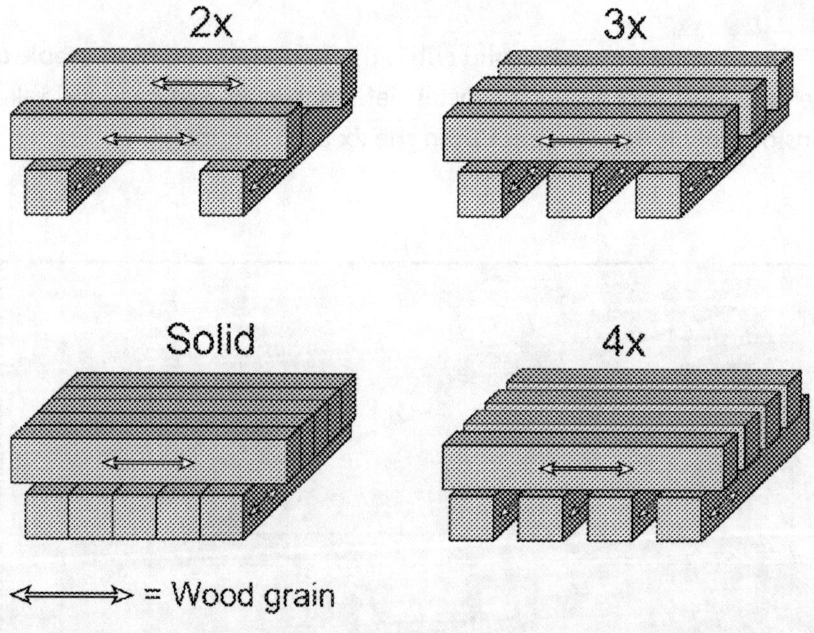

Fig. 6–42 The most common types of cribbing ensembles.

To achieve a crib's maximum load bearing capacity, the lumber must have full bearing on each other and must overlap the corners of the crib an amount as least equal to the width of the cribbing material itself. There may be times when space limitations and access problems dictate that you construct the crib on an angle.

The cribs' support capabilities depend entirely on the amount of surface contact between the tiers of the lumber. The more surface contact, the more weight the crib is capable of supporting. In Figure 6–43, on the left is a 2x-angled crib with a support capacity of more than 12 T. Full capacity is realized with an angled box crib as long as each member has full bearing on the other members, just like in a square crib. However, the stability of the angled crib is much less than the square crib. You have to keep the height-to-width ratio closer to 1½-to-1, instead of 3-to-1, which is normal.

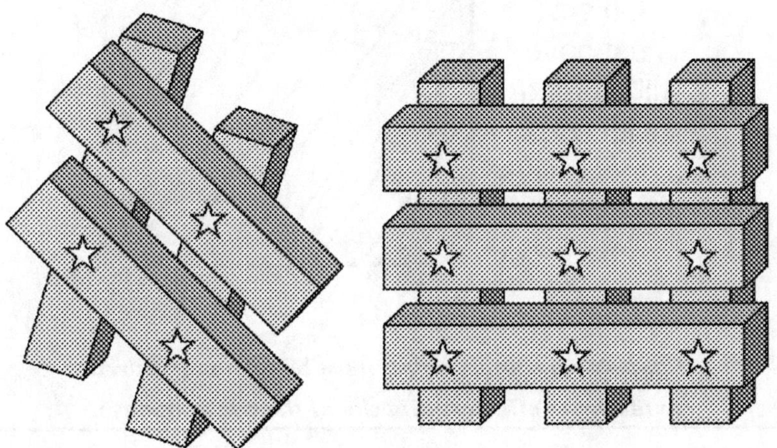

Fig. 6–43 The stability of the angled crib (l) is much less than the square crib (r).

The crib in the right side of Figure 6–43, a 3x setup, can support a load of more than 54,000 lb. A 2x crib erected of good quality construction grade 4x4s can withstand a load of more than 12 tons. A crib's capacity to support weight depends on its bearing points, the areas of contact from one tier of cribbing to the other. A 2x bearing crib has four bearing points regardless how high it is stacked. A 3x crib has nine bearing points. The 3x crib is generally the most efficient. It is more than twice as strong as a 2x crib and yet uses only one additional piece of lumber per tier. It takes very little time to install this additional piece of lumber, and the tradeoff is worth it.

It is very important for the cribbing to bear evenly under the load it is to support. Without the proper load distribution, the crib cannot support the anticipated weight. To reach the effective capacity of the strength of crib, all the contact points must be fully pressurized.

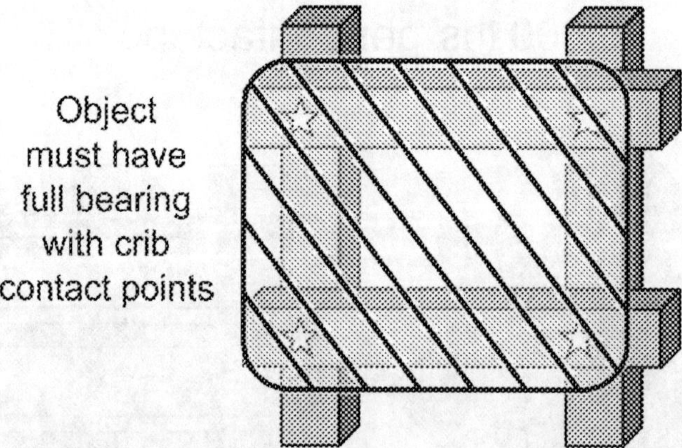

Fig. 6–44 All contact points must be fully pressurized to reach the effective capacity of the crib's strength.

The crib setup shown in Figure 6–45 illustrates what happens when there is uneven loading of a crib support system. Although it looks like it's spreading the load out, in fact, this is not the case. Each crib is only supporting one-fourth of its bearing capacity and in an unstable manner as well. Catching only one corner of the crib makes for an unbalanced load, resulting in a seriously unstable crib.

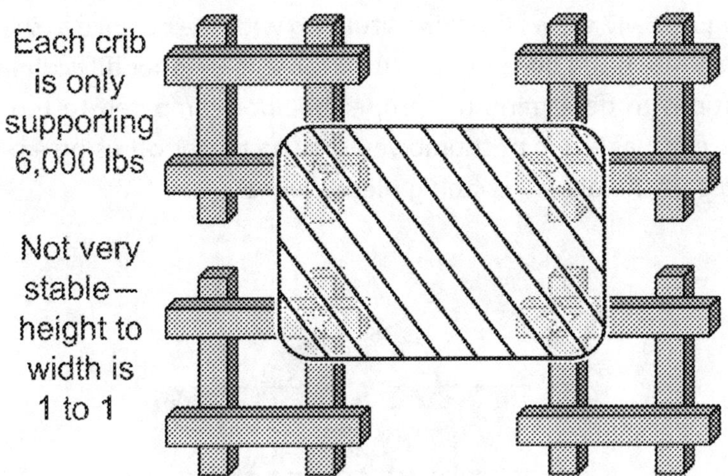

Fig. 6–45 Uneven loading of a crib support system.

When your crib is installed on an angle such as the parallelogram shown in Figure 6–46, the overall stability is a little shaky. In this case, you can still use the 3-to-1 ratio, but the width factor is going to be reduced. Of course, the longer the cribbing and the more severe the angle the height-to-width ratio varies. The inside, smaller square is the stable area of the crib. Notice that it is smaller than the parallelogram.

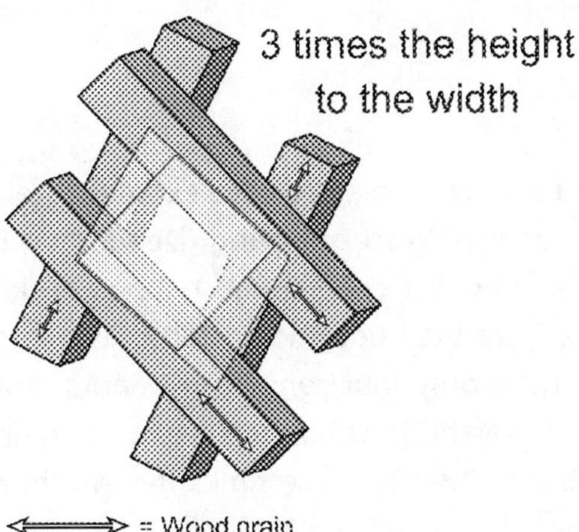

Fig. 6–46 The small, inner shaded square is the stable area of this parallelogram.

Figure 6–47 shows the same situation with the balance as shown in Figure 6–46, just a different setup. Because it is rather difficult in field conditions to determine the proper balance compared to the angle of the cribs, as a rule of thumb you may be better off using a ratio of 1½-to-1 of the overall cribbing piece's length.

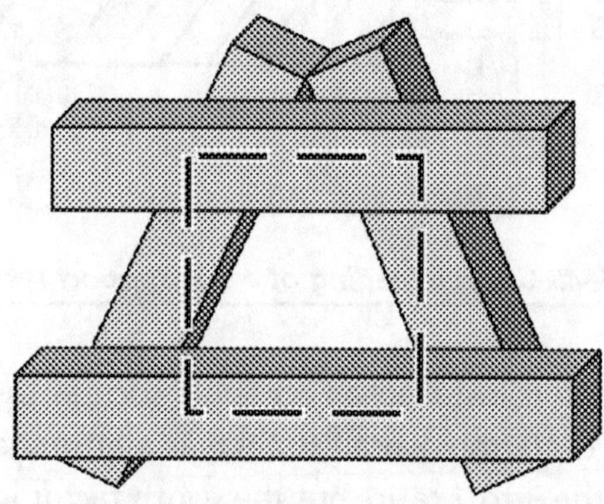

Fig. 6–47 It is rather difficult in field conditions to determine the proper balance compared to the angles.

Crib stability

Box cribbing on the whole provides excellent stability compared to many other types of rescue shoring. Depending on the type of cribbing being erected, the crib can reach a height up to three times its width before it starts to become unstable. The 2x crib is the least stable because it has only four contact or bearing points. Generally, the rule of thumb for the 2x crib in a collapse scenario is to have the height no more than two to three times the width. For example, if your cribbing material is 24 in. long, you really shouldn't erect the crib more than roughly 4–6 ft high. Any of the other crib types, 3x, 4x, etc., can be erected with ratios of three times the height to the width. For cribbing 24 in. long, do not exceed 6 ft in height. If you do exceed that height, the crib becomes potentially unstable and could easily fall over with only a slight lateral load against the sides of the crib.

For the most part, box cribbing is designed to withstand compression. However, when a crib is erected on a sloped surface, it must be able to handle lateral stress as well.

To get the full effectiveness of the cribbing you are erecting—which would be nice since we are using it to protect ourselves—keep the load positioned through the center of the crib. To accomplish this, keep the main part of the load in the center one-third of the crib.

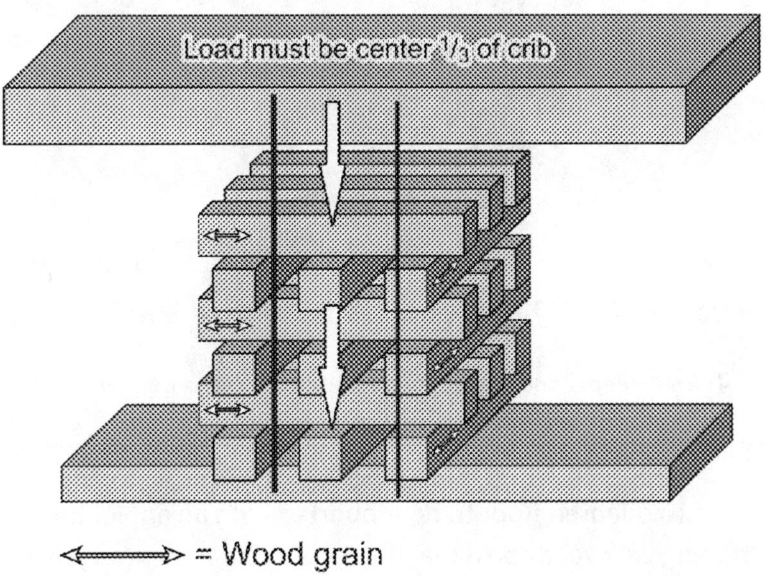

Fig. 6–48 Keep the load positioned through the center of the crib.

The lateral stability of the crib and the angle of the force of the load become issues as the slope of the crib increases. As you can see in Figure 6–49, the steeper the angle, the farther from the center of the crib the load will be. Unfortunately, what happens is that some of the contact points are taken out of the equation as the angle increases. That fact reduces the load capacity of the crib significantly, but it is hard to determine how much.

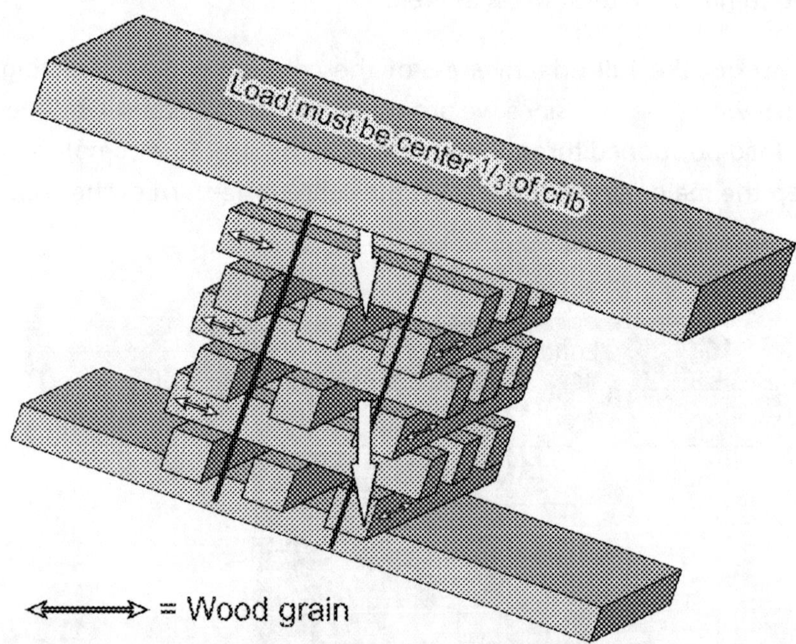

Fig. 6–49 The steeper the angle, the farther from the center of the crib the load will be.

When a collapsed floor to be supported is on an angle, try to erect the cribbing as level as possible. If necessary, shim or wedge the bottom first, trying to erect the crib approximately level. Then gradually slope the crib into the angled surface. A gradual slope is far more effective than just wedging the top of a crib. Finally tighten up the crib and anchor it to the top and bottom floor sections. In this way, you ensure that the crib is stable, and you are taking advantage of its full load-bearing capacity.

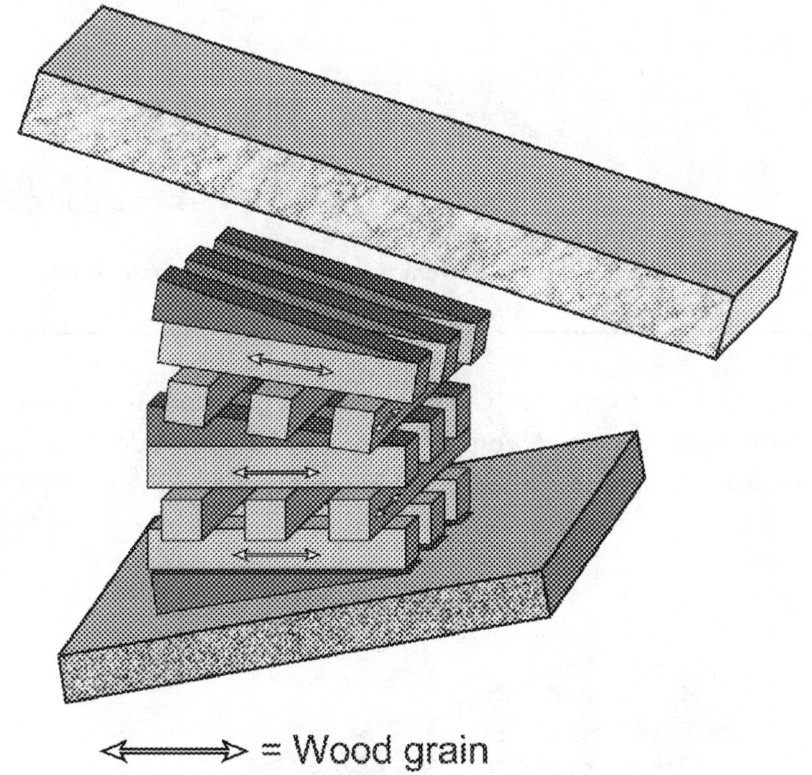

Fig. 6–50 Try to erect the cribbing as level as possible.

When assembling the crib, try to keep the bottom tier at right angles to the floor to distribute the weight. As you erect the crib, keep it as level as possible. Alternate each tier. Do not stack two consecutive levels in the same direction.

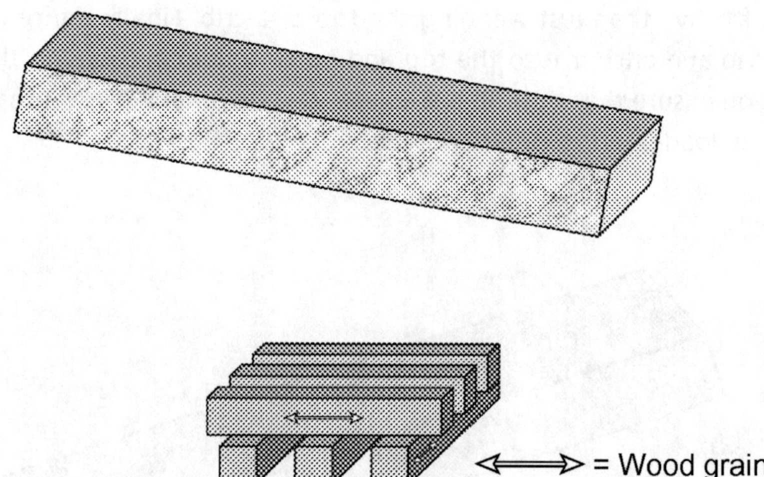

Fig. 6–51 Try to keep the bottom tier at right angles to the floor, alternating each tier.

As you start to approach an uneven floor, gradually slope the crib into the angle you need. You can do this by adding wedges that are the same length as the cribbing, 24 in. in the case of the example in Figure 6–52.

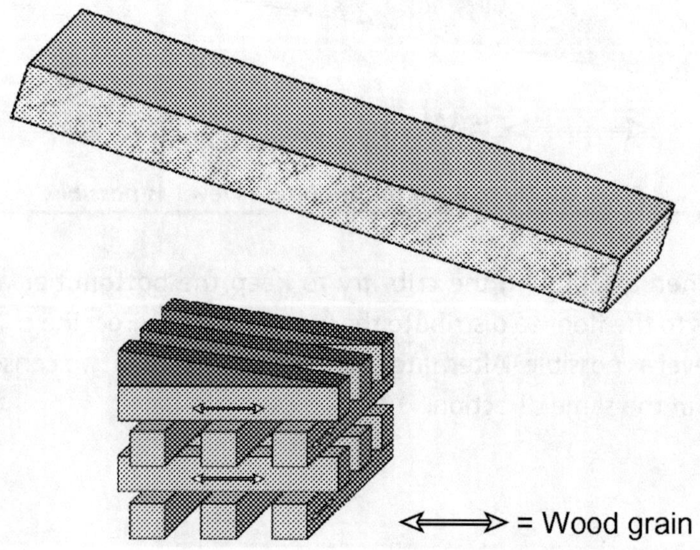

Fig. 6–52 Gradually slope the crib into the angle you need.

As you get closer to the floor, check the angle again. Ideally when you are approaching the point where the last two layers of cribbing fit, try to get those layers parallel to the floor.

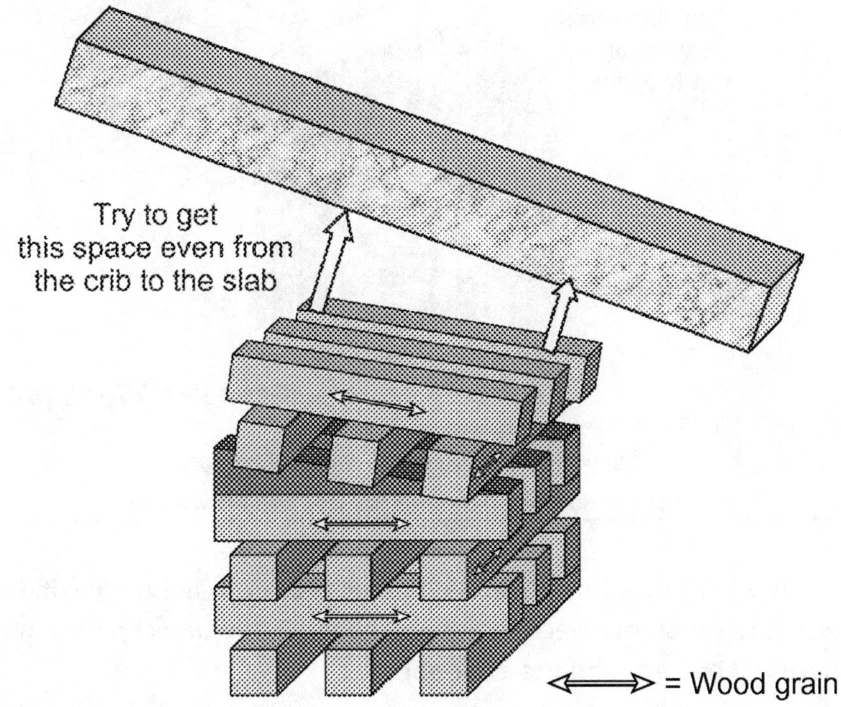

Fig. 6–53 As you get closer to the floor, check the angle again.

You can keep the distance between the floor and the cribbing even by inserting another set of wedges. The remaining space between the crib and the floor is now parallel, making the installation of the last pieces of the crib much easier to do.

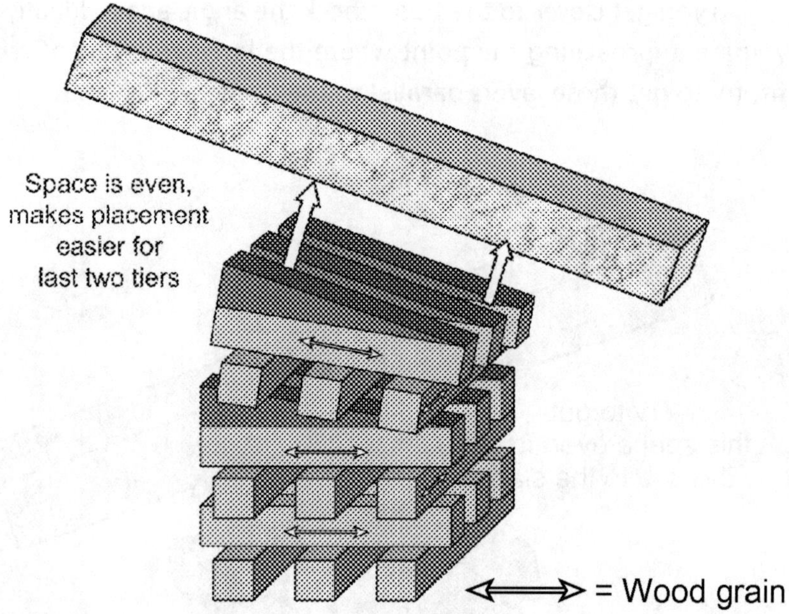

*Fig. 6–54 Inserting another set of wedges
can keep the distance between the floor and cribbing even.*

The last two layers are in, so it's much easier to finish the crib this way. You can shim if necessary. Remember, all the contact points must have full contact with the floor above.

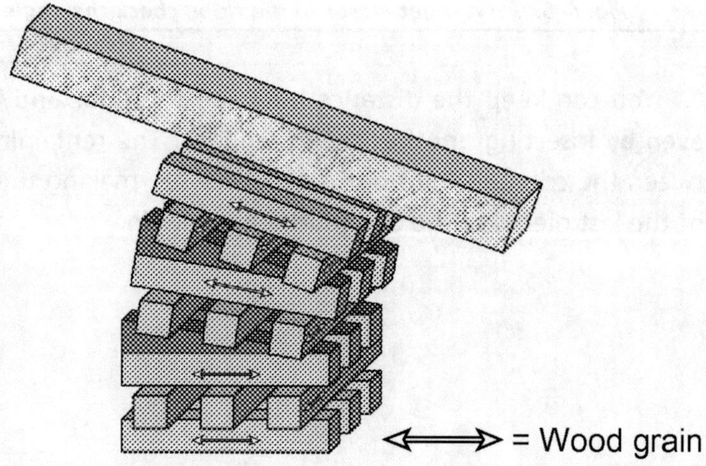

Fig. 6–55 All contact points must have full contact with the floor above.

The use of wedges at the top of the crib

In almost all scenarios, you need to shim or wedge from the top layer to the bottom of the floor. Here are some guidelines to follow.

The situation pictured in Figure 6–56 is unacceptable. You cannot stack more than two wedges on top of each other. They are much too unstable, and the center wedge is likely to be forced out if an impact load presses against the crib.

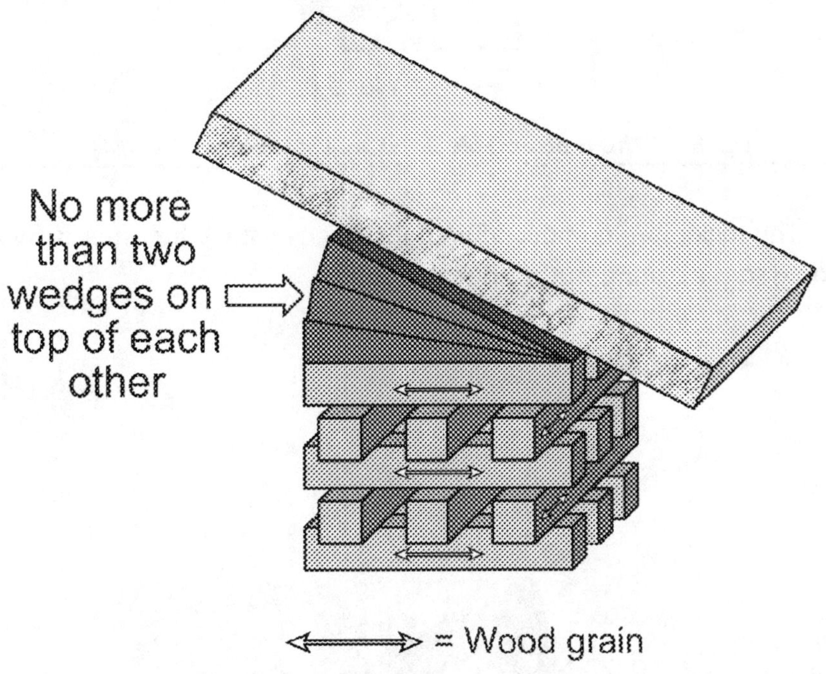

Fig. 6–56 This is not an acceptable crib.

Figure 6–57 shows a way to use wedges that would be acceptable if the area is too small for 4x4s. Place some 2x4s in between the two wedges at right angles to the wedges to help tie the top of the crib together and keep it more stable.

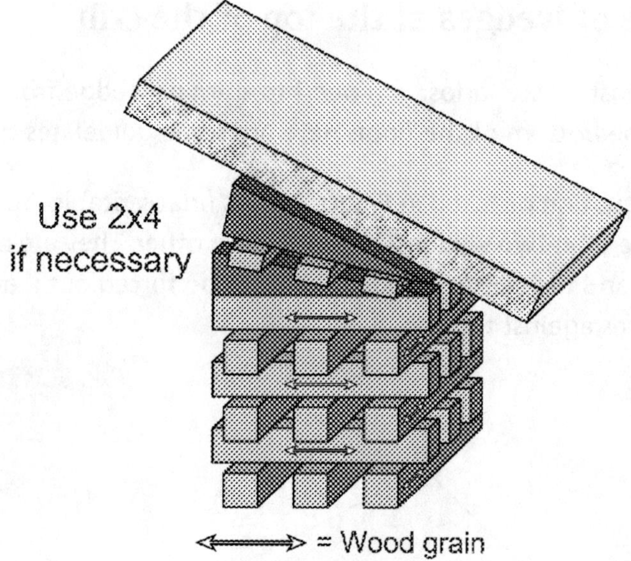

Fig. 6–57 This is acceptable if the area is too small for 4x4s.

The maximum number of wedges you can stack together is two. Doing so is all right, but it's not the best.

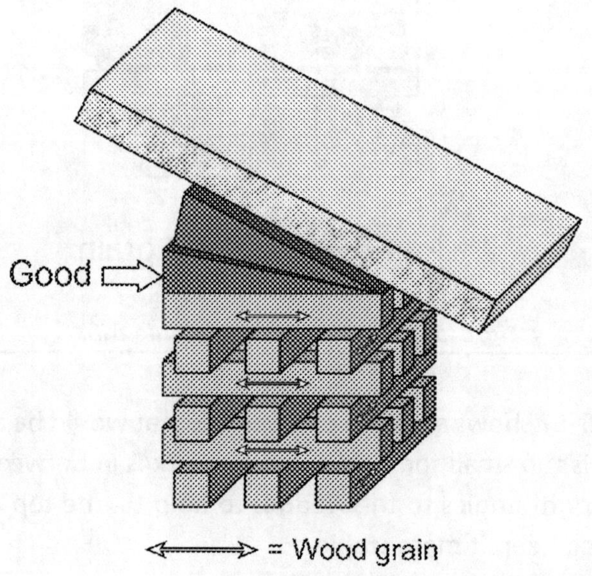

Fig. 6–58 The maximum number of wedges you can stack together is two.

Figure 6–59 shows a much better way to use wedges for the balance of the materials and the stability of the shore. Every element has an intersecting point, and the last tier is to tighten up with the installation of wedges.

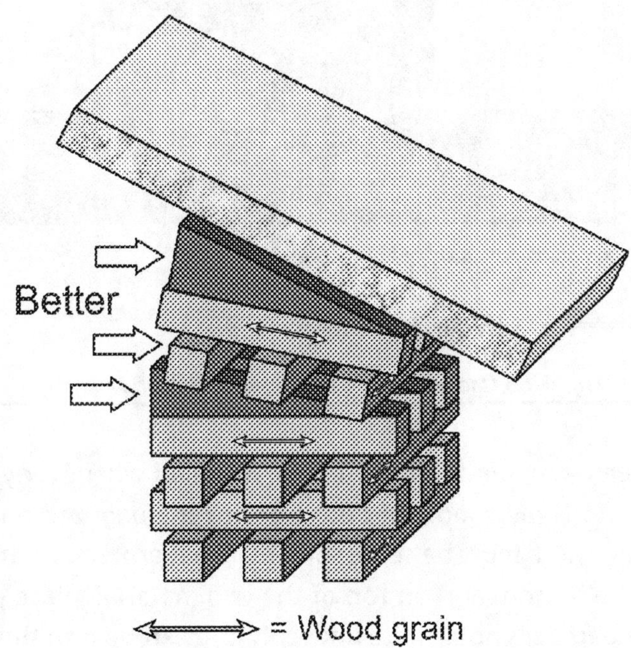

Fig. 6–59 A much better way to use wedges.

Cribbing and air bags

It is also possible to create a box crib and air bag ensemble. Use an air bag with a solid crib if you can. Do not use a 2x crib with an airbag. For safety reasons, an airbag and crib ensemble must be used with at least a 3x crib because all airbags lift from the center point of the bag out. The center of the bag must be solidly supported at all times.

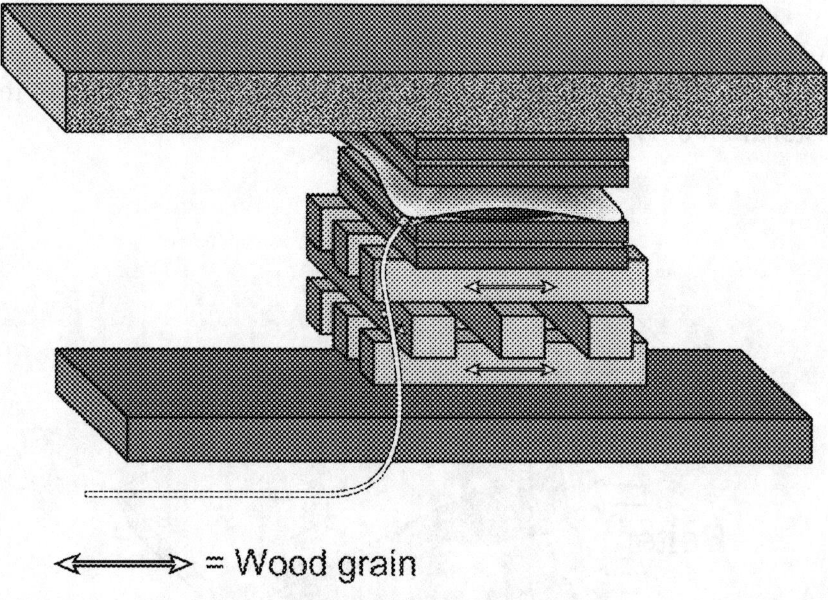

Fig. 6–60 Use an airbag with minimum of 3x crib.

Another safety item that should be installed whenever you erect an airbag crib is plywood. On both sides of the bag and on the top and bottom, there must be some solid material protecting the airbag itself. On the bottom and on top of the crib material, place plywood so that the bag does not inflate unevenly and extend into the spacing between the lumber. Not only would it detract from the lifting capacity of the airbag, but it would also put lateral pressures against the crib material, potentially making it unstable or forcing the cribbing material apart. The bag also should always be protected at the top. As it is forced into the element or debris, it lifts. Damage could result if the bag is not protected.

Always lift straight up when using the airbag crib; never lift on an angle! You could easily force the crib to move, causing a secondary collapse as well as possible crib failure.

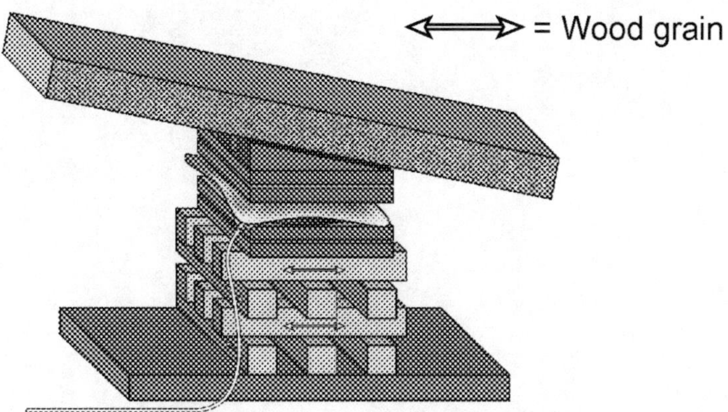

Fig. 6–61 Always lift straight up, never lift on an angle!

It is very important that the bag be padded out properly. It must be securely blocked up to any uneven surface, or the bag will inflate on an angle like a wedge. This causes the bag to apply a lateral load against the crib as well as to the floor above, possibly shifting one or the other, potentially causing a secondary collapse. Always lift straight up when using the airbag crib; never lift on an angle! You could easily force the crib to move, causing a secondary collapse as well as possible crib failure.

Figure 6–62 shows the result of such an application. The bag could fly out and cause severe injury and further collapse to occur. This is *not* the way to use an airbag with a crib.

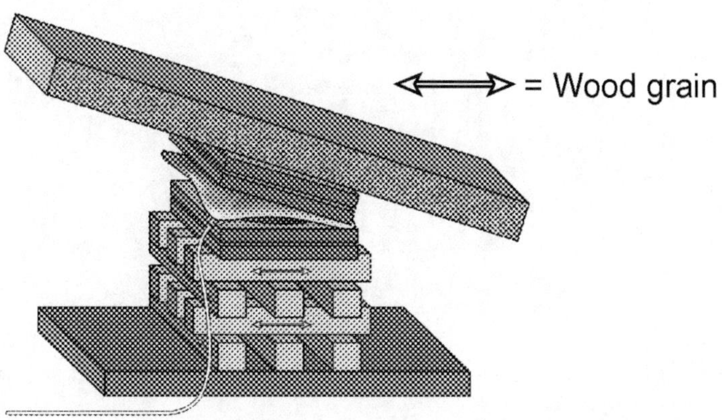

Fig. 6–62 Improper use of an airbag could cause severe injury and further collapse.

Tools and Equipment Utilized in Rescue Shoring

Hand Tools

The following is a list of some of the more common tools and equipment needed to successfully complete a typical rescue shoring operation. This list is by no means all you may need in a given situation, but it will cover most of the items you generally have to utilize. Later in this chapter, information about available tools in each category and the common uses for the tools is provided.

Typically, these tools are the ones in your cache and are utilized the most. Hand tools can be broken down into the following six basic categories:

- Measuring
- Hammering
- Digging
- Prying
- Mechanical
- Cutting

Every firefighter on your team must be proficient in the use of all of these tools in order to bring your operation to a safe and successful conclusion.

In many cases, especially in large structures with catastrophic collapse situations, it may be much easier and faster to place battery-operated tools into use before setting up electric equipment because they are portable and can be placed into operation rapidly. But remember to have backup, electrical equipment for these tools for any of your operations; the batteries don't last forever. Your team should try to utilize the 18- or 24-V type tools for more power and endurance. One manufacturer even has a line of tools including circular saws that utilize 19.2 V. Bear in mind, for long-term usage, electric or pneumatic tools are usually the best choices. Most people recognize that batteries can cut out early when tools are utilized in extremely tough environments such as collapse rescue scenarios. Extremely cold weather also has a tendency to draw the life out of the batteries a little faster than usual. Another factor that may affect the decision of the tools to use is the preference of your rescue team members. Your experienced mechanics and carpenters who are familiar with and utilize the tools on a regular basis have much better luck with the battery life because they know how to properly balance and handle the tool for optimum use.

There are two main types of batteries in use today, nickel-cadmium, (NiCad) and nickel-metal hydride, (NiMH). They both have their pros and cons. NiCad is good for high-drain applications, and NiMH is good for low-drain applications. As technology improves, the efficiency of the batteries and their operation time will increase. However, one major deciding factor that may come into play, especially with the advent of large and higher capacity batteries, is the issue of tool weight. Currently, some electric tools don't weigh any more than the cordless varieties; and at some point, it will be more efficient to just plug in the lighter electric tool.

The following sections of this chapter provide information about some of the more common cordless tools you may want to utilize for your rescue shoring operations. There are several other tools available not described that your team may consider using. The choice is up to the members. The ones discussed here are the ones most commonly utilized in any given situation. Remember, these items must be checked on a continual, rotating basis in order for the batteries to be constantly charged and ready for use at any time.

Measuring tools

Measuring tools can also be referred to as *layout tools*. The inclusion of these tools in your team's tool cache is a definite *must*. The erection of emergency rescue shoring cannot be accomplished without them.

Tape measures. There are numerous sizes and different styles of tape measures available to your rescue team today. These items are some of the most important and versatile devices your team members use in a shoring operations. They are durable, compact, and can quickly be brought into use at the scene. Of the many choices, the best suited for collapse rescue shoring is the 16 ft- and 25 ft-long tape measures. At least one tape should be included with each tool pouch setup. Ideally each member involved in the erection of any rescue shoring should possess a tape measure on the job.

Some of the more important features the tapes should have are easily retractable blades and a power-locking device to enable the blade to stay extended so that it can be set down without the blade retracting. The inclusion of both of these items makes operations with the tapes go quite a bit smoother and faster. The size and style of the retractable blade are also very important. The best type for rescue use is a tape measure with a 1-in. wide, slightly curved blade. This size and type of blade can be fed out approximately 7 ft past the tape while still holding its shape. This feature makes the tool an excellent item to use when measuring in and around hard-to-reach places, typically what you are dealing with collapse scenarios.

The blade also should have large, clear numbers on it—keep it simple. It should be single read only (just one set of numbers and slashes per tape). Tape measures with metric measurement and standard feet and inch graduations on the same face of the blade are too confusing to use and risk causing mistakes. The majority of the time, firefighters do not work in ideal conditions. Envision a collapse occurring at 3 A.M. in the pouring rain. Visibility certainly is not at optimum conditions.

These two tapes just described are used primarily for measuring lumber for your shores.

Another type of tape measure to use for measuring dimensions of rooms or buildings is a tape measure that reaches lengths of either 50 or 100 ft. These tapes are used primarily by builders or surveyors to get overall measurements. They can come in cloth or metal tape styles. For the purpose of shoring, metal tapes generally work better. They last longer under the typical conditions in which firefighters use them.

Laser measuring device. These devices are excellent for getting surprisingly accurate measurements without having to put yourself in physical jeopardy. A laser-measuring device is very handy for taking measurements that are out of your reach. They are battery operated, very portable, and can measure in inches and centimeters if needed. A laser-measuring device is excellent for use in extremely hazardous or unstable areas where you want to keep the presence of rescue personnel at a minimum.

Wood rulers. This group of tools generally consists of folding wood rulers with a one-way brass slide, enabling additional measurements up to 6 in. There are two general sixes, 6 ft and 8 ft. You can use either size; however, the 6-ft size is typically chosen due to its compact size and ease of handling. Many professional carpenters refuse to use anything but these types of rulers because the ruler measures more accurately than tape measures. They work extremely well when needing to measure between two solid points in an opening such as a doorway or window frame opening. Make sure your team purchases the carpenter's rule. A word of caution: there are wood rulers used by masonry contractors and are designed for the spacing of bricks. These rulers have graduations different from the carpenter's ruler and using them will throw your measurements way off.

Chalk lines. Also know as *snap lines*, chalk lines are used for the marking of straight lines on flooring, walls, plywood, and lumber. Use a chalk line for layout work just as carpenters do. When your team has to start a major operation and the use of numerous gusset plates is called for, this tool comes in handy. Most gusset plates used are 12x12. If you cut them from a 4-ft x 8-ft sheet of plywood, using the chalk line makes measuring the squares much faster and easier than other methods. The chalk box itself is hollow and made of metal in the shape roughly of a teardrop. It has a reel of string inside it with a hand crank affixed to it. Turning the handle retrieves the line after it has been played out. There is a slot with a sliding cover where the chalk is located in the tool. Your team should have two colors of chalk: red and blue. The chalk color to use depends on the application and the shade of the material being used.

Marking implements. The purpose of a marking implement is to identify and mark the lumber to cut for your rescue shoring operation. Some of the types of marking implements are carpenter's pencils, paint sticks, and lumber crayons, also known as *keel*. Use carpenter's pencils to lay out and mark lumber where it is to be cut and to identify the length of the piece. Carpenters pencils are heavy-duty pencils in the shape of a rectangle. There are several grades of hardness of the lead. Your team should decide on the grade to use. This will be strictly a personal preference for your shoring team members.

You can also utilize paint sticks and lumber crayons to identify lumber that has been cut. Lumber crayon comes in several sizes, styles, and colors; it's a good idea to have a least two colors available for your team. The most common colors available are yellow, red, and black. There are other colors, but these three generally cover any possibilities that your team might need. Normally, write the length of the lumber in inches on the item itself to prevent any problems in identifying a particular piece of lumber. Marking like this should be done at all times, especially when there are multiple pieces of lumber being sent into a collapsed structure. The shoring team should be able to easily identify a particular piece of lumber without resorting to measuring each piece. Marking lumber saves time and keeps operations less confusing.

Magic markers are also available and can generally be found without any problems. These come in numerous sizes, styles, and colors. They can also be utilized like the lumber crayon or the paint stick for marking the size of material on the item itself.

Squares. There are several types of squares your team needs to successfully complete an emergency rescue shoring operation. These are some of the most valuable and necessary tools in the operation. Without them, firefighters cannot properly cut the necessary square and angle cuts needed to complete their mission.

6-in. speed square. This is a very handy item that is compact enough to carry on your tool pouch and is easy to work with. It has numerous uses as angle finder, straight edge, right angle scribe, saw guide, and angle template, just to name a few utilized in rescue shoring. The speed square was developed for carpenters in the house-framing field. It was designed to give a quick and accurate means of laying out rafters for roof construction. The layout of rafters is based on a right triangle, just as raker shore systems are. For this reason, the speed square is an excellent tool for use in calculating and scribing angles for raker shores. There are several scales that are inscribed on the square:

- A protractor scale is measured in degrees.

- An inch square, graduated in ⅛ in.

- A rafter scale, defined by inch rise per foot of run.

The 6-in. speed square is in the shape of a right angle. Each side of the angle is 6¾ in. long. It has a base with a lip to place against the material and a pivot point where the square is shifted to get all the angles necessary for a rescue shoring operations.

Speed square with adjustable arm. As with many of the other tools, there are variations of the speed square. One version has an adjustable arm at the bottom instead of having a lip on the bottom of the square. The adjustable arm can be moved and locked in position when necessary. Another variation has three main components—the body, the adjustable arm, and the locking screw. The body usually is made of die-cast aluminum in the shape of a right triangle and is roughly 6¾x6¾. Cast into both sides of the square are three types of tables: inch scale, which is graduated every ⅛ in.; a protractor scale, which is graduated in degrees; and a rafter scale, which is graduated in inch rise per foot of run. The adjustable arm is also die cast aluminum. To use the arm, position its top edge with the desired degree or rise per foot run marking on the body. This gives an angle for marking and cutting a piece of lumber. When the arm is in the desired position, tighten the locking screw, also of aluminum, to lock the arm in that position.

12-in. speed square. This square is very similar to the 6-in. speed square but is 12 in. long and 12 in. wide. It has the same basic markings and is utilized in the same way to acquire the proper angles. It can be used for squaring up and as a saw guide for larger lumber.

Combination square. A combination square can be utilized for the same purposes as the ordinary try square. (The try square is a fixed right angle square with a 6- to 12-in. blade and a 4- to 8-in. body. It is used strictly to square up a piece of lumber or mark right angles on a board.) However a combination square differs from the try square in that the head (or body) can be made to slide along the blade and clamp at any desired location. The sliding of the head is accomplished by means of a central groove in the blade that acts as a guide, enabling the head to travel anywhere along the blade. This groove is etched into the entire length of the square, allowing the blade to be completely withdrawn and used separately as a ruler if the need arises.

Also located in the head of the square is a spirit level that enables the tool to be utilized as a simple level. Since the blade is easily moved in the head, the combination square makes a good marking gauge. To accomplish this, set the head where you want it on the lumber and clamp it. The entire combination square can then be slid along the edge of the lumber just like an ordinary gauge. For marking convenience, most of these types of squares also have scribes placed in the head of the square. For layout purposes, the combination square may be used to scribe lines at miter angles as well as at right angles. It can do this because one edge of the square head is at a 45° angle. This tool is good for use as a gauge or marking right angles when lumber has to be cut square.

2-ft framing square. In almost all building construction work, the steel square, also known as a *framing* or *rafter square*, is an invaluable tool for measuring and determining angles. It is most commonly called a framing square because of its various scales and tables, which are used extensively in house framing and building construction.

The steel square consists of a tongue and body (or *blade*, as it is also called). The tongue is the shorter and narrower part, usually 16 in. long and 1½ in. wide. The body generally is 24 in. long and 2 in. wide. The point at which the tongue and body meet on the outside edge of the square is called the *heel*. The face of the square is the side on which the manufacturer's name is stamped. It is also the side that is visible to you when you hold the body in your left hand and the tongue in your right hand. The back of the square is the side opposite the face. As with any tool, it pays to buy a good quality product. The best squares have their tables and scales etched into the metal, making them the most wear resistant. A square with the tables painted on does not wear well; the numbers quickly wear off with use, rendering the square ineffective as a measuring device.

A good quality square has seven scales and tables: a rafter or framing table, an Essex table, a brace table, an octagon scale, a diagonal scale, octagon scale, and 100ths scale.

Rafter table. A rafter table is always found on the face of the square. It is normally used to determine the lengths of common, hip, valley, and jack rafters and the angles at which they must be cut to fit properly to the ridge board and top plate for roof framing. A rafter table is invaluable not only for determining the lengths of rake shores but also for laying out the angles needed for these shores. The table consists of six lines of figures whose uses are scribed on the left end of the body.

- First line – Lengths of common rafters per foot run

- Second line – Lengths of hip and valley rafters per foot run

- Third line – Length of the first jack rafter and the differences in length of the adjoining jack rafters at 16 in. on center

- Fourth line – Length of the first jack rafter and the differences in length of adjoining jack rafters, spaced at 24-in. intervals

- Fifth line – Figure to be used to determine the edge bevel for the side cuts of jack rafters

- Sixth line – Number of the edge bevel for side cuts of hip and valley rafters

Essex table. This table is always found on the back of the square on the body. It provides the board measure in feet and 12ths of a foot of boards 1 in. thick and of common lengths and widths. This table is usually not needed for rescue work.

Brace table. Found along the center of the back of the tongue, a brace table is used to give you the lengths from 24 to 60 in. of common braces, where the rise and run are equal, forming a 45° angle. The brace table can be very useful when installing short rakes or additional bracing.

Octagon scale. The octagon scale is located along the center of the face of the tongue and is used for laying out a figure with eight sides on a square piece of lumber. The graduations are generally a series of dots located $5/24$ in. apart and numbered at every fifth dot. This scale is not necessary for rescue shoring.

100ths scale. Located on the back of the tongue near the heel between the brace and Essex tables, the 100ths scale consists of 1 in. divided into 100 parts. Underneath the scale, 1 in. is divided into $1/16$ths, making it easier to convert decimals to more commonly used fractions. This scale is especially useful for dealing with rafters and braces whose lengths are given in 100ths. Use this scale when you are determining the lengths of bracing and raker shoring.

Inch scale. Located on both the body and tongue along the inside and outside edges of the square, this scale measures inches graduated in $1/8$ths, $1/10$ths, $1/12$ths, and $1/16$ths. On the square's face, the outside edge of the body and tongue is graduated in sixteenths. The inside edge of both sides of the face is graduated in $1/8$ths of an inch. On the back of the body, the outside edge is graduated in $1/12$ths, while the inside edge is divided into $1/16$ths. The back of the tongue on the outside edge is also divided into $1/12$ths, and the inside edge is graduated in $1/10$ths of an inch. These scales are used in measuring and laying out work to precise dimensions.

Many squares include a diagonal scale, the object of which is to give minute measurements without having the graduations so close together that they are hard to work with and to read. This scale is rarely used in rescue work.

T-bevel square. This tool is small and compact; it has a movable arm that also slides along its center. It is used primarily to copy specific, existing angles. The *T*-bevel square is quick and easy to use—and it's almost fireman proof. Just set up the main body and slide the angle section in position. Tighten the thumbscrew that holds the movable arm to lock the arm in position.

If the cutting station is close enough, you can bring the bevel square to the cutting station and transfer your angle. If the cutting station is not that easily accessible, you can transfer the angle to a block of scrap wood and give it to a runner to take it to the cutting station. If neither one of those options is a good one, you can scribe the angle on a scrap piece of material and lay a speed square or framing square on the angle and measure the degrees from there. Relay the angle to the cutting station. The cutting team cuts the angle you ask for and relays the angled piece to you via runner. This tool is excellent for a quick, no-nonsense angle size-up, just what you need when installing a rescue shoring.

A square. This is a newly developed collapsible triangle that is designed for precision layouts. When unfolded, it is four times bigger than a framing square, thus almost guaranteeing accuracy. It has an unfolded and locked dimension of a 3 ft x 4 ft x 5 ft right triangle. The positive locking mechanism ensures accuracy by locking the tool only when it is in the correct position. It folds up nicely and can be stored in a relatively small area. It is excellent for forming right angles and perfect for use when preassembling your raker shore systems as well as any other systems that have to be square.

Levels. The most common levels are probably the 2- and 4-ft level, the 6-in. torpedo, and the post level. Use the 2- and 4-ft level to check if walls, buildings, and shores are plumb or how far out of plumb they are. The 6-in. level is a good tool for determining if each element erected is plumb. The post level can be placed on the edge of any vertical post, and it will plumb up that post in both directions at the same time.

The various styles of levels available to your team are enough to confuse anyone. There are wood, metal, box beam, magnetic, lighted, and combinations of all of these. The first thing your team should be buying is heavy-duty or contractor-grade tools. These tools last a lot longer than some of the other homeowner-grade tools. This is a *must*, and the abuse these tools take in rescue situations is extensive. Some firefighters prefer the wooden type with brass edges and open hand-grips in the center of the level. This type of level stands up to quite a bit of punishment and is an extremely accurate tool. Don't forget, when stored in your apparatus, tools are subjected to quite a bit of pounding and vibrations. For this reason alone, your levels should be the type with the embedded vials. The adjustable vial-type invariably loosens up and has to be adjusted with every use.

Another type is the aluminum, box beam constructed type. It is heavy-duty and durable. All metal but with no interior handles, it has to be gripped along the outside, which normally isn't a problem, but the handles are a definite advantage. There are some of the smaller torpedo-type levels that even have lights. Something that is very handy in confined areas and at night or where the light is not the best is a magnetized level. These levels can cling to steel, leaving room for hands-free operation with the level in place.

There are also other types of levels on the market that can detect if any movement has occurred after an item has been installed. These levels can be anchored to a structural item and read on a continuous basis. A digital readout indicates if the element has shifted.

45° template. When your team starts to erect its shoring items, especially at night or in bad weather, the use of templates greatly speeds your operation and helps to make potentially confusing situations mistake-free. The use of templates can eliminate several opportunities for mistakes. Using templates also helps the team be able to concentrate on other items and not be too concerned with doing the math or physically making the angles with the tools at hand. The 45° template is used for your average raker shore angle especially for the solid sole raker shore. The 45° angle is the angle of choice. There are several different styles of templates, and they can be made out of wood or metal.

60/30° template. This template is utilized mainly in the construction of a split sole raker when it is being erected on soil. It is recommended that this angle be used because it directs more of the pressure into the soil at a steeper angle; therefore, the shore is less likely to slide laterally against soft soil conditions. If the shore were to slip laterally, it most certainly would fail. By placing the raker at a steeper angle, you can alleviate that possibility. The top of the raker has a 60° bevel, and the bottom angle has a 30°-angle bevel. This is important if the base of the raker is being anchored into something substantial. The use of these templates can speed up your layout operation and takes the guesswork out of some of your measuring.

Hammering tools

Framing hammers. One of the most common and versatile tools in the rescue arsenal is the framing hammer. There are several types of hammers available—a vast assortment indeed. However, this book addresses a few that will serve your needs very effectively. To nail up the rescue shoring, use what is commonly referred to as a framing hammer. If members on your team have carpentry skills and experience, then by all means use their experience to pick the type and size that works well for them. If team members do not have carpentry experience, here are some guidelines to use when selecting a framing hammer.

The hammer should be the straight-claw, also called *rip claw*, variety. This style works much better than the curved claw type for working in structural collapses. The next issue is the hammer's weight.

It has to be heavy enough to drive 16d nails yet not too heavy to cause arm fatigue quickly. The ideal weights of most hammers are from 22 to 24 oz. This is the weight of the metal head as well as the handle.

For leverage purposes, your hammers should be no less than 15 in. long. For hammers with longer handles (up to 18 in. long) the head can be 20 or 22 oz. The additional leverage of the longer handle compensates for the lighter head. However, firefighters who have not used a longer handle hammer extensively should stick to the 15-in. size.

Another important item is the handle material. Most hammer handles longer than 15 in. are generally wood or steel shank. Years ago, the only type available was a wooden handle, generally hickory or ash. Over time, several other types have become available, especially in the last 20 years. Wood is definitely losing popularity; steel, fiberglass, graphite, jacketed I-beam, and jacketed graphite are now all commonly available. There are pros and cons with all types, and opinions are varied as to which type is better. In any event, your team should choose the type it works with most comfortably. Many experienced carpenters still like the wooden handle. The jacketed graphite handle and the jacketed I-beam handle are also good quality. As far as steel handles, if they are solid steel, generally I-beam construction, they also are good for shoring work. If your team is less familiar working with tools, fiberglass or wood handles are not a wise choice. The firefighters have a tendency to break the handles. With the other three types, the durability of the tool will serve your team well. These hammers are used for almost anything, from clearing debris from floors and ceiling, pulling nails, and breaking away of light masonry to splitting thin pieces of lumber. It is a very useful tool and must be included in the tool cache. There should be at least one framing hammer for each tool pouch in the cache.

Hammer drills. These tools are also known as lump hammers. They generally range in size from 3 to 5 lb. There are no claw ends on these tools. They have driving heads on both ends, which makes them excellent to use when just a little more weight than a framing hammer is needed. In collapse operations, hammer drills are commonly used for driving large pins, wedges, and stakes and to break up light sections of masonry or concrete. It is a very handy tool when you need just a little more power to accomplish the task.

½-in hammer drill (cordless). With numerous manufacturers and tool sizes available, your team has to make its own decisions on the brand of ½-in hammer drill to purchase. For your purposes, the ½-in. chuck will work the best. The most practical size batteries to consider are the 18-, 19.2-, and 24-V varieties. The revolutions per minute and the blows per minute vary slightly among all types. Pick the one that is most suited for your team's particular needs based on the structures in your response district.

Sledgehammers. The next step up from the hand-drilling hammer is the sledgehammer. It is a much larger tool and is utilized for driving stakes and 1-in. pins and for breaking up sections of masonry or concrete. Generally the weight of a sledgehammer can be 10 to 12 lb. There are larger sizes, but using them effectively requires larger, stronger individuals. The tool length of sledgehammers tends to be roughly 32–35 in. Those lengths give the tool plenty of leverage.

Rotary hammer. A rotary hammer is used to drill holes, drive pins, break up masonry, or to hammer drill. It is a versatile tool necessary to keep in your tool cache. There are several styles and sizes available, including electric, pneumatic, hydraulic, and gas operated. There are also various drill and hammer bits for the tool. Have your team try out several models and make its own decision as to the type to use. One use for this tool is to drill holes into the street or sidewalk to accept 1-in. steel pins to anchor the blocking for raker shores. This is the ideal tool for that particular job.

⅞-in rotary hammer (cordless). These tools have rapidly developed in the cordless market led by a few manufacturers' developing of this type of tool. This tool utilizes 24- and now 36-V battery packs for maximum power and staying power. For a quick application, this tool may very well fit the bill. Normally used to drill holes for anchor bolts in concrete, it is also used for stitch drilling to remove sections of concrete.

Power actuated hammer. This tool is necessary to anchor down specific pieces of lumber to masonry surfaces, possibly to walls or sidewalks or the street itself. As with any other type of tool, there are various brands and models. For the best results when erecting rescue shoring, get a heavy-duty model. There are several strength sizes as well as pin sizes available. Have a range of sizes for both. There are two types of pins: those used for steel and those used for concrete. The pins for steel use have a fluted shaft, and the pins for concrete have a smooth shaft.

Power nailers. It is extremely advantageous for your rescue team to utilize, purchase and be proficient in the use of the various types of power nailing systems. They are not difficult to set up or learn to implement. With a little practice, your team members can become quite skilled in their use.

Gas fuel nailing systems. There are two types of gas fuel nailing systems available to your rescue team today. Both are quite compact. One runs on a battery and a methylacetylene-propadiene (MAPP) gas cylinder; the other runs on a gas cylinder and an internal ignition source. They are very good for initial deployment and operations. The system can be stored close to the firefighters and brought off the apparatus with the initial tool complement. The advantage of these tools is their quick deployment. They only need the fuel and batteries loaded and are ready for work. Easily transportable, they can be put to work in less than a minute. Both types can set approximately 1,200 fasteners in place before having to be refueled, which is quite a bit of work accomplished with little effort. These tools don't cause any vibration, perfect for what is needed in rescue shoring operations.

Pneumatic nailing system. Another type of power nailing system is the pneumatic nailing system. This is a commercial system that has been around for several years, and there are numerous styles and sizes. Two main types are the coil nailer and the stick nailer. For shoring purposes, generally the stick nailer is fine. Typically it is smaller than a coil nailer, making it easier to use in close quarters. A pneumatic nailing system can fasten all kinds of nails as well as staples. However, you only need the ones that fasten 8d and 16d nails. With the proper air supply and hose set up, you can almost use this tool indefinitely. Its only draw is the equipment needed to initially set it up. Many times in a collapse scenario the room to work and the access to the building is not good to say the least. You will need an air supply, generally a compressor of some type, and a set of air hoses. Where in the operation the compressor is to be setup depends on the number of hoses you need. Sometimes, if several hoses are needed, the air has to be constantly regulated. The real benefit of using these tools is the speed with which they are able to nail items and their ability to anchor those items without any significant vibrations. For these two reasons, using the pneumatic or fuel-powered nailers is recommended for your rescue shoring operations.

Cutting tools

Utility knife. A utility knife is a basic tool that comes in handy to cut many building materials encountered as debris: sheetrock, ceiling tiles, carpeting, paneling, and several other items. It is especially useful when you have to sharpen your carpenter pencils. When you get a utility knife, make sure for safety reasons that it has a retractable blade. You don't want to have any accidents while you are trying to erect your shores. Make sure you have plenty of blades with the tool. The items you are cutting determine how quickly the blades become dull. When installing these items in your tool cache, be certain there are several spare blades inside each knife. When the blade does become dull, you can easily change it right on the spot without much effort.

Handsaws. Another item that is not used all the time but should be on hand is a handsaw. Used to cut small things quickly when electricity is not readily available, handsaws are lightweight and portable—ideal for use in some close quarters—and should be included in an initial shoring kit. Handsaws come in various sizes and types, but there are two main types: the cross cut and the ripsaw. The one most often utilized in a collapse rescue scenario is the crosscut type. Among the things to look at when determining the type of handsaw to use are the length of the saw and the number of points of the saw. The *points* of the saw are the points on the saw blade. The number of teeth per inch or per saw blade length is important. For example, a 10-point saw has 10 teeth per inch, and an 8-point has 8 teeth per inch. For shoring purposes, the 8- and 9-points are generally sufficient.

Chisels. There are two types of chisels useful in most collapse shoring operations. The situation dictates the type to utilize. If there is a wood floor and ceiling scenario, then the wood chisel is the obvious choice. There are a few styles and numerous widths of chisel blades. Choose the ones with which your personnel are most comfortable. Generally, it is a matter of personal preference; some workman like the wider blade tools, and others are more comfortable with the narrower blade chisels. These chisels generally have wood, plastic, or fiberglass handles. All of them are fine. Again select the type your troops like best. If you are working with some type of masonry, whether it is cement block, brick, or even concrete, the solid metal, cold chisel should be utilized. The blade widths are approximately the same as the wood chisels. Both of these chisels are good for scraping or chipping away small pieces of wood or masonry from floors or ceilings where many times it is necessary for these little items to be removed in order to install your rescue shoring.

Cutters. Cutters are used to cut wires, cables, or sheet metal. The cutting tools used most often are the ones used for cutting wire, BX (a type of electrical cable covering) cable, and thin sheet metal. Side cutters, aircraft shears, and cutting dykes are the most common types.

Reciprocating saw. Also known as a *sawzall*, a reciprocating saw is an excellent tool for making quick adjustment cuts when shores are in position. This saw also can be used to cut out protruding elements that may be in the way of the shores. This is a very handy tool to have for all collapse rescue operations. Reciprocating saws can be powered by battery, electricity, gas, or compressed air. There are also several styles. Your team should make sure it gets a heavy-duty, variable speed model. It is much easier to start cuts in lumber with a reciprocating saw. The style your team picks should be decided after testing the different saws. The blades are made to cut wood or metal; some blades can be used to cut both. The best type of blade to purchase for collapse rescue work is the bi-metal type. It comes with various sizes, tooth styles, and blade thickness.

Reciprocating saw (cordless) variable speed. These tools are good for cutting protruding pieces of lumber, electric cables, bolts, nails, and the like. They range in size from 14 to 18 to 24 V. Your team should choose whichever type it likes the best; however, you should have at least two of these saws and extra batteries in order to help keep the tool in continuous use.

Power miter box. A bit of a luxury item, a power miter box is a very good tool to have, especially if you are working in a long-term operation with numerous shores to be erected. There are several makes and models. A 14-in. blade should be the minimum you purchase. It can cut some of the larger material if necessary. The blades also vary, but whichever one you choose, be certain it has a carbide tip; it will last longer. The box can make very accurate cuts at any angle with minimal layout required. The angles are right on the box table, and most models have stops at the more common angle cuts, such as 90°, 45°, 22.5°, etc.

Chainsaws. A chainsaw is an excellent tool for cutting any shoring material in a remote area. The most common types are gas- and electric-operated saws. The electric is excellent for shoring station uses. It doesn't make a lot of noise, gives off no fumes, and handles rather well. There are various bar lengths. The 14 in. is a good length for cutting normal size shoring lumber such as 4x4s, 4x6s, and 6x6s. When using these saws for cutting shoring, make sure you don't use carbide tips. They make a cut too sloppy for shores. Another use for the saws is to cut more than one piece of lumber at a time.

Circular saws. The most common type of saw shoring teams use is the circular saw. There are two main types: the worm drive and standard drive. The worm drive is a little more powerful than the standard drive but needs more electrical power. Sometimes, not always, this is an advantage when we are operating off site, using portable electrical generators. The two most common sizes of circular saw are the 7¼ -in. and the 10¼ -in. saws. The 7¼-in. saw is a little easier to handle and is good for cutting 2-in. dimensional lumber; the 10¼-in. saw can cut the 4x4s commonly utilize in one pass.

6-in., 6½-in. circular saws (cordless). There are at least eight or nine manufacturers that produce 6-in. and 6½-in. circular saws. There are smaller size blades available; however, for shoring work they really are not efficient enough. The two sizes mentioned last a little longer than saws with smaller blades and can cut 2-in. material numerous times with one charge. As with any battery-run tool, you need to maintain spare batteries to keep the tool running effectively.

Prying tools

Cat's paw. A small hand-held nail puller, a cat's paw gets its name from the shape of the pulling head. It is excellent for pulling out nails with heads embedded deep into the material. They are made of steel and generally are about 12 in. long. Occasionally you will have to pull apart some of the material that has been previously assembled. This tool is perfect for the job. By angling the fork of the tool just in front of the embedded nail head and striking the tool with a hammer, the fork embeds into the lumber under the nail's head. When the fork of the tool surrounds the nail head, you then can pry up the nail and remove it.

Pinch bar. Also called a crow bar, a pinch bar comes in various lengths: 18 in., 24 in., 36 in., and 48 in. It is a good idea to have on hand a few pinch bars of different sizes. Use the tool to pull large sections of material apart, pull larger nails out of lumber, or as a lever to move specific items. These are handy items to have that don't take up a lot of room in your tool cache.

Wrecking bars. A wrecking bar is very good to have when you have to lift debris. Two of the sizes commonly used are the 5-ft length and the 3-ft length. The 5-ft wrecking bar can be used to lift some substantially heavy items when there is a proper fulcrum available. Remember, when utilizing this type of tool, the more leverage you acquire the better, more efficient, and easier the lift will be. A wrecking bar can also be used to pry material apart from a standing position, such as separating floor planking from their anchor beams. In a masonry situation, the bar can also be used to break up brick and block and even chisel out small pieces of concrete if necessary.

The smaller bar is for similar uses in a more limited and confined environment. It can perform similar functions on a smaller scale, but there is less leverage available with the smaller length bar. These are both very versatile tools to have in your team's cache.

Wrenches and pliers. Tightening and unscrewing items may require wrenches and pliers. As usual, there are numerous types and sizes of these tools that your personnel can choose. The most common type of wrench needed is an adjustable wrench—two sizes are generally enough, such as 6 in. and 12 in. For larger jobs, 12-in., 24-in., or even larger pipe wrenches can be carried on the apparatus. These wrenches are excellent for unscrewing gas or plumbing pipes that may be in your way.

Digging tools

Shovels. There are numerous types available. To cover all the bases, have on hand a pointed and flat blade type. An entrenching tool may also come in handy. It is a good idea to have both styles of shovels in both short D-handle and long-handle types. The main use of these tools is to clean the area where the shore is to be installed. In almost all occasions, there is debris in the way that must be cleaned away. Typically, the area should be roughly 4 ft wide and 4 ft longer than the size of your shoring systems. In most situations, the smaller handled shovels serve your rescue team much better that the longer handle variety. The small entrenching tool can be good for cleaning out those hard-to-reach areas or where access is limited.

Mechanical tools

Ladders. In many cases, the use of folding ladders is called for, especially when the ceiling height is over 8 ft. Ladder sizes can run the gamut from the small 24-in. high step stool to the 8-ft high folding ladder. Although these ladders are constructed of several types of materials, including wood, metal and fiberglass, the better the quality of the material and workmanship, the better the ladder functions and stands up to typical abuse. Many firefighters prefer a good quality fiberglass ladder. It stands up well and is very sturdy while being relatively lightweight. It's a good idea to keep at least three different sizes of folding ladders on the apparatus; for instance, 2, 4, and 6 ft heights are pretty common sizes. Of course the response area of your rescue team pretty much dictates the size ladders you require.

Sawhorses. A *must* to have in order for your operation to be done properly, saw horses hold the lumber used at the cutting station. The lumber should be placed on top of these saw horses. Doing so makes it safer and easier to cut and to handle the lumber for your shoring operation. There are several sizes and styles of these saw horses. Select the type that best meets your team's needs and preference. However, whichever style you choose, the sawhorses should be a low-profile type and collapsible if at all possible. Your tool cache should have several sets of these saw horses. Make sure you purchase the heavy-duty style so that it can withstand the abuse that the firefighters place on it.

Rescue struts. Another type of pneumatic device is the air shore or rescue strut. These shores are constructed of aluminum alloy and are 3½ in. or 3 in. in diameter. They are used for stabilizing collapse debris and shoring purposes. When entering into a collapsed building, it is imperative to observe as many safety precautions as possible. The more stable you can make the remaining structure, the safer your operations will be. Rescue shoring in collapsed structures is an inherently risky business, and you must do everything possible to put the safety factors on your side. One of the fastest ways to accomplish this is through the use of specially designed support systems.

Several different types of mechanical and pneumatic shoring systems available today have been adapted for use in rescue operations. Rescue struts are particularly versatile and can be used in various building collapse situations as well as for several other shoring and rescue operations. This factor is a major plus for departments with a limited budget. This system affords your rescue team a choice of securing devices: an acme-threaded strut and a self-locking strut. The self-locking and the acme-thread struts have been tested by an independent company and have an axial crush strength of more than 50,000 lb. These struts have an axial working load capacity of 20,000 lb, making them excellent choices for initial safety shoring and for working in tight void areas.

The *activation force* is the amount of pressure needed to raise the shaft of the rescue strut. As the activation force increases, the support force of the strut increases proportionally. For example, if a pressure of 50 psi is exerted against the shaft, the strut is capable of exerting 245 lbs of force. If a pressure of 350 psi is placed on the strut, the strut will exert a force of 1700 lb. These shores have limited lifting capabilities and generally should be used only as stabilizing tools, especially in collapse operations where any additional movement could be hazardous to your rescue personnel.

Self-locking strut. A shoring system can contain self-locking struts of three sizes: 24½–36½ in. (12 in. stroke); 36½–58 in. (21½ in. stroke); and 55½–91 in. (35½ in. stroke). *Stroke* means the strut expands to the stated length. Each strut is made of 3-in. aircraft aluminum alloy tube with a solid 2½-in. aircraft aluminum alloy moveable-grooved shaft. These struts normally are extended by hand for building collapse operations but can be activated by air, carbon dioxide, or nitrogen if the need arises.

The distinctive feature of the self-locking strut is that it locks automatically in an extended position. Its special locking feature, a double row ball-lock coupling, does not require a member to manually lock it in place, and there are no safety locking pins to install by hand, providing a greater inherent safety factor. The hands-free locking feature allows the rescue team to extend and lock the strut from a remote location if necessary.

Taking down and repositioning the strut can be accomplished more safely than the same operation with some other types of struts. The locking mechanism can be released by removing the load pressure and pulling a release ring. If for some reason the load shifts or if further collapse occurs, the rescue team member need only let go of the release ring; and the strut immediately locks in place again, stopping the debris from shifting further.

Acme thread strut. An acme thread strut system (a system is composed of a group of struts and components) contains three sizes of acme-thread struts: 24¾–36½ in. (1 1¾ stroke); 36½–58½ in. (22 in. stroke); and 58–90 in. (32 in. stroke). These struts are constructed of the same aircraft aluminum alloy as the self-locking strut and also consist of either 3½ in. or 3-in. aircraft aluminum alloy tube with a 3 in. or 2½-in. solid aircraft aluminum alloy acme threaded shaft. Like the self-locking variety, they can be extended manually or from a pressure source such as air, nitrogen, and carbon dioxide. However, locking the strut is a manual operation.

The distinctive feature of the acme thread strut is that this collapse rescue support system allows for extremely soft placement in most collapse rescue operations. It can be brought to and gently tightened at any point under a load. The strut is secured in place with a large nut, which extends with the shaft that the user manually screws down and locks against the tube. Taking down the acme-threaded strut can be accomplished as safely as taking down the self-locking strut. Remove the load pressure and manually twist the acme nut. The locking nut releases. If debris shifts while the strut is being lowered, the firefighter simply lets go of the acme nut, and the shore locks in place, resupporting the load.

Strut extensions. The system also contains three rigid strut extensions, 12, 24, and 36 in. long. They are constructed of 3-in. aircraft aluminum alloy tubing. These extensions were designed to allow rescue personnel to add length to either strut type—self-locking or acme thread struts—multiplying support capabilities and applications in building collapse situations. Each strut can be used as a rigid support device if the need arises. The larger strut extensions made of 3½-in. aluminum alloy tubing are 24, 48, and 72 in. long.

Low clearance supports. The low clearance support system is a series of four solid extensions constructed of the same material as the acme thread and self-locking strut shafts in 1, 3, 5, and 7 in. lengths. This part of the system is designed for rescue work in very close quarters. These supports and the extensions can be used with a variety of bases.

Bases and fittings. Thanks to a specially designed adapter called the extension converter, any of the systems' bases and end plates may be attached to either end of the strut extensions to create a strong and rigid support device. Numerous bases and connectors are available for the rescue strut systems previously discussed, making them very versatile systems. They can be used for building collapse, trench rescue, and automobile extrication, just to name a few of the operations for which they are suited.

3-in. standard base. This is a simple 3-in. diameter cap 2 in. high with a ¼ in. pull-and-twist locking pin. Covering both the strut base and the end of the shaft, the standard base protects the strut ends during simple bracing operations.

Rigid base. A 6x6-in. rigid base with a nonskid, grooved surface was developed to provide greater stability than a standard base. This base works well on solid surfaces and when the shore is utilized at right angles to the object that is bracing.

Swivel base. A fitting with a 6x6-in. square base can swivel 20° in any direction. The swivel adds approximately 3½ in. to the length of the strut. A swivel base was developed for cases in which the items to be braced are not in direct alignment. This is a very versatile base that can be used in many different situations.

Cone point base. The cone base basically is the standard base with a ¾-in. pointed cone in its center. It is primarily for holding the struts at a slight angle against smooth surfaces such as the sheet metal skin of an automobile.

Spring-loaded connector. A fitting developed especially for use with the self-locking strut, the spring-loaded connector keeps the strut compressed even if slight movement occurs. The spring activates at roughly 200 lb and has a travel length of approximately ½ in.

Threaded adjustable connector. The threaded adjustable connector is designed for use in situations in which any forceful movement can be dangerous as it can in building collapse-shoring operations. Manufactured from a 1¾-in. threaded aluminum shaft, it can be finely adjusted up to ½ in. It is excellent to use with the self-locking strut in collapse operations where supporting the collapse load is the primary function.

Hinged base plate. Designed for use with the rescue strut system, the hinged base plate rotates 90° in one direction, making the struts available as an initial safety rake shore at 45°.

4x4-in. channel base. This bracket, specially designed for shoring operations, can be used on the base of the strut or the end of the shaft. It will lock on to a section of 4x4 to which it can be anchored with nails or screws. The bracket is 6 in. long, 2¾ in. high and 3½ in. wide and is designed to fit snugly to the shoring lumber.

6x6-in. channel base. Designed for larger lumber operations, 6x6-in. channel base is used in trench rescue and in larger buildings where more substantial shoring may be necessary.

V base. A 3-in. x 3-in. cylindrical fitting with a 90° *V* base approximately 1 in. deep in its center, this base is used for stabilizing anything with an angle or corner such as a tractor trailer leaning after an accident or a beam with a square edge in a building collapse.

T plate. This *T*-shaped fitting has a lip and a return that can be used to hang the strut for hands-free operation, necessary in a trench cave in where the whalers have to be re-braced before personnel can enter the excavation.

Hydraulic ram. A recent addition to the strut system is a 10-T hydraulic ram with a 4-in. lifting stroke and a separate power pack with a 6-ft hose. The ram can be utilized in several collapse situations as well as in a retrieval system for shores under pressure. A coupling on the bottom of the ram accepts any of the rescue struts, and a fitting on the top accepts all bases and connectors. A hydraulic ram is small enough to be easily maneuvered in a collapse void and used as a temporary support for damaged or unstable structural elements. In special cases, it may be able to gently lift certain collapse debris that is pinning a trapped victim, freeing the victim from the rubble.

Ellis shores. A shoring method that incorporates the use of lumber and a set of clamps to form vertical bracing is called an Ellis shore. This system enables you to slide two pieces of lumber together and lock them in position with two specially designed clamps.

These clamps come in four sizes for use on 2x4s, 3x4s, 4x4s, and 4x6s. In rescue shoring, you use the 4x4 size almost exclusively. The clamps are generally painted red and are made out of ½-in. round rod and formed into a rectangular collar. With the collar are two heavy-duty malleable castings. Both castings are serrated on the flat surface for firm gripping of the lumber when the shore is pressurized. The clamps are placed 12 in. apart and can be used many times over. An Ellis jack is used to pressurize the two pieces of lumber and then the clamps are nailed in place. The manufacturer recommends that the shores don't support more than 6000 lb each. As with any mechanical system, it has its advantages and disadvantages. Hydraulic shores are definite options that your team can utilize in a collapse rescue operation.

Miscellaneous tools/equipment

Tool pouch. Another item that is almost a *must* is the tool pouch or, known by its other common name, nail apron. It will keep the necessary hand tools that you use the most in one place and easily accessible. Don't be cheap; buy the good stuff. It will last much longer that the lower grade pouches.

There are several different types of pouches. Use one that suits your team's needs, generally the one that holds tools as well as nails. The better quality aprons will have anywhere from 8 to 11 pockets, including hammer loops on both sides. You can put several sizes of nails in the pockets as well as most of the small hand tools needed to erect the shoring. The typical tool pouch should have a speed square, 6-in. level, hammer, tape, utility knife, carpenter's pencil, lumber crayon, nail puller, 8d and 16d nails. You can also, if you like, have nail pouches with both size nails prefilled and ready to go. Each shoring position has unique needs, and additional tools can be included into the pouches assigned to those positions.

Steel pickets. Steel pickets are 1 in.-thick steel stakes used mainly to anchor raker shores or their anchor blocks. Use at least two of these behind each raker shore. Depending on the soil conditions, the stakes can be anywhere from 36 to 48 in. in length. They should have a point in the center of the stake so that as it is driven into the ground, it remains straight. On the top of the stake, it is a good idea to have a cap slightly larger then the pin size. This cap makes driving the stakes with a sledgehammer easy. Normally these stakes, or pickets, are made of cold, rolled steel stock and can be purchased this way. If for some reason you can't seem to locate this design, you can substitute steel reinforcing rod as a viable alternative. Although technically not as strong as the cold, rolled steel, steel reinforcing rods work well for shoring. The rebar should be at least 1 in. thick. Number 8 bar would be the technical term for that size. Rebar slightly larger is fine too. These pickets are also good as anchors for your interior sloped floor-shoring systems.

Anchors. Another method of securing your different types of shores, especially the solid sole rakers, is to use anchor bolts. There are numerous styles of bolts and several different methods used to secure them. One style particularly useful is the *Hilti* HSL heavy-duty anchor. This anchor comes in various lengths and thickness. The ones that work especially well are 5–6 in. long and roughly 1 in. thick. There is another type of heavy-duty anchor that is almost fireman proof. It is the heavy-duty anchor that has an indicator cap installed on the head of the bolt. A red cap is set at a predetermined torque; and when the bolt is tightened with a wrench, the cap breaks off at the specified torque. Both of these anchors are the sleeve-expansion type.

Another type of anchor that works well is the Kwik bolt type. It comes in roughly the same sizes as the Hilti anchors and has a wedge expansion system that sets it in the predrilled holes. One thing that absolutely must be present for the anchors to work is the proper hole set up. The hole must be the exact diameter required by the manufacturer's specifications. It cannot be off in the slightest; if it is, the anchor will most likely pull out. After drilling the hole the predetermined depth and thickness, you must clean it out. There cannot be any debris or powder in the hole to interfere with the proper setting of the anchor. When that has been accomplished, tap the bolt in with a hammer and sink it flush with the washers. Then tighten it with a wrench to the proper torque. These anchors can be quickly put into place and are ready for use immediately after setting.

Nails. Generally speaking in all shoring operations, use two sizes of nails: the 8d nail and the 16d nail. There are numerous sizes and style of nails; however, at most rescue-shoring operations limit the nail supply to these two sizes of nails. The common nail is the one utilized on most occasions. You can use duplex nails, also known as double headed nails or green sinkers. These nails have a rosin coating on them. The 8d nail is 2½ in. long, and a 16d nail is 3½ in. long. Use the 8d nails to nail plywood together or nail plywood to dimensional lumber. Use the 16d nail to connect only dimensional lumber together regardless whether face nailing or toenailing the lumber together. Your shoring team must make sure that there is plenty of material on scene. You should have at least 100 lb of each type of nail in your tool cache at all times.

Nail caddy. It is always a good idea to make sure that you have enough nails on hand to complete any shoring operations. The easiest way to accomplish this is to make up a box, or caddy, out of plywood. The most common nail sizes used are the 8d and 16d common. Although used for general work, rather than for shoring purposes, 10d nails also can be used. Make up the caddy with three bays and put different sizes of nails in each bay. If you are going to employ the use of several caddies then you can dedicate one caddy to each nail size and label accordingly. The typical caddy can be roughly 20 in. long, 7 in. wide, and 12 in. high. It has three 6x6-wide compartments, and is constructed of ½-in. thick plywood. You can use ¾-in. pipe as the handle. This can make for a compact and easily portable caddy that can be stored almost anywhere on your apparatus and will not be in the way at an operation.

Tool Assignments

A tool assignment is a list of some of the more common items needed at a typical rescue shoring operation. This is by no means a complete list of all the tools you may need, for each situation is different and unique unto itself. The following information provides tools most likely utilized at all rescue-shoring operations.

Officer

The officer's position must be flexible, and it must be a supervisory one if at all possible. He should be staying at least one step ahead of the team in order for the operation to run smoothly. He needs the following:

- Radio

- Personal Protective Equipment (PPE)

- Shoring manuals

- Safety checklist

- Shoring size-up checklist

- Paint stick

- Laser pointer

- Tape measure

Shoring firefighter

The firefighter should have the following at his disposal:

- 22- to 24-oz straight claw hammer, steel or graphite handle recommended for durability
- Leather tool apron
- 16-ft tape measure with a 1-in. thick blade
- Carpenters pencils
- Utility knife
- 18-in. pry bar
- 6-in. torpedo level
- 4-in. level
- 3-lb lump hammer
- 8d nails
- 16d nails
- Nail caddy
- Pneumatic nailer
- Pneumatic nails
- Fuel-driven automatic nailer
- Spare battery
- Spare fuel
- Flat screw driver
- Shovels
- Wedges
- Shims
- Short hand saw
- PPE

- Radio

- Nail puller

- Shoring book

- Light

- Hacksaw

- Adjustable wrench

- Snips

- Battery-operated sawzall

Measuring firefighter

The main job of the measuring firefighter is to concentrate on measuring all of the shore's elements. Therefore, this team member needs the following tools:

- 2-ft. framing square

- Carpenter's pencils

- Paint stick

- Marker

- 16-ft or 25-ft tape measure

- Light

- Nail puller

- Tool apron

- Hammer

- Utility knife

- Shovels

- Paper and pencil

- Speed square

- 4-ft and 2-ft levels
- Laser measuring device
- *T*-bevel square
- PPE
- Radio

Note: The levels are strictly an option; most shoring can be done by eye.

Layout firefighter

This firefighter's main job is to concentrate on measuring all the shore's elements. Therefore, the tools needed are as follows:

- A 16- or 18-ft tape measure
- 2-ft framing square
- Carpenter's pencils
- Paint stick
- Light
- Tool apron
- Hammer
- Utility knife
- Felt tip marker
- Flat shovel
- Speed square
- Angle templates
- *T*-bevel square
- PPE
- Radio
- Note pad

Cutting firefighter

It is imperative that the cutting firefighters have the following:

- All the PPE required by the manufacturer of whatever saw being used

- A saw capable of cutting 4x4s in one pass, which includes a 10¼-in. circular

- Beam cutters

- Chainsaws, gas or electric (the electric would be better because it is easier to handle and quieter)

- Spare blades and other saw accessories

- Hammer

- Utility knife

- Speed square

- Tape

- Cutting guide

- Radio

Tool and equipment firefighter

The tool assignment for the tool and equipment firefighter can be flexible. His main responsibility is to track on paper the location of all the tools in the team's cache. This person should at least have a radio, PPE, and a tool pouch with the items common to other team members.

Tools to Order for a Shoring Operation

Hand Tools

16-ft and 25-ft tape measures with 1-in. wide blades

50-ft and 100-ft steel tapes

Battery-operated laser measuring devices

6-ft carpenter measuring rules

Chalk lines

5-lb chalk, red

5-lb chalk, blue

22- to 24-oz straight claw framing hammers with steel or graphite handles

10-lb sledge hammers

3-lb sledge hammers

5-lb sledge hammers, non-spark

10.5-lb sledge hammers, non-spark

Leather carpenter's tool pouches

6-ft speed squares

12-in speed squares

Combination squares

2-ft carpenter's framing squares with rafter tables

T-bevel squares

4-ft levels, wooden

4-ft smart levels

2-ft levels, wooden or aluminum box beam

2-ft smart levels

6-in. levels, torpedo type

Post levels

Laser levels

Utility knives with retractable blade

Carpenter's pencils

Lumber crayons, red, yellow, and black

Cat's paws nail pullers

Wonder bars

36-in. pinch bars

5-ft wrecking bars

3-ft wrecking bars

26-in., 8-point, crosscut hand saws

15-in., 9-point hand saws

12 in.-blade, 24-point hacksaws

Jab saws with blades

Carpenter's hatchets

9-piece set of wood chisels

Cold chisels, full sets

Shovels, short handle, D-handle, square point

Shovels, short handle, D-handle, round point

Shovels, long handle, square point

Shovels, long handle, round point

Shovels, entrenching type

18-ft "Little Giant"-type ladders

24-in. pipe wrenches

12-in. pipe wrenches

9-in. wire cutters

30-in., casehardened bolt cutters

Saw horses, Metal Mule or equivalent

1 in. x 4 ft steel pickets

50-lb box of 8d common nails

50-lb box of 16d common nails

50-lb box of 10d common nails

45° template

60° template

Electrical Tools

7¼-in., worm drive circular saws

10¼-in. circular saws

24-tooth, 7¼-in. carbide tip circular saw blades

40-tooth, carbide tip circular saw blades, 10¼ in.

Heavy-duty, ½-in. hammer drills with bits

³⁄₈-in., cordless drills and extra batteries

Heavy duty, variable speed reciprocating saws

1½-in. rotary hammers with bits

14-in. miter saw and stand with blade

Initial Shoring Tool Box

16- or 25-ft tape measures

50-ft steel tapes

Laser measuring device

Chalk line

Straight claw framing hammer

Utility knife

Cat's paw

Tool pouch

Respiratory protection

Eye protection

Carpenter's pencil

T-bevel square

Speed square

6-in. level

Glossary

These are some of the items and definitions that you may encounter when your team or department decides to get involved with technical rescue or emergency rescue shoring operations that require familiarity with construction terms. This list is not exhaustive and is provided to enable your personnel to become familiar with the more common items and terminology associated with shoring activities.

30° – The lowest angle that can be utilized in the fabrication of raker shores. Although this is an excellent angle to place shores, in most cases the lumber will not be long enough to establish the angle.

45° – The preferred angle of use for solid-sole raker shore due to the efficiency of the cuts and the ease of construction.

60° – The steepest angle that should be utilized in raker shore operations. When installing a split-sole raker in the ground, this should be the angle of choice.

Adjustable Square – A try square whose arm is at a right angle to the handle and can be moved to form either a *T* or an *L* shape; can be utilized for laying out angles or scribing lines; a simple tool generally used to mark right-angle cuts in sections of dimensional lumber.

Air Compressor – A machine that can draw in outside air at atmospheric pressure, compress the air to pressures higher than the normal atmosphere, and deliver it at a rate great enough to operate pneumatic tools and equipment. An air compressor can be either fuel- or electric-driven. Generally the ones used by firefighters are fuel-driven, normally by gasoline.

Air-dried Lumber – Wood that has been dried by exposure to air under natural conditions. The lumber usually does not have moisture content greater than 24%.

Allowable Load – The load that induces the maximum allowable unit stress at a critical section of a structural bearing element. By calculating the load that has been applied during and after the collapse, it is possible to determine if any structural integrity remains in that element.

Anchor Bolts – A steel bolt usually fixed in a building's structure with its threaded portion projecting; used to secure frameworks, timbers, machinery, and the like that can be used in a variety of situations in rescue shoring.

Arch – A curved type of construction that spans an opening; usually consists of wedge-shape blocks or a curved structural member supported at both ends. Arches vary in shape from semi-elliptical to the acutely pointed type. Failure of any element in an arch may cause total arch failure. This type of construction, although not commonly used recently, can cause serious problems for any rescue team. Overloaded arches are an extremely dangerous situation; in a collapse scenario, any arch must be examined for structural integrity.

Axial Force – The action of compression or tension along the length of a structural member, usually expressed in pounds; a load applied to the center of a structural member such as a column or strut.

Axial Load – The resultant longitudinal, internal component of force that acts perpendicular to the cross-section of a structural member and at its center producing uniform stress throughout the element.

Beam – A structural element that sustains transverse loading and develops internal forces of bending and shear in resisting loads; also called a *girder* if large scale, a *joist* if small scale or closely spaced in sets, a *rafter* if used for a roof; one of the most common structural elements encountered.

Beams – Horizontal or inclined load carrying structural members, supported on two or more points; a structural member whose prime purpose is to carry transverse loads, typically floors.

Bearing Wall – A wall that supports a vertical load other than its own weight; the main bearing component of any unframed building. These walls may be located anywhere in the structure but generally are the exterior walls.

Bending – A combination of two states of stress, compression and tension, in different fibers of the same structural element; turning action that causes change in the curvature of a linear element; one of the most common state of stressing seen in collapse operations.

Bevel Square – A carpenter's tool similar to a *T* square but having a blade on one end that can be adjusted to any angle; excellent tool for transferring preexisting angles to other pieces of lumber.

Board Foot – In lumber a unit of measure equivalent to a board that is 1 ft square and 1 in. thick.

Boards – This is a broad definition that is usually given to dimensional lumber that is 2 in. or more wide and 1½ in. thick. This would include 2x3s, 2x4s, 2x8s and the like.

Bottom Plate – See sole plate.

Brace – A metal or wood member used to stiffen or support a structure or a shore; a strut that supports or fixes another member in position or ties together several members at once; a very important element in a rescue-shoring situation.

Bracing – Usually refers to the resistance to movements caused by lateral forces or by the effects of buckling. This is one of the most important elements in keeping rescue-shoring effective against secondary collapse potential. All shoring systems must be properly braced to protect the rescuers as well as the victims.

Buckling – The collapse or destruction of a structural element in the form of a sudden sideways deflection. Buckling is very common in columns or any slender structural element subject to an overload in the form of compression. It is a common occurrence in many collapsed buildings with columns used for support, such as any framed structure.

Cantilever Beam – A beam that is generally supported only at one end. This can be a very unstable element in a partial collapse situation. This type of beam has to be examined thoroughly before it can be moved.

Carpenter's Level – A tool or instrument used by carpenters and mechanics to determine an even horizontal or vertical line. It is used in leveling and plumbing up shoring elements during the erection of a shoring system. It can also be utilized to determine how far out of plumb existing walls are in a partially collapsed or leaning structure. There should be several sizes of levels in a rescuer's rescue tool cache.

Center Brace – A strip of wood, usually 1x6 or ¾-in. plywood, nailed in the center of a vertical shore or laced post shore to add additional stability. A center brace can also be 2-in. dimensional lumber; also called *midpoint bracing*.

Chain Saw – A saw powered by gas, electricity, or hydraulics, usually hand held for cutting wood or concrete. It is used extensively for cutting shoring lumber and collapsed debris and can cut multiple pieces of lumber in one pass.

Checks (Lumber) – Splits or cracks in a board, generally caused by drying or seasoning too rapidly; a separation of the wood naturally occurring across or through the growth rings. This condition reduces the strength of the lumber.

Circular Saw – A power-operated saw, normally driven by electricity and having a circular steel blade with different shapes and numbers of teeth along the perimeter of the blade. Common blade sizes are 7¼ in., 8½ in., and 10¼ in. To cut a 4x4 in one pass, you need the 10¼ in. saw, the most effective size saw for cutting 4x4s because you would have to rotate the wood and cut twice if you used a small diameter blade.

Cleat – A small block of wood usually 2 in. thick nailed on a shoring member or a surface to stop another member from moving or sliding; a very important part of various shoring systems. If a cleat is not properly installed or nailed, the shore will not be effective and may fail.

CMU – An engineering and construction term; an abbreviation for concrete masonry unit; just plain, regular concrete block.

Collapse – The origin of this word comes from the Latin word *collabi*, describing the sudden and rapid failure of a structural member or structure due to a variety of possible forces. Any building or structural element in a building has the potential to collapse.

Column – In standard building construction, a relatively long, slender structural element under compression, usually vertical in nature and supporting a load that acts in the direction of its longitudinal axis. This element has a high collapse potential, and if it fails, other items supported by it also collapse. A column is a primary structural element that has to be examined immediately in almost all situations.

Combination Square – An adjustable carpenter's tool, consisting of a steel rule that slides through an adjustable head; can be used as a try square, marking gauge and straight edge.

Combined Load – Two or more different types of loads impacting on a structure at the same time such as dead load, live load, and wind load. Many times after a collapse, these combined loads affect numerous structural elements, causing major secondary collapse danger for the rescuers.

Combined Stress – The combination of axial loading and bending stresses acting on a structural member simultaneously. Frequently impacting collapsed columns, this additional stress affects the bearing capacity of the columns and has to be carefully examined before any material is removed from around the columns.

Common Nail – A cut or wire-made, low-carbon-steel nail having a slender, plain shank and a medium, diamond tip; generally used in framing and utility work; can be used to assemble and anchor rescue shoring systems. The two most commonly utilized nails in rescue shoring are the 8d and the 16d.

Compound Beam – A built-up, rectangular beam composed of smaller timbers over which planks or plywood sheets are nailed on each side. The entire unit is then joined by bolting all the elements together.

Compression – The force applied to a structural member that has a compressive or pushing effect on the member and its end connections; the state of stress in which the particles of the material are pushed one against the other, generally causing overall shortening of objects in the direction of its action. Compression is one of the most common states of stress in structural collapse as well as rescue shoring.

Compressive Strength – The maximum compressive stress force that a specific material can handle before that element reaches it failure point. Knowing the working compressive strength of all shoring material is a *must* during the erection of rescue shoring.

Compressive Stress – The amount of stress that will resist the shortening effect of an external compressive force.

Concentrated Load – One of the most dangerous situations encountered in a structural collapse scenario. A concentrated load acts on a small, localized area of a structure and is a load that is applied unequally over an area. It is the opposite of a distributed load. In rescue shoring, concentrated load must be determined and redistributed via shoring throughout the structure to good ground or other structural elements that can handle the additional load.

Connection – The union or joining of two or more distinct elements. In a structure, the connection itself often becomes a separate entity. In a structural collapse situation, it is imperative that the connection points involved in the collapse area are all checked for damage. Connection points are generally the weakest point of any building component.

Continuous Beam – A simple beam extending over three or more supports and evenly supporting a load.

Cribbing – Framework constructed of timbers or steel to provide support for material above it. In urban search and rescue, it is the assembly of multilayers of dimensional lumber used to support and stabilize damaged structural elements or specific items. Cribbing provides one of the simplest methods of supporting unstable items in collapse operations.

Cripple Stud – A structural element shorter than a standard stud; as a stud in a short wall or above a door or window opening.

Cripple Wall – A wall that is shorter than the surrounding wall sections; can also be a wall section on top of another wall, in which case the joints of the two walls become an issue in a collapse situation because they generally are less stable than the larger walls in a structure.

Cross Brace – Any system of bracing in which there are intersecting diagonals. Normally constructed of 2x6s, a cross brace is a very important component of rescue shoring. Without the bracing in certain situations, a lateral load against shoring could cause premature failure.

Cross Grain – The grain in the lumber not parallel to the axis of the member. This characteristic is undesirable as it lessens the strength of the wood.

Curtain Wall – A non-load-bearing wall built between a series of exterior columns and beams. This wall is supported entirely by the frame of the building, rather than being self-supporting. A curtain wall is an item that must be identified as soon as possible in a collapse situation.

Dead Load – The overall weight of a structure, including the weight of any fixtures or equipment permanently mounted within, such as a vault or HVAC equipment.

Dead Shore – An upright series of timbers used as a support of a dead load during structural instability; another name for the vertical shore.

Deck – The flooring of a building or other structure or an open platform, generally resting on some type of beam setup.

Deflection – Generally refers to the lateral movement of a structure caused by loads; the amount of deformation of a member; the displacement of a structural member as a result of loads applied to it such as vertical sag of a beam or the lateral sway of a tower. Visible deflection of any members in a collapse scenario is a cause for concern.

Deformation – Any change in shape, including shortening, lengthening, twisting, buckling, or expanding of a structural element without breaching the continuity of its parts. The obvious deformation of any structural element in a collapse is cause for concern and must be investigated at all times.

Design Loads – Value of a load calculated as the total dead and live loads that a structural member is designed to support. Any element that is forced to exceed its designed load must be examined and resupported if necessary.

Diagonal Brace – An inclined structural member in compression or tension, usually installed to stabilize a framework against horizontal forces. This item is used in almost all types of shoring, usually as a tension element for additional stability. A diagonal brace is a very important piece of rescue shoring and must be attached to the shores properly in order for them to be fully efficient.

Direct Stress – Application of only compressive or tensile stress, resulting in the bending or shearing of an element. Material will support substantial amounts of direct stress; however, in collapse scenarios, this is rarely the case.

Distributed Load – A load that is applied relatively equally over a given area. Almost all elements in a building are designed to support distributed loads. However, that same load may become concentrated, causing a problem for a rescue forces. This is due to the fact that a smaller number of structural elements may be trying to support that heavier load.

Door Shore – A series of uprights and lateral sections of lumber used to stabilize a door opening or wall breach for access by rescue personnel; generally constructed of 4x4s with a series of wedges.

Double-headed Nail – A nail having two heads, one above the other, used for temporary work and generally utilized for the erection of scaffolding or concrete forms. The upper head is driven with a hammer until the lower head bears on the surface into which the nail is driven. The space between the upper and lower head is used to withdraw the nail. Double-headed nails are excellent to use for training purposes because pulling them from lumber is easier than pulling single-headed nails.

Double T-Shore – A temporary, initial, safety shore designed to be portable and lightweight. It consists of a header, two posts, gusset plates, wedges, and a sole plate and is generally constructed of 4x4 material.

Dry Rot – The deterioration of wood caused by fermentation and chemical breakdown when attacked by fungus, giving the wood a white hue. It is actually caused by continual contact with moisture. On most occasions it can be readily seen with the naked eye.

Ductile – The strain behavior that results from the plastic yielding of materials or connections. To be significant, the plastic strain prior to failure should be considerably more than the elastic strain up to the point of plastic yield.

Duplex Nail – Another name for the double-headed nail, the technical term associated with the nail.

Dynamic Load – Any load in a non-static or moveable state; a load that changes location or value rapidly such as the wind or a moving live load. An elevator is an example of a dynamic load, as is an escalator full of people.

Earthquake Load – The total force exerted on a structure by the actions of an earthquake; can be a rolling motion, lateral force, or an up-and-down motion, depending on how far from the epicenter of the earthquake the structure is.

Eccentric – The direction of a load on an element that does not have the same centerline of bearing through the length or width of that element.

Eccentric Load – A load normally on a column that is non-symmetric with respect to the central axis of the column, thereby causing a bending action. When this situation occurs, it drastically takes away from the strength of the column. In a collapse scenario, eccentric load can be a very dangerous condition.

Elastic – Two aspects of stress-strain behavior. The first is a constant stress-strain proportionality or constant modulus of elasticity, as represented by a straight-line form of a stress-strain graph. The second is the limit within which all the strain is recoverable, that is, there is no permanent deformation.

Elasticity – The property of a body that causes it to tend to return to its original shape after deformation from stretching, torsion, or compression. Basically, when an item is deformed and returns to its original shape after the overload is removed, it still has strength.

Face Nailing – The direction of nailing on which the nails are driven perpendicular to the face of the material. In rescue shoring, almost all bracing and all gusset plates are face nailed.

Failure – The condition of becoming incapable of a particular function. It may have partial as well as total implications on a structure. For example, a single connection may fail, but the structure might not collapse because of its ability to redistribute the load.

Fatigue – A structural failure resulting from a load applied and removed or reversed repeatedly through a large number of cycles. Fatigue can be a major cause of collapse. When collapse occurs because of fatigue, the remains of the entire building have to be examined for potential secondary collapse problems.

Fishplate – A piece of wood or metal used to fasten the ends of two members with nails or bolts; generally utilized to stop the ends of the elements from moving or rotating. In a structural collapse situation, a fishplate has to be checked for continued integrity.

Flitch Plate – A steel plate normally sandwiched between two or more pieces of structural lumber and bolted together; normally used to place more rigidity into the section making it more efficient. On occasion, a flitch plate can be placed on the outside face of the material.

Flying-Raker Shore – An initial safety shore erected in place before a series of raker shores are to be assembled; used to make the area safer for rescue operating personnel. It is a temporary shore and can be taken down and reused after the main shoring system has been installed.

Flying Shore – A larger type of shoring system placed between two buildings for the temporary support between two adjacent walls, generally for use above the first floor. A flying shore can be a rather complicated setup.

Force – An effort that tends to change the shapes or the state of motion of an object. In collapse scenarios, the force applied to a specific element or connection point determines whether it holds together. The cause of the collapse must be known in order to determine how much force has actually been applied to the structure.

Framing Square – A right-angle tool that is 24 in. long and 2 in. wide on one end and 16 in. long and 1½ in. wide on the other, forming a right angle. This tool has several scales and tables that a carpenter can utilize for many functions, among them calculating and determining the angles necessary for the erection of raker shores.

Girder – A major horizontal member used to carry a series of beams or a large load; a beam that supports other beams, generally at right angles to those beams.

Grades of Lumber – General classification of lumber according to the strength and utility of the different species of lumber.

Green Lumber – Freshly sawed or unseasoned lumber that has not been dried. The use of this material should be avoided in rescue shoring operations. It has a tendency to split and twist, and the excessive water content detracts from the strength of the material. Also it is rather difficult to work with and plays havoc with tools.

Gusset Plate – Normally a section of ¾-in. plywood, generally a 12x12 utilized to hold two pieces of lumber together so that they cannot move or be dislodged. In raker shores, it is a primary item; in interior shoring, they are a secondary item.

Hardwood – Classification of lumber from broad-leafed deciduous trees, heavy and close-grained. Oak and maple are two common types of hardwood. Although very strong, hardwood is not generally suited for rescue shoring. It is usually heavy and splits and twists with long-term storage.

Header – Upper horizontal cross member between the jambs that forms the top of a window or door frame and provides structural support for construction above; designation given to the top member in the vertical, window, door, sloped floor, and laced post shores.

Horizontal Brace – Structural element utilized for the main horizontal legs in the truss support system anchoring a series of raker shores together. A horizontal brace is also found in laced posts and vertical shores more than 9 ft in height.

Horizontal Shore – Any shore erected in the interior or exterior of a structure and supporting stable or unstable walls; generally erected with either two or three horizontal struts, depending on the amount of damage and the height of the walls.

I-beam – A common name for the American standard beam because of its resemblance to the letter *I*. The flanges of the beam are normally smaller than the web and are beveled for support strength.

Impact Load – The dynamic effect on a structure of a forcible, momentary contact of another moving body; basically, one item driven into another such as collapsed debris falling in a structure; one of the most dangerous and unpredictable loads encountered in a collapse situation.

Initial Shoring – The installation of any temporary safety shoring before the main shoring activity begins; generally consists of a lighter, less complicated item that can be reused.

Joist – One of a series of parallel beams of timber, steel, or reinforced concrete used to support floor and ceiling loads and supported in turn by larger beams, girders, or bearing walls. The widest dimension of the joists is vertically oriented.

Kiln-dried Lumber – Wood that has been seasoned in a special chamber by artificial heat. This is generally the best material to use for shoring purposes because it is less likely to warp, split, check or crack when stored.

Laced-post Shore – Also called a shoring tower, it is one of the strongest and most stable shores used to stabilize heavy concentrated loads; can be used as a safe haven area in an earthquake situation.

Lateral – An orientation, meaning to the side or from the side with reference to the vertical direction of the gravity forces, wind, earthquakes, and the like; often used in reference to something that is perpendicular to a major axis or direction.

Lateral Bracing – The horizontal or vertical bracing for a wall, beam, or structural member. A very critical part of emergency building shores, lateral bracing adds stability and enables the shoring material to more efficiently handle loads generated against them.

Level – In construction, the term refers to the position of a line or plane when that line is parallel to the surface of still water.

Lintel – A horizontal structural member placed over an opening in a wall such as a door or window to carry superimposed loads; can be wood, metal, or masonry.

Live Load – Any load that is not of a permanent nature; generally refers to any load other than the dead load on a structure; can include people, snow, water, and temporary loads.

Load – Term used to define a force or systems of forces exerted on or carried by a structure or part of a structure; the active force or combination of forces exerted on a structure.

Midpoint Brace – Used in several types of shores as additional lateral support when the posts reach a designated height; usually made of a 1x6 piece of plywood or 2x6 dimensional lumber. The idea is to stop the deflection of the posts when they are used at a specific height. Also can be called center bracing.

Member – In building structures, this term describes one of the distinct elements of an assemblage.

Moment – Action tending to produce turning or rotation; product of a force times a lever arm, giving a unit of force time and distance. Bending moment causes curvature; tensional moment causes twisting.

Mudsill – A plank or timber laid directly in mud or soft soil, used to help distribute the weight of an object above it by displacing more contact with the earth.

Nail – A straight, slender piece of metal, for the most part made of steel, pointed and having a head. Nails are normally driven in place with a hammer and used to anchor two or more pieces of lumber together. There are numerous styles and sizes of nails available today. In rescue shoring, firefighters use two basic types, common and duplex, and two basic sizes, 8d and 16d.

Neutral Axis – An imaginary line in a beam, shaft, or other member, subjected to bending where there is no tension or compression and where no deformation has taken place.

Nominal Size – The dimensions of sawed lumber before it is dried or surfaced.

Penny – A unit denoting the length of a nail; also an indication of the shank and head diameter. The higher the penny number, the larger the diameter and the longer the shank of the nail.

Pitch – The slope or angle of the raker shore when measuring the angle with a carpenter's framing square. A point 9 ft high and 12 ft back gives a pitch of 9 over 12.

Plumb – Exactly vertical, at right angles to a level line; exactly 90° from that line. The item is level in a vertical plane when it is plumb. Any posts in interior shoring generally should be plumb to effectively handle the load.

Plywood – Structural wood made of three or more layers of veneer, usually an odd number; generally laid with the grain of adjoining plies at right angles to each other and glued together. There are various types and thickness available. In rescue shoring, the most common type used is ¾-in. plywood.

Post – The vertical members in the window, door, vertical, and laced post shores. These are the main elements that transfer the damaged loads to good bearing.

Primary Structural Members – Usually refers to the various elements that support the main structure of the building and/or other structural members. Normally columns, arches, beams, girders, and bearing walls are considered primary structural elements.

Punching Shear – The punching of a hole through a base by a heavily loaded column as a result of failure of the base. This event can easily happen in rescue shoring if the loads are not properly distributed.

Raker Shore – A specific type of rescue shoring used to stabilize leaning walls of a building. The strength of the raker shore is in the use of an inclined member. There are three distinct types of raker shore: the solid-sole, split-sole, and flying raker.

Rescue Shoring – The erection of a series of timbers and bracing to stabilize walls or floors in an attempt to arrest any further movement or collapse of unstable structures. It is strictly a temporary measure, used to provide a degree of safety for rescue personnel operating in unstable structures. It is not used to restore structural elements to their original positions or shapes.

Rescue Struts – Pneumatic shores that can quickly be set up in a collapse situation and are easily and manually adjusted. They can withstand pressures greater than normal wood in the right application.

Safe Load – The load on a structure that does not produce stresses in excess of the designed allowable stresses. In a collapse scenario, the remains of the building are generally subjected to more than the safe load allowed by the design of the building.

Screw – A fastener with an externally threaded shaft that has slightly better holding power than a nail.

Shake – A defect, usually a split or crack in wood, resulting from damage during growth or unequal shrinkage during drying of the material. Shake can detract from the strength of the member if it is severe.

Shear – A deformation in which parallel planes slide relative to each other so as to remain parallel. A force that is lateral (perpendicular) to the major axis of a structure or a force that involves a slipping effect as opposed to a push-pull effect. Wind and earthquake forces are sometimes visualized as shear effects on a building. This is due to the fact that they are forces perpendicular to the major vertical axis of the structure.

Shear Strength – The maximum amount of shear stress that a material is capable of handling before permanent failure occurs.

Shims – Thin pieces of wood or metal, usually tapered, that are inserted under one member to adjust its height or to fill in an open area.

Shore Tower – A substantial shoring setup used when a large concentration of weight is located in one spot above; another name for the laced post shore.

Shoring – The application of materials, normally wood, temporarily supporting a damaged structure; the temporary support of structures during construction, demolition, alteration, renovation, etc. in order to provide the stability that will protect property as well as construction crews and the public.

Sill – A horizontal timber at the bottom of a wood structure that rests on the foundation or ground.

Sleeper – Any long, horizontal beam on the ground that distributes a load from the members above it. In rescue shoring, sleeper generally describes any lumber placed under a shoring system to distribute the weight evenly throughout the ground around the shore.

Softwood – Lumber from trees with a needle- or scale-like leaf. Douglas fir and Hem fir are two of these types.

Sole Plate – The designation given to the bottom member of any shores that rests on the floor or ground.

Solid-sole Raker – One type of raker shore normally used on concrete or hard surfaces; named because the sole plate is a solid piece of material, generally a 4x4.

Split – Separation of wood due to the tearing apart of the wood cells. It can detract greatly from the strength of lumber.

Split-sole Raker – Another type of raker shore that is normally used when soft ground is encountered; named because the sole is two 2x6s lapped against the side of the raker.

Static Load – Any load placed on a structure that doesn't change in magnitude or position with time; a load that remains constant and is applied slowly.

Strain – A failure in the shape or form of a body or material that is subjected to an external force. Permanent deformation resulting from stress is usually measured as a percentage of deformation, sometimes called unit strain or unit deformation.

Stress – Internal forces set up at a point in an elastic material by the action of external forces; the mechanism of force within the material of a structure, visualized as a pressure effect, tension, compression, or shear effect on the surface of a unit of the material and quantified in units of force per unit area. Allowable, permissible, or working stress refers to a stress limit that is used in stress design methods. Ultimate stress refers to the maximum stress that is developed just prior to failures of the material.

Stringer – A long horizontal member that supports a floor or deck.

Strut – A brace or any piece of material that resists thrust in the direction of its own length; it may be vertical, diagonal, or horizontal.

Stud – An upright post or support; normally one in a series of vertical structural members that act as the supporting elements in a wall or partition.

T-Shore – A shore normally erected of 4x4s in the shape of a *T* and used mainly as an initial safety shore.

Tensile Strength – The resistance of a material to rupture when subject to tension; the maximum tensile stress that a material can sustain.

Tension – The force exerted on a structural member that has the effect of either pulling apart or elongating the structural member. The resulting action produces straightening effects and elongation.

Timber – Dimensional lumber that is 5 in. or more in its least dimension. This would include 6x6s, 6x8s, 8x8s and the like.

Toenail – A common method of anchoring two sections of lumber together by driving the nails on a slant or an angle; utilized quite frequently in all shoring operations.

Torsion – The twisting of a structural member about its longitudinal axis by two equal torques at opposite ends of the member. Lateral loads produce torsion on a building when they tend to twist it about its vertical axis.

Torsional Load – A load creating a force that is offset from the shear center of a structural element and causes a twisting of that element.

Transit – A surveying instrument used for measuring and laying out of horizontal and vertical angles, distances, directions, and differences in elevations; excellent for determining the slightest movement in a collapsed structure from a safe distance.

Ultimate Stress – The maximum amount of stress that a material can stand before it physically breaks apart. All structural elements have a measurable amount of ultimate stress. In building design, all loads placed on any element are calculated in such a way as to never reach ultimate stress. However in structural collapse, some items can reach their ultimate stress level.

Uniformed Load – A load that is equally distributed over a given length of a structural member. Ideally in building design, all structural elements are uniformly loaded. In collapse rescue operations, the structural elements that are normally uniformly loaded can suddenly become eccentrically loaded, causing them to shift or fail.

Vertical Shore – Sometimes know as a *dead shore*; generally used to support any loads from floors above in a damaged structure to specific structural elements, such as girders; the most common type of shore erected in collapsed buildings.

Wall – A vertical, planar building element. Foundation walls are those that are partly or totally below ground. Bearing walls are used to carry vertical loads in direct compression. Shear walls are those used to brace a structure against horizontal forces due to wind or seismic shock.

Wall Plate – The first members erected against damage walls that are to be stabilized; used in raker shoring and horizontal shoring.

Wedges – Pieces of wood or other material, thick at one end and tapering to a thin edge at the other; used in pairs to take up space between supporting elements or to apply pressure against two shoring elements. Wedges are generally six times as long as they are thick.

Window Shore – Any shore that supports an unstable window opening through the use of lumber and wedges; generally constructed of 4x4s.

Working Stress – The unit stress that has shown to be safe for a specific material, while maintaining a proper degree of safety against structural failure. For safety reasons, all structural elements are designed to accept specific loads only up to a working stress.

Index

D

O

P–Q

R

Y–Z